Environmental biology of European cyprinids

Developments in environmental biology of fishes 13

Series Editor
EUGENE K. BALON

Environmental biology of European cyprinids

Papers from the workshop on 'The Environmental Biology of Cyprinids' held at the University of Salzburg, Austria, in September 1989

Guest Editors:
WOLFGANG WIESER, FRITZ SCHIEMER,
ALFRED GOLDSCHMIDT & KURT KOTRSCHAL

Reprinted from *Environmental biology of fishes* 33 (1-2), 1992
with addition of species and subject index

Kluwer Academic Publishers
Dordrecht / Boston / London

Library of Congress Cataloging-in-Publication Data

Environmental biology of European cyprinids : papers from the workshop on "The environmental biology of cyprinids" held at the University of Salzburg, Austria, in September 1989 / guest editors, Wolfgang Wieser ... [et al.].
p. cm. -- (Developments in environmental biology of fishes ; 13)
"Reprinted from Environmental biology of fishes 33 (1-2), 1992, with addition of species and subject index."
ISBN 0-7923-1484-0 (acid free paper)
1. Cyprinidae--Europe--Congresses. 2. Cyprinidae--Europe--Ecology--Congresses. I. Wieser, Wolfgang, 1924- . II. Series.
QL638.C94E58 1992
597'.52--dc20 91-37264

ISBN 0-7923-1484-0

Published by Kluwer Academic Publishers,
P.O. Box 17, 3300 AA Dordrecht, The Netherlands.

Kluwer Academic Publishers incorporates
the publishing programmes of
Martinus Nijhoff, Dr W. Junk, D. Reidel and MTP Press.

Sold and distributed in the U.S.A. and Canada
by Kluwer Academic Publishers,
101 Philip Drive, Norwell, MA 02061, U.S.A.

In all countries, sold and distributed
by Kluwer Academic Publishers Group,
P.O. Box 322, 3300 AH Dordrecht, The Netherlands.

Contents

Before, during, inbetween and after the oral presentations at the workshop 'Environmental Biology of Cyprinids' in Salzburg: a – Eugene Balon and Wolfgang Wieser before the first lecture; b – Kurt Kotrschal and Fritz Schiemer in the lecture theater; c – Alexandra Sänger with Alfred Goldschmidt at the registration desk; d – Kurt Kotrschal explaining something to Wim Hoogenboezem; e – Monica Zaunreiter admires a story by Reiner Berg before dinner at the local restaurant, while (f) Jan Osse and Wolfgang Wieser still talk science.

Jan Osse and Babu Ali enjoy drinks at the reception in the poster exhibit (a), and so do Tony Pitcher and Fritz Schiemer (b); c – Frans Sibbing delivering his lecture and later (d) listens to somebody elses presentation sitting behind Wolfgang Wieser, Eugene Balon and Jan Osse; in the evening talks continue in the nearby restaurant with (e) Renate Siegmund, (f) Babu Ali, Walter Bock and Hans Winkler, and (g) Julius Ojwang-Okwor.

The participants of the workshop on 'Environmental Biology of Cyprinids' at the garden side of the new Zoology building of the University of Salzburg, September 1989.

Environmental Biology of Fishes **33**: 9–12, 1992.

Introduction to the environmental biology of European cyprinids

Wolfgang Wieser
Institut für Zoologie, Universität Innsbruck, Technikerstrasse 25, A-6020 Innsbruck, Austria

The hope of achieving the same degree of predictability in describing life processes as achieved by physicists in the description of non-living processes, has always been a motivating force for biologists. Now, striving for predictability certainly is a measure of good science, but the description of complex, dynamic, and evolving systems requires concepts quite different from those underlying the description of physico-chemical processes.

Take the example of the temperature relationship of a poikilothermic animal, let us say a cyprinid fish: the effect of environmental temperature on, for example, the rate of oxygen consumption of the members of a homogeneous age group belonging to one species, can be established quite rigorously. There will be more variability than in the response of a chemical reaction to temperature, but by monitoring the behavior of the animals and controlling the conditions of the experiment, most of the variation can be accounted for and, in consequence, normalized. The result will be a quantitative relationship of the type of Krogh's 'normal curve' (Krogh 1914), which allows probabilistic predictions as to the metabolic responses of a particular organism to a change of temperature. However, if before carrying out this test the fish had been acclimated to a low temperature, the resulting curve would look different. Once this is recognized, acclimation temperature becomes a parameter in the functional relationship between temperature and metabolic rate. If instead of laboratory animals we use fish freshly caught in our favourite lake, we discover that season and photoperiod also profoundly influence the shape of the reaction/temperature curve (Koch & Wieser 1983). The same is true of the state of nutrition, or of the conditions prevailing in the environment during a sensitive period of embryonic development (Kinne 1960, Brett 1979). A new dimension of variability is introduced when different populations of the same species (Lavin & McPhail 1987), or different individuals of the same population (Garland et al. 1990) are compared, and so on. In this way new parameters are added to a seemingly simple relationship until the latter is revealed as an experimentally sound but quite abstract function, with little predictive power regarding the responses of a population in the wild.

Generalizations of the *functional response* type contain information about operational characteristics of a given biological function, like metabolic rate (Krogh 1914) or feeding rate (Holling 1965), but they provide little ecological perspective. To achieve this perspective, functional relationships have to be expanded into what has been called the *reaction norms* of specific functions, the sum of which constitutes the *reaction norm of the species* (or population) in question. This concept, introduced into ecology by Woltereck (1909), has been resurrected by Stearns & Koella (1986) in an attempt to model the evolution of phenotypic plasticity of life-history traits. 'A norm of reaction represents the range of potential phenotypes that a single genotype could develop if exposed to a specified range of environmental conditions' (Stearns & Koella 1986, paraphrasing Woltereck 1909). Evolutionary biologists tend to assume that an animal chooses (or is forced into) that phenotype which under a given set of circumstances promises to maximize fitness. Stearns & Koella, for example, constructed models for the reaction norm of age and size at maturity, assuming that the members of a population would choose that path which under the prevailing conditions maximizes 'r', the

Malthusian parameter. Similar models could be constructed for any organismal function that lends itself to quantification by a human observer.

The reaction norm of a species is an evolutionary concept introduced to account for the phenotypic flexibility of a given genotype (Futuyma 1986). It is also a challenge to the experimental biologist since in order to understand (or predict) the role of a given species in an ecosystem its degree of phenotypic flexibility has to be matched by the degree of sophistication of the questions asked and the complexity of the experiments performed. In consequence, the one-dimensional curves in two-dimensional grids which have become a way of describing the properties of dynamic events, from chemical reactions to feeding responses of animals, turn out to be less and less useful tools for the description of ecological relationships. This is so because in ecology, in contrast to physics and chemistry, boundary conditions cannot be ignored or standardized, but have to be accepted as variable components of the system to be described.

Communities are structured by the accidents of evolutionary history and by the rules of biological interactions, mainly competition and predator-prey relationships. Within the ecosystem a species may play a significant role as a keystone predator and at the same time as a major prey item for higher trophic levels (Winfield et al.). Two species occupying different niches on the same trophic level may obey a set of fairly well defined rules, but these rules may be altered drastically by the introduction into the system of a third species occupying a different trophic level (Lammens et al.). These and other relationships, predictable or unpredictable, between the players of an ecological game, depend on, and are constrained by, the reaction norms of the players themselves; thus, if we want to understand the rules, the ecological potential of each participant has to be taken into account. This is not always appreciated by the designers of ecological games played in laboratories. It is the task of the environmental biologist to complicate matters by defining new boundaries, introducing new rules, and even dragging new players from behind the wings where they have been pulling invisible strings. One of the major driving forces in such an endeavour must be the *comparative approach,* since, as in evolution, the performance of the individual player can be measured only in relation to the performances of its relatives, competitors and opponents (which in the realm of community ecology is intended to mean the prey or the predator). Furthermore, the 'range of potential phenotypes' into which a single genotype might develop calls for an equivalent range of interventions by which the players are forced to show their hands. Some of the phenotypic expressions of genotypes are quite flexible, others are more rigid. Thus a 'range of flexibilities' determines the direction, magnitude, and rate of adaptive processes. According to what Schoener (1986) called a 'mechanistic approach' to community ecology, models of community organization should more often and more seriously take into account this range of flexibilities embodied in the behavioral, morphological, and physiological reaction norms of constituent species (Dill 1987, Persson & Diehl 1990).

This special issue of 'Environmental Biology of Fishes' illustrates the idea that the construction of mechanistic models of the roles played by different species in an evolving ecological game requires the combined efforts of experts specializing in different areas of biology.

The major question asked by by the organizers of a long-term research project on the ecology and ecophysiology of Austrian cyprinids was to what extent the spatial and temporal patterns of distribution of the 30 or so cyprinid species occurring in Austrian lakes and streams (Schiemer 1988, Schiemer & Spindler 1989) can be correlated with species-specific structural and functional features. Following this deterministic trail it was attempted to define differences in these features between species known to be essential in structuring European freshwater communities by both top-down and bottom-up effects. The same approach formed the basis of a workshop held in Salzburg in September 1989, in which geographic restrictions were relaxed so as to include the cyprinid fauna of other European countries. In the present volume, containing a selection of the key-note contributions at the Salzburg meeting, restrictions have been relaxed even further to include one paper (Huber & Rylander) on a New World group of cyprinids because it neatly supplements a paper on the same subject by Kotrschal & Palzenberger.

Most of the papers comprising this volume are devoted to what has been recognized above as one of the major driving forces in environmental biology, i.e. the comparative approach. Comparisons can be made between individuals solving problems by means of *genotypic* styles (Wanzenböck; Keckeis & Schiemer; Uiblein; Winkler & Orellana; Wieser et al; Sänger; van den Berg et al.; Gomahr et al.; Kotrschal & Palzenberger; Huber & Rylander), or between individuals solving problems by means of *phenotypic* styles (Hoogenboezem et al.; Wanzenböck; Winkler & Orellana; Koch et al.; van den Berg et al.). These styles, be they characteristic of different genotypes (documenting the principle of *variability*) or of different phenotypes (documenting the principle of *flexibility*) ought to be analyzed on all levels of biological organization, from molecules to organisms (Koch et al.). On the organismic level, genotypic variability and phenotypic flexibility are expressed in a variety of physiological functions: sensory (Gomahr et al.), locomotory (Sänger; Kaufmann & Wieser), integrative (Kotrschal & Palzenberger; Huber & Rylander), metabolic (Wieser et al.), behavioral (Uiblein; Keckeis & Schiemer; Winkler & Orellana) and developmental (Wanzenböck; van den Berg et al.).

Within the framework of a mechanistic program such concerted efforts should be helpful in defining the capacity of populations to deal with both predictable and unpredictable features of their habitats. In theory at least, this would contribute to a better understanding of complex events and processes in the environment, like changes of the fish fauna in lakes and streams due to eutrophication (Lammens et al.) or river engineering (Balon), diversity of habitat use by groups of species (Copp), adaptations to extreme environments (Koch et al.; Penttinen & Holopainen), and the multitude of competitive and prey-predator interactions which have not been addressed in this collection of studies, but constitute one of the major topics of present-day community ecology (e.g. Ebenman & Persson 1988, Persson & Diehl 1990).

It is hoped that the mechanistic approach, or – to use an old-fashioned terminology meaning the same thing – the study with modern techniques of the autecology and environmental biology of individuals and populations, will shed more light on the meaning of the word 'adaptation', because in dim light it is difficult to distinguish between structures serving static functions and those of purely aesthetic value in the vault of San Marco (Gould & Lewontin 1979).

Acknowledgments

The editors of this special issue of *Environmental Biology of Fishes* wish to record their thanks to the 'Fonds zur Förderung der Wissenschaftlichen Forschung in Österreich' which for many years has given financial support to the 'Austrian Cyprinid Project' (no. S-35/01-04). Thanks are also due to the organizers of the Salzburg workshop on the Environmental Biology of Cyprinids in September 1989, to E.K. Balon for his editorial efforts in putting this issue into shape, and to Joy Wieser who spent much time in improving the language of some of the reports included in the present volume.

References cited

(The references cited without year of publication refer to the contributions in this volume).

Brett, J.R. 1979. Environmental factors and growth. pp. 599–677. *In:* W.S. Hoar, D.J. Randall & J.R. Brett (ed.) Fish Physiology, vol.8, Academic Press, New York.

Dill, L.M. 1987. Animal decision making and its ecological consequences: the future of aquatic ecology and behaviour. Can. J. Zool. 65: 803–811.

Ebenman, B. & L. Persson. 1988 (ed.). Size-structured populations: ecology and evolution. Springer Verlag, Berlin. 284 pp.

Futuyma, D.J. 1986. Evolutionary biology. 2nd edition. Sinauer Associates, Sunderland. 600 pp.

Garland, T., A.F. Bennett & C.B. Daniels. 1990. Heritability of locomotor performance and its correlates in a natural population. Experientia 46: 530–533.

Gould, S.J. & R.C. Lewontin. 1979. The spandrels of San Marco and the Panglossian paradigm: a critique of the adaptationist programme. pp. 147–164. *In:* J. Maynard Smith & R.

Holliday (ed.) The Evolution of Adaptation by Natural Selection, The Royal Society, London.
Holling, C.S. 1965. The functional response of predators to prey density and its role in mimicry and population regulation. Mem. Entomol. Soc. Can. 45: 1–60.
Kinne, O. 1960. Growth, food intake, and food conversion in a euryplastic fish exposed to different temperatures and salinities. Physiol. Zool. 33: 288–317.
Koch, F. & W.Wieser. 1983. Partitioning of energy in fish: can reduction of swimming activity compensate for the cost of production? J. exp. Biol. 107: 141–146.
Krogh, A. 1914. The quantative relation between temperature and standard metabolism in animals. Intern. Z. physik.-chem. Biol. 1: 491–508.
Lavin, P.A. & J.D. McPhail. 1987. Morphological divergence and the organization of trophic characters among lacustrine populations of threespine stickleback (*Gasterosteus aculeatus*). Can. J. Fish. Aquat. Sci. 44: 1820–1829.
Persson, L. & S. Diehl. 1990. Mechanistic individual-based approaches in the population/community ecology of fish. Ann. Zool. Fennici 27: 165–182.
Schiemer, F. 1988. Gefährdete Cypriniden – Indikatoren für die ökologische Intaktheit von Flußsystemen. Natur und Landschaft 63: 370–373.
Schiemer, F. & T. Spindler. 1989. Endangered fish species of the Danube river in Austria. Regulated Rivers: Research & Management 4: 397–407.
Schoener, T.W. 1986. Mechanistic approaches to community ecology: a new reductionism? Amer. Zool. 26: 81–106.
Stearns, S.C. & J.C. Koella. 1986. The evolution of phenotypic plasticity in life-history traits: predictions of reaction norms for age and size at maturity. Evolution 40: 893–913.
Woltereck, R. 1909. Weitere experimentelle Untersuchungen über Artveränderung, speziell über das Wesen quantitativer Artunterschiede bei Daphniden. Verh. Deutsch. Zool. Ges. 1909: 110–172.

Environmental Biology of Fishes **33**: 13–21, 1992.

A model for switching between particulate-feeding and filter-feeding in the common bream, *Abramis brama*

Wim Hoogenboezem[1,2], Eddy H.R.R. Lammens[1], Yvette van Vugt[2] & Jan W.M. Osse[2]
[1]*Limnological Institute, De Akkers 47, 8536 VD Oosterzee, The Netherlands*
[2]*Department of Experimental Animal Morphology and Cell Biology, Agricultural University, Marijkeweg 40, 6709 PG Wageningen, The Netherlands*

Received 2.2.1990 Accepted 3.11.1990

Key words: Snap frequency, Zooplankton density, Buccal cavity, Swimming speed, Energy costs, Cyprinidae

Synopsis

A model has been developed to describe the process of switching between particulate- and filter-feeding in common bream, *Abramis brama*, in relation to fish size and zooplankton density. The model assumes that the encounter rate of fish and zooplankton is determined by the density of zooplankton and the swimming speed of fish. However, if zooplankton density is so high as to allow at least one prey to be engulfed per random snap, the encounter rate is determined by the volume of the buccal cavity and by zooplankton density, but is independent of swimming speed. The snapping frequency will be maximal at the time of switching, decreasing with increasing zooplankton density because of the extra time needed for intra-oral prey handling. The model predicted switching from particulate- to filter-feeding only for bream > 15 cm standard length at zooplankton densities $< 500\,l^{-1}$. The snap frequency of six size classes of bream (7.5, 10.4, 12.5, 15, 24 and 29.5 cm) was measured at varying densities of *Daphnia*. The model predictions for snap frequencies of all size classes corresponded to the highest values observed. The average of the observed snap frequencies was only 50% of the predicted values, probably because the calculated average distance between prey animals assumed an ideal swimming route of the fish and error-free vision for particulate-feeding, and the handling time was ignored.

Introduction

The abundance of common bream, *Abramis brama* (L.), has strongly increased in The Netherlands due to eutrophication. An additional food resource for all size-classes of bream is provided by the abundant planktonic crustaceans (Lammens et al. 1987, Löffler 1984). Lammens (1985) observed two feeding modes in bream: particulate-feeding (PF), i.e. visual selection of prey which are swallowed one by one, and filter-feeding (FF), i.e. several prey being engulfed by a series of undirected snaps (Jansen 1976, Lammens 1985, Lammens & Hoogenboezem 1991). In choosing its diet, bream is confronted with alternatives at least at three successive steps: (1) the place of foraging, i.e. benthic or pelagic, (2) the mode of feeding, i.e. particulate- or filter-feeding and (3) the size and species of food organisms (Lammens & Hoogenboezem 1991). At each step the choice is probably determined by some optimality principle in which the energetic constraints are most important (Pyke 1984), although the presence of predators may also exert an influence (Werner & Hall 1988). In the present study we are mainly interested in the second decision, that is, the switch between particulate- and filter-feeding of zooplanktivorous bream.

The switchover from particulate-feeding to filter-feeding has been observed in several other zooplanktivorous fish species (O'Connell 1972, Jansen 1976, Holanov & Tash 1978, Crowder & Binkowski 1983, Crowder 1985, Gibson & Ezzi 1985, Lazzaro 1987, Ehlinger 1989, Sibbing 1991) and was found to be determined by fish-size, density and size distribution of prey, and light conditions.

In this paper we develop a model to predict the switchover from particulate- to filter-feeding in bream, using buccal volume, swimming speed and inter-prey distance as the most important variables. The model also describes maximum feeding rates (snap frequencies) under ideal conditions, that is, when vision and prey-handling are not limiting and when prey consists of only one species with little variation in size.

Material and methods

Snap frequency

Feeding experiments were performed with six bream (7.5, 10.4, 12.5, 15, 24, 29.5 cm standard length (S.L.) obtained from Tjeukemeer, a shallow eutrophic lake in the north of The Netherlands. Because social interactions may influence feeding behavior in the experiments, only one fish was used per trial. Prior to the experiments, the fish were kept in the laboratory for at least one month; before the experiment started the fish had been starved for 24–48 hours. Zooplankton density was zero when the experiment started and was raised every 2 or 4 minutes by adding known amounts of zooplankton using perforated plastic tubes (2 mm holes) to ensure a uniform distribution of zooplankton (Fig. 1); this was repeated 15–30 times. A continuous flow of water through the system ensured that the fish were not conditioned to the addition of zooplankton. A 50 μm plankton filter prevented the escape of prey through the outlet. For small fish (7.5–12.5 cm S.L.) an aquarium of $25 \times 25 \times 10\,cm^3$, for bream $\geqslant 15$ cm an aquarium of $70 \times 30 \times 48\,cm^3$ was used. The entire setup was surrounded by black plastic to reduce external disturbance of the fish during the experiment. Zooplankton was obtained from lakes and ponds and consisted predominantly of *Daphnia hyalina* (mean length 1.14 mm) and *D. pulex* (mean length 1.29 mm). The density of the zooplankton was corrected for feeding losses, which amounted to 10–20% at the beginning of the experiments in the small aquaria but to less than 5% in the 100 l tanks.

Apart from the switching experiments in which the zooplankton density was raised, additional experiments were carried out in order to determine the effect of satiation and light on feeding mode and snap frequency (SF). Satiation was tested in an 8 hour experiment with a constant prey density of 300–400 l^{-1} *D. hyalina* (mean length 1.22 mm) and a 24 cm bream. SF was determined during the whole period. To study the influence of light, the decrease of zooplankton was determined when a density of approximately 300 l^{-1} was offered to three small bream (11–12 cm) or to one 29.5 cm bream in complete darkness. At the beginning and the end of the experiment (2 hours) zooplankton density was determined by taking ten 0.5 litre samples using a perspex tube. The zooplankton was counted and measured.

Video recordings, with time registration, were used to analyse the number of snaps in relation to zooplankton density and swimming speed of foraging bream. Time and distance were measured when a fish swam parallel and close to the front window of the aquarium. This was necessary in order to reduce projection errors. We used relatively small aquaria to facilitate focussing on the mouth of the fish and the recording of separate snaps. Consequently, swimming speed was measured over short distances (20–50 cm) and short periods (3–10, occasionally 25 seconds).

Buccal volume

Buccal volume was measured according to the ellipse-method used by Drost & van den Boogaart (1986). Simultaneous dorsoventral and lateral ciné-recordings were made of four foraging bream (8.0,

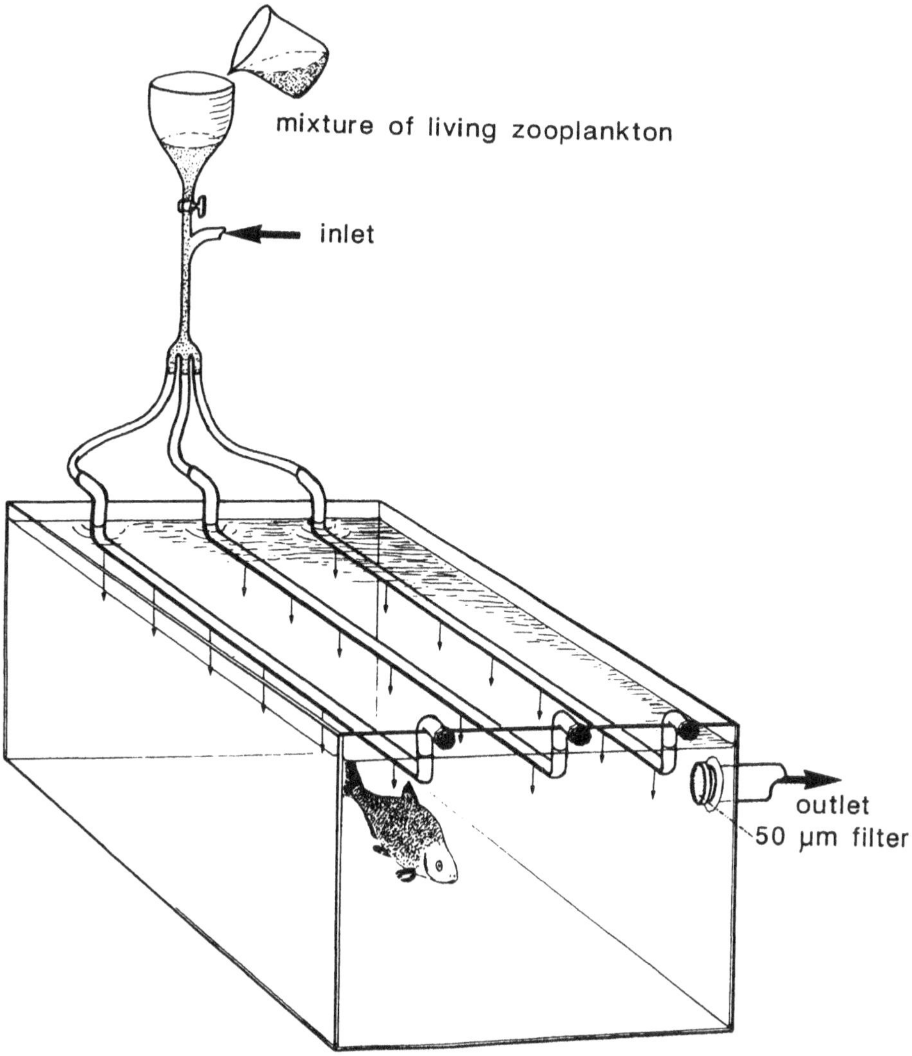

Fig. 1. Aquarium setup. Zooplankton is added to the tank via three PVC tubes through 2 mm holes. Prey density can be increased to a known level by adding zooplankton through the funnel; a continuous water current prevents conditioning of the fish to the added zooplankton.

23.3, 28.5, and 33.8 cm S.L.) Differences between maximal adduction and abduction of opercula were measured using enlarged film frames. Maximal adduction is measured using the last film frame prior to the snap, when no head expansion occurred. For maximal abduction the last film frame before the opening of the opercula was used. The difference between these volume changes is assumed to be the volume of water intake during the snap.

The fish were trained for several weeks to feed under light conditions generated by six 500 W lamps, necessary for high speed filming. A Teledyne D.B.M. 54 camera, with Angenieux zoom objective 10 × 12A, was used at film speeds of 48, 100 or 200 frames s^{-1}, on Kodak Double X negative film (200 ASA).

Development of the model

The snap frequency of particulate feeders is assumed to be determined by the encounter rate of

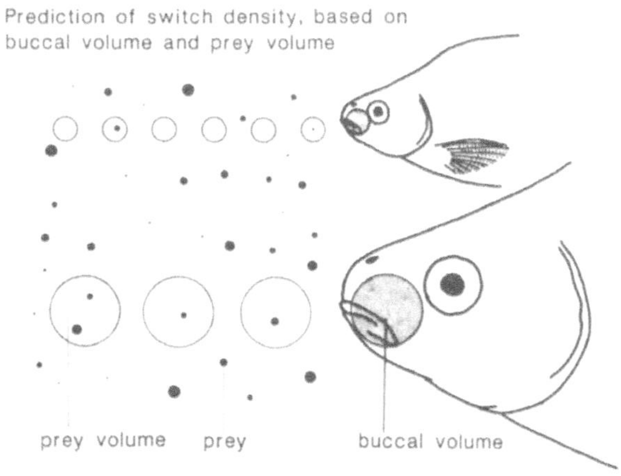

Fig. 2. Relation between buccal volume (cross-hatched circle), volume of water processed (open circle) and possible feeding mode. The buccal volume of small fish (top) is too small for them to engulf one prey at each random snap and therefore particulate-feeding will be most likely. Large bream (bottom) can ingest at least one prey per random snap and filter-feeding is most likely.

the zooplankton, and thus predominantly by swimming speed and zooplankton density, when handling time and visibility are not limiting. However, when each random snap contains at least one item of prey, the fish is assumed to switch to filter-feeding with a maximal snapping frequency which will be constant when handling is not limiting.

Particulate feeding. – The model assumes that the encounter rate is largely determined by the average distance between prey animals and by the swimming speed of the fish. Snap frequency can be predicted if prey are visible and randomly distributed, and if the fish finds the optimal swimming route between prey animals, snapping at every non-evasive prey encountered.

The mean inter-prey distance (cm) is calculated from the zooplankton density (d in $N\,l^{-1}$). The average volume around each prey item is considered to be a cube with a side length of A cm:

$$A^3 = 1000 * d^{-1} \qquad \text{(ml)}.$$

Each cube is surrounded by 26 similar cubes and therefore the average inter-prey distance D is

$$D = (6A + 8A\sqrt{3} + 12A\sqrt{2}) * (26^{-1}) \qquad \text{(cm)}.$$

Swimming from prey to prey at a mean speed of V $cm\,s^{-1}$ the fish will encounter $V*D^{-1}$ prey items per second. If every encountered prey is ingested, then:

$$\text{mean snap frequency} = V*D^{-1} \qquad (\text{snaps}\ s^{-1}).$$

Filter feeding. – It is assumed that if each random snap is successful, that is, when the average volume surrounding each prey is equal to or smaller than the buccal volume, snap frequency will be maximal (Fig. 2). When zooplankton density is much higher than 1 per buccal volume the time needed for intra-oral handling of the prey may increase as well, slowing down the actual snap frequency.

Results

Observed snap frequencies

Maximal snap frequencies (SF) varied from $0.5\,s^{-1}$ in 7.5 cm bream up to $2\,s^{-1}$ in 24 cm bream. In the small length classes SF increased in proportion to zooplankton density, whereas in the large length classes maximum SF was reached at a zooplankton density of $<100\,l^{-1}$, remaining constant or gradually decreasing thereafter (Fig. 3). In large bream the variation of SF at zooplankton densities $>50\,l^{-1}$ was lower than in small bream. When foraging, the smaller fish swam at relatively low speeds of 1–2 $cm\,s^{-1}$; they frequently stopped feeding and rested at the bottom of the tank or displayed bursts of activity. The larger fish swam at 5–7 $cm\,s^{-1}$ and in a much more regular manner. Only the larger bream (24 and 29.5 cm) had a buccal volume sufficiently large (9.5–16 ml, Fig. 4) to catch prey with random snaps at zooplankton densities $\geqslant 100\,l^{-1}$, and were usually filter-feeding. In large bream, particulate-feeding was observed only when large and conspicuous *Daphnia pulex* were present at low densities. However, the buccal volume of the smaller bream (0.3–2 ml, Fig. 4) allowed filtering only at densities $\geqslant 500\,l^{-1}$. Due to the more irregular feeding behaviour of small bream it was often difficult to observe which feeding mode was used. Usually, particulate-feeding fish made small up-

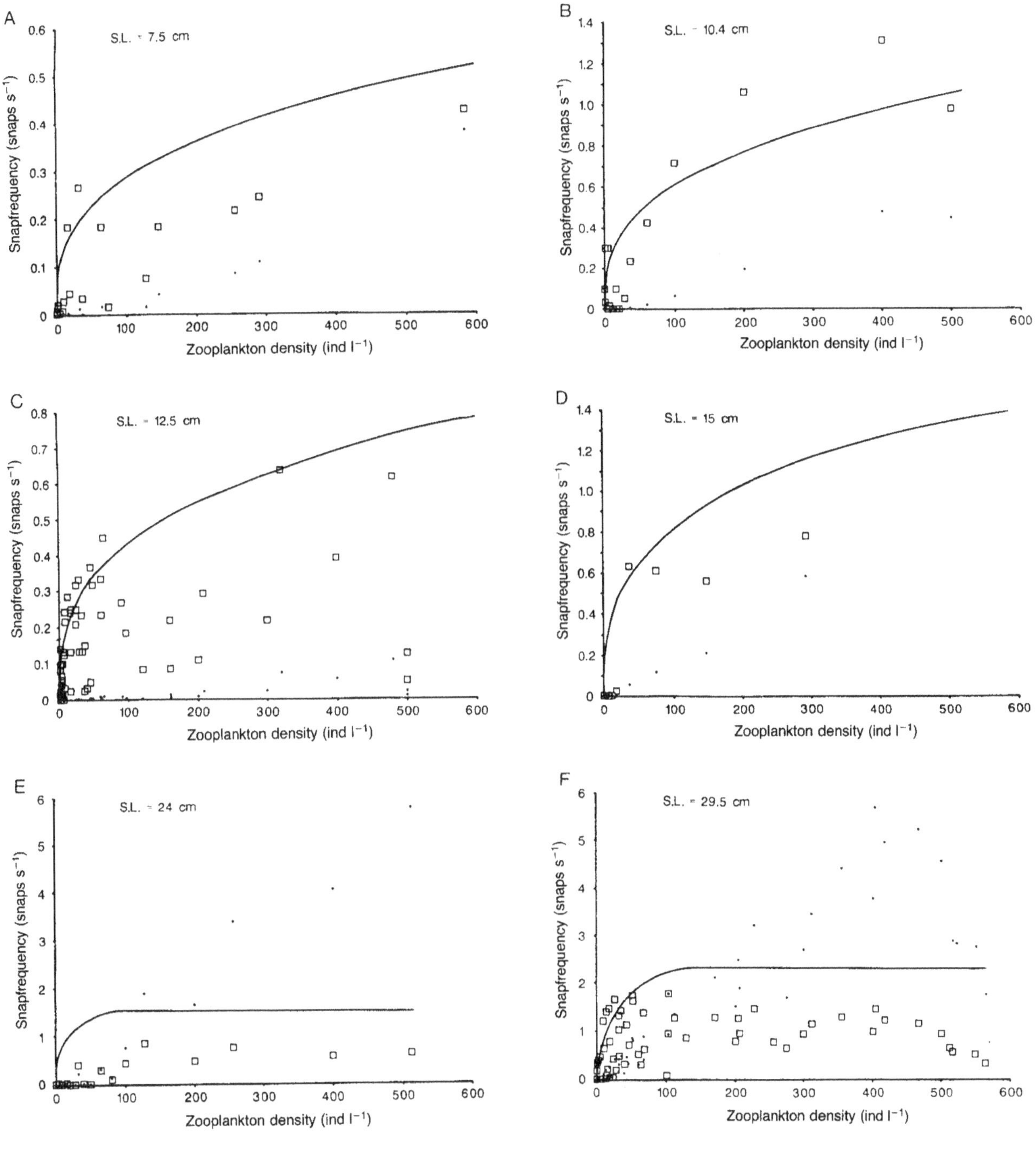

Fig. 3. Observed (open squares) and predicted (line) snap frequencies of six size-classes of bream (S.L. = standard length), at different *Daphnia* densities. Also indicated is the calculated number of prey items caught (dots) assuming strictly random snaps. The plateau of the curve in E and F indicates filter-feeding, with density-independent snapping rate.

ward movements with the head, which filter-feeding fish did not do.

In the 8 hour experiment, the 24 cm bream foraged at a high and constant SF (1.27 ± 0.33 snaps s^{-1}; $n = 21$ observations) for 7 hours. This was not significantly different from the values found in the

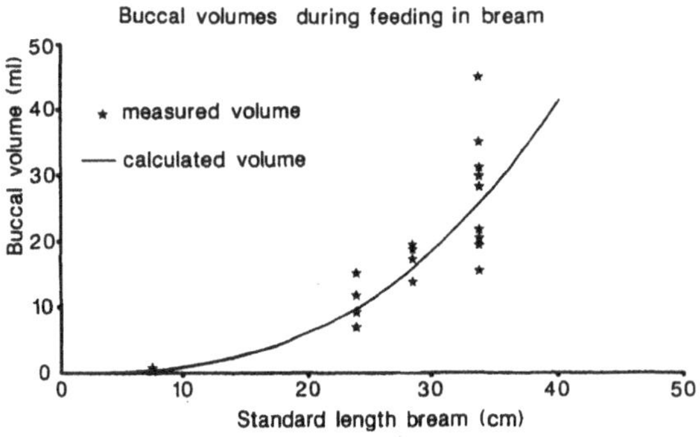

Fig. 4. Buccal volume as measured during feeding in several size-classes of bream and calculated for all bream < 40 cm S.L.

switching experiment for 24 cm bream at zooplankton densities of 300–400 l^{-1} (Fig. 3E). Thereafter the feeding rate dropped distinctly (0.46 ± 0.23 snap s^{-1}; n = 6) and the fish seemed satiated.

In complete darkness, one large bream (29.5 cm) reduced the zooplankton density from 262 to 158 l^{-1} in two hours in a 100 l tank (Fig. 5B), whereas three small bream (mean length 11.7 cm), under identical conditions, were not as efficient (Fig. 5A).

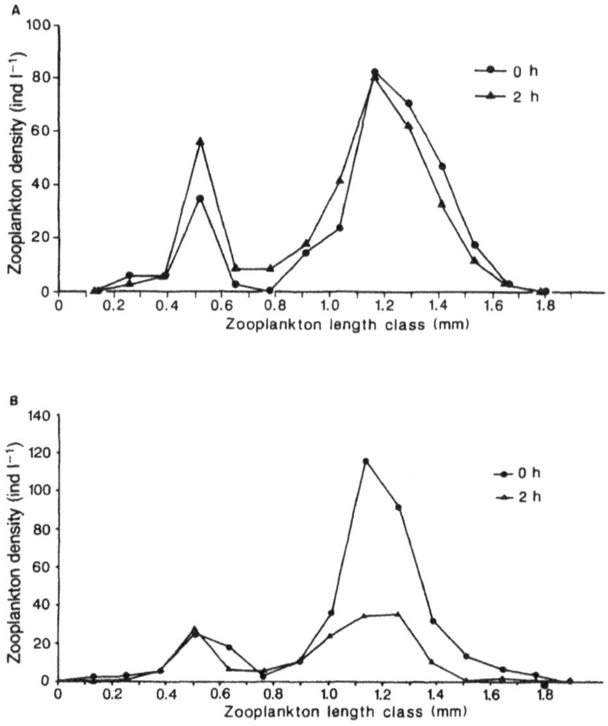

Fig. 5. Size distribution and density of zooplankton at the beginning and end of a 2 hour feeding experiment in total darkness with small (A) and large (B) bream.

Predicted snap frequencies

The average distance between prey objects, the swimming speed, and the buccal volume of the fish were the most important variables of the model, whereas other possible variables were kept constant. The model predicted that large (24 and 29.5 cm) bream would switch to filter feeding at ca. 100 and 60 zooplankters l^{-1}, beyond which density they would maintain a constant or even decreasing SF, whereas smaller bream would not switch at all. The general shape of the model predictions corresponded quite well with the observed values, but the predicted SF was almost twice that actually observed (Fig. 3). The SF of the filter-feeding mode was derived from the maximum SF of the particulate-feeding mode. When the distance between prey objects is so small that only handling time but not swimming speed is limiting, SF will remain constant or decrease. This was confirmed by the observed values. It seemed, however, that in large bream particulate-feeding was restricted to very low densities (24 cm bream) or was not employed at all (29.5 cm). For the small bream the model predicted filter-feeding at zooplankton densities > 500 l^{-1}, the feeding experiments giving no evidence for filter-feeding at lower zooplankton densities. The model predictions corresponded to the maximum SF observed. Particularly the 7.5 cm bream showed large deviations at high zooplankton densities.

Bream of 11.7 cm S.L. did not feed in darkness, whereas 29.5 cm bream did (Fig. 5). If the small bream were able to forage as particulate feeders their SF would have been ca. 1 s^{-1}, corresponding to 7200*3 snaps in 2 hours and a decrease of zooplankton density by about 216 l^{-1}. With the filter-feeding mode, three random snaps were needed for one prey which would have led to a decrease of 72 l^{-1} with the same SF, and to 142 l^{-1} with the theoretically derived SF. Large bream did forage successfully in darkness, with a somewhat higher SF than was found in the switching experiment.

Discussion

Bream are able to feed on zooplankton during their whole lifespan (0.5 to 50 cm S.L.). Selectivity and feeding mode change in relation to the size of the fish and the density and size composition of the zooplankton (Lammens 1985, Lammens et al. 1987). Eutrophication in Dutch inland waters has stimulated the development of zooplankton up to densities of several hundreds per litre and has made zooplanktivorous feeding important for the formerly benthivorous fish (Löffler 1984, Lammens 1986). The branchial sieve plays an important part in filter-feeding of larger bream (Hoogenboezem et al. 1990, Hoogenboezem et al. 1991. We have tried to model the switchover from particulate- to filter-feeding and to determine the feeding intensity in relation to fish size and zooplankton density. Our feeding model differs from other feeding models as it describes two alternative feeding modes and the intensity of feeding, taking into account the size of the fish; however, it ignores feeding selectivity, which is the most important part of most feeding models (Werner & Hall 1974, Eggers 1977, O'Brien et al. 1976, Confer & Blades 1975, Wright & O'Brien 1984).

An important variable in our model is the volume of the buccal cavity. Most fish have to suck in some water surrounding the prey and this amount of water corresponds to, or is somewhat larger than, the volume created by head expansion (van Leeuwen 1984, Drost & van den Boogaart 1986). Our hypothesis was that if the average volume of water surrounding one individual prey is less than the buccal volume created by head expansion, the most profitable feeding mode will be by random snaps. At a density of 1 prey l^{-1} a 15 cm bream, in order to catch one prey object, can choose between 50 random snaps as a filter-feeder and 1 snap as a particulate-feeder. In this case the choice is quite easy. However, with increasing density it will become more profitable at some point to begin filter-feeding. In our model we used an arbitrary switching point of 1 prey per 1 snap, because we ignored the cost of searching at these high zooplankton densities. This proportion may be 1 to 2 or 1 to 3, if the cost of searching is higher than expected or if the cost of snapping is low. However, for bream exceeding 20 cm in length the prey is so small that possibly the fish has problems locating it. The 29.5 cm bream did not, or only sporadically, feed at low zooplankton densities (fig. 3F), although it ought to be rewarding to be a particulate-feeder under these conditions. The 24 cm bream approached its maximum SF at a zooplankton density of ca. 30 l^{-1}, which means that in the filter-feeding mode it takes 3 snaps for one prey. It is possible that for these fish particulate-feeding on small particles creates more problems than filter-feeding so that they prefer to do 3 blind snaps for 1 prey, instead of 1 'particulate' snap.

Although we realised that prey handling might be a limiting factor it could not be quantified. The handling of the prey occurs during 'closed protrusion movements' when the prey collected in the branchial sieve is detached and swallowed (Sibbing 1988). The closed protrusion movements are hard to quantify as they are difficult to distinguish from normal feeding movements when the mouth is opened and the upper jaw protruded. During benthic feeding these closed protrusion movements are clearly visible and occupy a large part of foraging time (Lammens et al. 1987) Ignoring of the handling time was probably largely responsible for the exaggerated model predictions. Another point regarding the buccal cavity is that the calculated amount of water sucked in is a minimal estimate and that the real value may be higher because some water can flow through the buccal cavity when mouth and opercula are still open, especially during swimming (Muller & Osse 1984). In foraging trout the amount of water passing through the mouth cavity with each strike far exceeded the volume of the buccal cavity (van Leeuwen 1984). This implies that the real amount of water moved is probably higher, and therefore filter-feeding can start at a lower density, than predicted from the calculated buccal volume.

The model also predicted a higher SF for the particulate-feeding mode than that observed. The model assumed a more or less ideal fish which sees all prey objects, forages continuously, and is not limited by satiation or resting pauses. We assumed that swimming speed was constant during foraging and also that the swimming speed was not related to zooplankton density or other factors. However, a fish does not swim from prey to prey but follows a more or less straight line, so that the inter-prey distance experienced by the fish will be greater than the actual distance. At these high zooplankton densities it is probably much more profitable to swim in a straight line and to snap any prey encountered than to change direction continuously in order to take the shortest distance between prey items. At densities from 10 to 500 l^{-1} the distance between prey items varies from 5 to 0.5 cm, which is much less than the length of the fish. As it seems almost impossible to determine the real encounter rate we therefore assumed an ideal fish finding an ideal swimming route. The swimming speed of *D. hyalina* and *D. pulex* is negligible compared to that of fish. Drenner et al. (1978) showed that cladocerans, contrary to copepods and *Chaoborus* larvae, do not have strong evasive abilities. Thus our model, which does not take into account evasive behavior of the prey, can only be used to predict the interactions between fish and cladocerans.

The main variables of our model allow reasonably good qualitative predictions of the switching behavior in relation to the size of fish and the density of food organisms. The model is certainly not perfect, some variables, such as handling time, the actual distance between prey, and the real amount of water passing the buccal volume per snap, have been ignored. By comparing observations with predicted SF we became more aware of the possible importance of these variables. If it were possible to determine the real amount of water per snap, the switching point could be estimated more accurately, and if it were possible to estimate the 'real' distance between prey the prediction of the number of snaps would also be more accurate.

Acknowledgements

We thank Koos Vijverberg, Nand Sibbing, Ramesh Gulati, Onno van Tongeren, and Wolfgang Wieser for their critical remarks on the manuscript. We are also grateful to Peter McGillavry and Bert van den Bergh for their assistance during the feeding experiments and to Jos van den Boogaart, who prepared the figures. The investigation was supported by the Foundation for Fundamental Biological Research (BION), subsidized by The Netherlands Organization for Scientific Research (NWO), project 811-437-241B.

References cited

Confer, J.L. & P.I. Blades. 1975. Omnivorous zooplankton and planktivorous fish. Limnol. Oceanogr. 20: 571–579.

Crowder, L.B. & F.P. Binkowski. 1983. Foraging behaviors and the interaction of alewife, *Alosa pseudoharengus* and bloater, *Coregonus hoyi*. Env. Biol. Fish. 8: 105–113.

Crowder, L.B. 1985. Optimal foraging and feeding mode shifts in fishes. Env. Biol. Fish. 12: 57–62.

Drenner, R.W., J.R. Strickler & W.J. O'Brien. 1978. Capture probability: the role of zooplankter escape in the selective feeding of zooplanktivorous fish. J. Fish. Res. Board. Can. 35: 1370–1373.

Drost, M.R. & J.G.M. van den Boogaart. 1986. A simple method for measuring the changing volume of small biological objects, illustrated by studies of suction feeding by fish larvae and of shrinkage due to histological fixation. J. Zool., Lond. (A) 209: 239–249.

Eggers, D.M. 1977. The nature of prey selection by planktivorous fish. Ecology 58: 46–59.

Ehlinger, T.J. 1989. Foraging modes switches in the golden shiner (*Notemigonus crysoleucas*). Can. J. Fish. Aquat. Sci. 46: 1250–1254.

Gibson, R.N. & I.A. Ezzi. 1985. Effect of particle concentration on filter- and particulate-feeding in herring *Clupea harengus*. Mar. Biol. 88, 109–116.

Holanov, S.H. & J.C. Tash. 1978. Particulate and filter feeding in threadfin shad, *Dorosoma petense*, at different light intensities. J. Fish Biol. 13: 619–625.

Hoogenboezem, W., F.A. Sibbing, J.W.M. Osse, J.G.M. van den Boogaart, E.H.R.R. Lammens & A. Terlouw. 1990. X-ray measurements of gill-arch movements in filter-feeding bream, *Abramis brama* (Cyprinidae). J. Fish Biol. 36: 47–58.

Hoogenboezem, W., J.G.M. van den Boogaart, F.A. Sibbing, E.H.R.R. Lammens, A. Terlouw & J.W.M. Osse. 1991. A new model of particle retention and branchial sieve adjustment in filter-feeding bream (*Abramis brama* (L.) Cyprinidae). Can J. Fish. Aquat. Sci. (in press).

Jansen, J. 1976. Feeding modes and prey selection in alewife (*Alosa pseudoharengus*). J. Fish. Res. Board. Can. 33: 1972–1975.

Lammens, E.H.R.R. 1985. A test of a model for planktivorous filter-feeding by bream, *Abramis brama*. Env. Biol. Fish. 13: 288–296.

Lammens, E.H.R.R. 1986. Interactions between fishes and the structure of fish communities in Dutch shallow, eutrophic lakes. Ph. D. Thesis, Agricultural University, Wageningen. 100 pp.

Lammens, E.H.R.R. & W. Hoogenboezem. 1991. Diets and feeding behaviour. pp. 353–376. *In*: I.J. Winfield & J.S. Nelson (ed.) Cyprinid Fishes: Systematics, Biology and Exploitation, Chapman & Hall, London.

Lammens, E.H.R.R., J. Geursen & P.J. McGillavry. 1987. Diet shifts, feeding efficiency and coexistence of bream (*Abramis brama*), roach (*Rutilus rutilus*), and white bream (*Blicca bjoerkna*) in hypertrophic lakes. Proc. V. Cong. Eur. Ichthyol. 153–162.

Lazzaro, X. 1987. A review of planktivorous fishes: their evolution, feeding behaviours, selectivities and impacts. Hydrobiologia 146: 97–167.

Leeuwen, J.L. van. 1984. A quantitative study of flow in prey capture by rainbow trout *Salmo gairdneri*, with general considerations of the actinopterygian feeding mechanism. Trans Zool. Soc., Lond. 37: 171–227.

Löffler, H. 1984. Zur ökologie des Brachsen (*Abramis brama* (L.)) im Bodensee. Schweiz. Z. Hydrol. 46(1) 147–162.

Muller, M. & J.W.M. Osse. 1984. Hydrodynamics of suction feeding in fish. Trans. Zool. Soc. Lond. 37: 51–135.

O'Brien, W.J., N.A. Slade & G.L. Vinyard. 1976. Apparent size as the determinant of prey selection by bluegill sunfish (*Lepomis macrochirus*). Ecology 57: 1304–1310.

O'Connell, C.P. 1972. The interrelation of biting and filtering in the feeding activity of the northern anchovy (*Engraulis mordax*). J. Fish. Res. Board Can. 29: 285–293.

Pyke, G.H. 1984. Optimal foraging theory: a critical review. Ann. Rev. Ecol. Syst. 15: 523–75.

Sibbing, F.A. 1988. Specializations and limitations in the utilization of food resources by the carp. *Cyprinus carpio*: a study of oral processing. Env. Biol. Fish. 22: 161–178.

Sibbing, F.A. 1991. Food capture and oral processing. pp. 377–412. *In*: I.J. Winfield & J.S. Nelson (ed.) Cyprinid Fishes: Systematics, Biology and Exploitation, Chapman & Hall, London.

Werner, E.E. & D.J. Hall. 1974. Optimal foraging and the size selection of prey the bluegill sunfish (*Lepomis macrochirus*). Ecology 55: 1042–1052.

Werner, E.E. & D.J. Hall. 1988. Ontogenetic habitat shifts in blue-gill: the foraging rate – predation risk trade-off. Ecology 69: 1352–1366.

Wright, D.I. & W.J. O'Brien. 1984. The development and field test of a tactical model of the planktivorous feeding of white crappie (*Pomoxis annularis*). Ecol. Monogr. 54: 65–98.

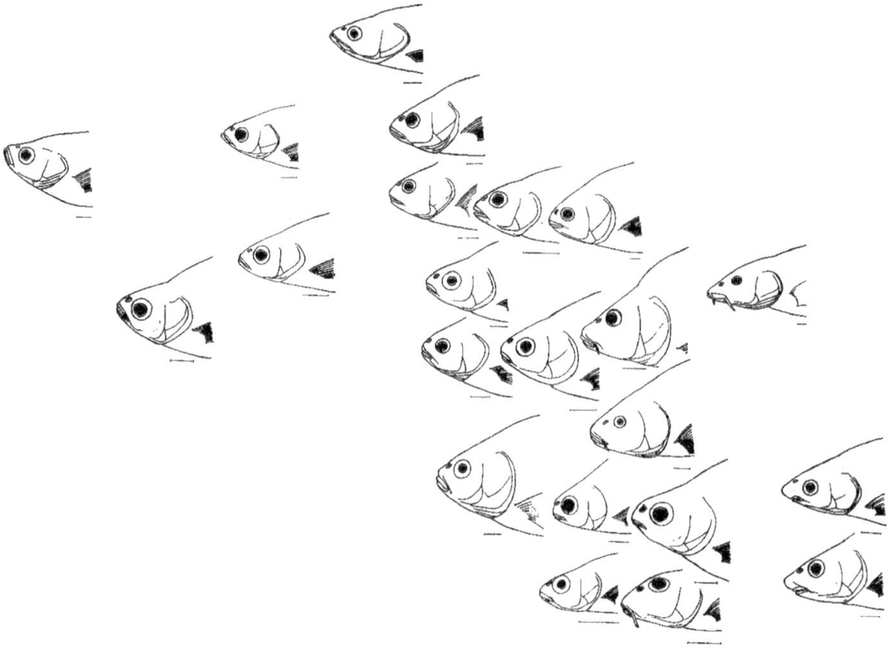

Environmental Biology of Fishes **33**: 23–32, 1992.

Ontogeny of prey attack behaviour in larvae and juveniles of three European cyprinids

Josef Wanzenböck
Institute of Zoology, University of Vienna, Althanstrasse 14, A-1090 Wien, Austria

Received 15.1.1990 Accepted 25.9.1990

Key words: Optimal foraging, Handling time, Pursuit time, Capture efficiency, *Rutilus rutilus, Alburnus alburnus, Abramis ballerus*

Synopsis

The ontogenetic change in time costs of prey attacks as well as the change in capture efficiency for representative cladoceran and cyclopoid prey was investigated in roach, *Rutilus rutilus,* bleak, *Alburnus alburnus,* and blue bream, *Abramis ballerus.* Video recordings were used for measuring the timing of attacks, whereas capture efficiencies were determined by direct observation. Decreases in the time cost of attacks reflect the decreasing importance of prey fixation during growth of the fish. No differences in capture efficiencies were found among the three cyprinid species, indicating that attack behaviour is unlikely to function as a basic mechanism leading to differences in prey selectivity among the investigated species. Increasing capture efficiency during early development may lead to increasing selectivity for cyclopoid prey in the field.

Introduction

Zooplanktivorous fish became favoured subjects of foraging models after their impact on zooplankton (Hrbáček et al. 1961) and on whole lake ecosystems has been described (see Lazzaro 1987 & Northcote 1988 for reviews). In order to explain the phenomenon of size selectivity several authors proposed optimal foraging models which implied that fish predators maximize their net energy gain by using a size selective strategy (Confer et al. 1978, Werner & Hall 1974, O'Brien et al. 1976). To test and improve those models a detailed knowledge of the prey capture process is essential. The present work intends to contribute to this knowledge by analysing the prey attack behaviour of planktivorous fish larvae and juveniles.

Prey capture behaviour of fish larvae and juveniles was originally investigated to determine critical food densities for the recruitment of economically important freshwater (Braum 1964) and marine (Rosenthal & Hempel 1970, Blaxter & Staines 1971, Houde & Schekter 1980) species. Modelling of feeding rates requires information on prey localization capability, time cost of prey handling, and capture success for various prey types. All these components are highly dynamic because of the rapid morphological and behavioural changes in developing fish larvae (Balon 1985). This paper concentrates on the development of the time cost of prey attacks as well as on the dynamics of capture success for representative cladoceran and cyclopoid prey in 0+ cyprinids. Quantitative data on the time cost of prey attacks are rare in the literature and completely lacking for the ontogeny of larvae and juvenile fish. The aim of this study was thus to give a quantitative description of the changes of time cost and its influence on capture efficiencies.

Available data on prey capture efficiencies of fish larvae are reviewed in Drost (1987). In many studies prey types are not specified and capture success is only investigated over a short period of larval development (maximally until two weeks after first feeding). In the present study two well defined prey types and a longer time period (until three months after hatching) were chosen. Differences in capture efficiency between species might provide a basic mechanism for differences in prey selection (Winfield et al. 1983). This paper expands the information of a previous study on the ontogeny of prey detection capacities (Wanzenböck & Schiemer 1989).

0 + cyprinids were chosen because of their ecological importance in many freshwater ecosystems. Young-of-the-year are a major factor affecting abundance patterns of littoral zooplankton (Whiteside 1988) and may even influence reproductive patterns of adult fish (Cryer et al. 1986). Roach, *Rutilus rutilus* (L.), and bleak, *Alburnus alburnus* (L.) were chosen because of their wide distribution and their dominance in many European water bodies (Lelek 1987). They commonly occur together, often at high abundances, and show distinct differences in feeding biology as adults (Prejs 1976). The roach can be characterized as an omnivorous fish with a wide food spectrum including detritus, plant material, zoobenthos, and zooplankton. The bleak is more specialized in surface and zooplankton feeding. Nevertheless, the larvae and juveniles of both species feed almost exclusively on zooplankton (Cryer et al. 1986, Mark et al. 1987, Keckeis et al. 1988, Schiemer et al. 1989). It was hypothesized that possible differences in prey capture behaviour may lead to niche separation in 0 + cyprinids during early ontogeny. In order to test this hypothesis more extensively the blue bream, *Abramis ballerus* (L.), a zooplankton feeder in the larval and juvenile period (Panova 1966) as well as in the adult period (Schiemer & Spindler 1989), was included in the interspecific comparison.

Materials and methods

Fish larvae were obtained by laboratory fertilization of eggs from adults from a Danube backwater near Vienna. Stocks of fish larvae and juveniles were kept in circular pools (60 l) with constant water flow and 16 h light period. Food, consisting of natural zooplankton and *Artemia* sp. nauplii, was provided in excess.

Time analysis of prey attack behaviour was carried out using video recordings. For a detailed description of the experimental setup, video system, and preparation of the experimental fish, see Wanzenböck & Schiemer (1989). The video recordings showed the position of both the fish's snout and the prey with an accuracy of ± 1 mm. The minimum measureable time interval was 0.02 second(s) (i.e. one video field). Pursuit time was defined as the interval from first reaction (determined by a sudden change in swimming direction and/or by sudden acceleration) until the fish snapped at the prey (i.e., when the prey entered the fish's mouth and was no longer visible). 'Fixation' was defined as an interruption of an approach in front of the prey and maintenance of position for at least three consecutive frames (0.06 s). Fixation occurred 2 to 4 mm in front of the prey. The interval between first reaction and beginning of fixation was defined as 'approach time', basically equivalent to the time required to overcome the reactive distance. The interval between the beginning of fixation and snapping was termed 'fixation time'. The average approach speed was calculated by dividing reactive distance by approach time. Twenty eight attacks of large bleak were analysed in greater detail. The position of the fish during search, approach, and after prey capture was determined on every video field to yield data on swimming speeds on a finer time scale. To calculate mean values of the swimming speeds data were pooled on a relative time scale because pursuit times of the 28 attacks were different.

The prey used in these experiments was *Ceriodaphnia reticulata* Sars and *Eucyclops serrulatus* Sars from laboratory cultures. Two distinct size classes were produced by sequential sieving through a set of sieves (500, 300, and 200 μm). The two size

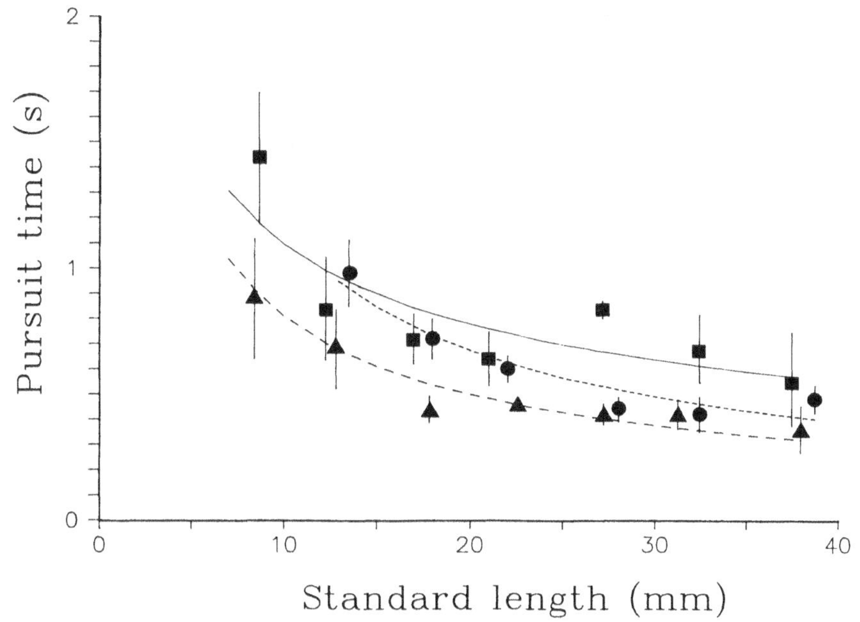

Fig. 1. Pursuit time versus standard length of fish. Symbols represent medians within each size category, vertical bars indicate ± 2 standard error (SE). Species indicated: Squares = roach, circles = blue bream, triangles = bleak. Curves are drawn according to the following functions derived from log-transformed data: roach *Rutilus rutilus* (solid line) $y = 3.43 \times^{-0.5}$, n = 567; blue bream *Abramis ballerus* (short-dashed line) $y = 7.13 \times^{-0.8}$, n = 606; and bleak *Alburnus alburnus* (long-dashed line) $y = 4.05 \times^{-0.7}$, n = 536.

classes had a mean length of 0.49 mm (± 0.053 SD, n = 100) and 0.65 mm (± 0.109 SD, n = 100), respectively.

Capture efficiency experiments were carried out in white plastic vessels. Vessel dimensions, water volumes, and prey densities varied with fish size (Table 1). Light was provided by two fluorescent tubes (18 W each) mounted 1 m above the vessels. Light intensity at the water surface was 1200 lux. At the beginning of an experiment three fish (starved for 24 h) were introduced and allowed to acclimate for 30 min. The appropriate number of prey was then added and the behaviour of one randomly choosen fish observed and monitored with the help of a tape recorder. Capture efficiency was expressed as the percentage of snaps that were successful. Only attacks in which prey were at least 1 cm off the side walls or the bottom of the vessels were included. Prey were either *Ceriodaphnia reticulata* or *Eucyclops serrulatus* of the same size classes as described above.

Experiments were carried out at weekly intervals during the first three months after hatching. Water temperature was 20° C (± 1° C). All experiments were done between 0900 and 1700 h.

Table 1. Dimensions of experimental tanks, water volumes, water depth (for both types of experiments) and prey densities (in experiments for the determination of capture efficiencies) used with different-sized fish (all three species).

Fish		Tanks			
Age (d)	Standard length (mm)	Diameter (cm)	Volume (ml)	Depth (cm)	Density (ind.l^{-1})
10–16	8–12	15	150	0.8	500
17–44	12–20	20	250	0.9	500
45–60	20–25	25	1000	2.0	100
61–90	25–30	30	3000	4.2	50

Results

Intraspecific ontogenetic changes

Total pursuit times obtained with different prey

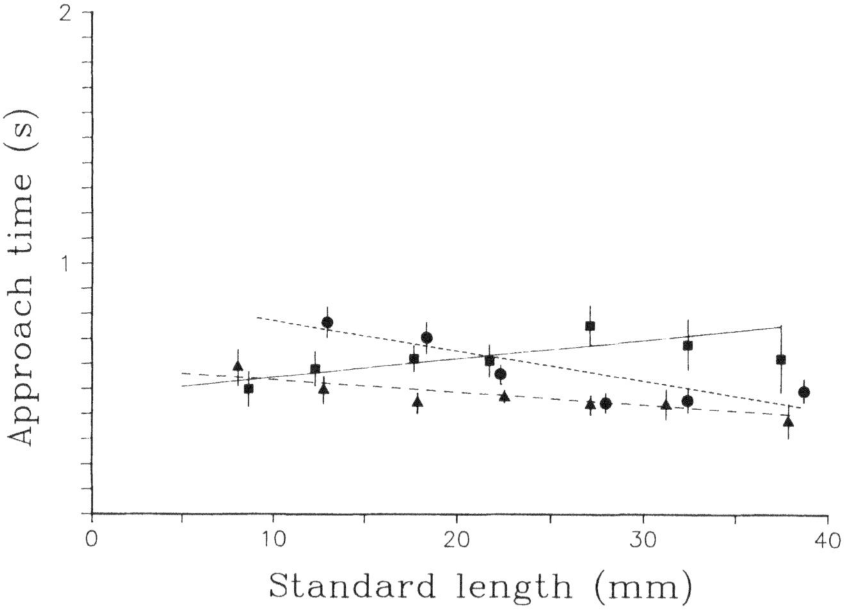

Fig. 2. Approach times versus standard length. Symbols represent mean values within each size class, vertical bars indicate ± 2SE. Species indicated as in Figure 1. Regression equations: roach (solid line) $y = 0.47 + 0.007\times$, $n = 570$; blue bream (short-dashed line) $y = 0.89 - 0.012\times$, $n = 595$; and bleak (long-dashed line) $y = 0.59 - 0.005\times$, $n = 535$.

size classes were pooled because no significant differences were found within each length class of fish (Mann-Whitney U-test, $p > 0.05$). The relationship between total pursuit time and fish standard length can be described by negative power functions in all three species (Fig. 1). The data were logarithmically transformed and linearity was tested (Table 2) by analysis of variance (Linder & Berchtold 1982).

One component of total pursuit time is approach time. No significant differences were found in approach time with different prey sizes (Mann-Whitney U-test, $p > 0.05$). Therefore data obtained with different prey sizes were also pooled. The relationship between approach time and standard length can be described by linear models (Fig. 2, Table 2). Although the slopes of the regression lines were significantly different from zero, differences in approach time were minimal.

The other component of total pursuit time is fixation time. Fixation was apparent in more than 70% of all attacks in the smallest fish size class (Fig. 3). This percentage dropped dramatically with fish size in all species (regression analysis of logarithmically transformed data, F-test, $p < 0.05$). Fixa-

Table 2. F-statistics for analysis of variance (linearity) and of covariance (comparisons of slopes and intercepts) of regression models relating different time values to standard length of fish. Slopes and intercepts are compared between species. Values above the diagonal refer to slopes, below the diagonal to intercepts. Values marked ** = significant ($p < 0.05$), n.s. = not significant ($p > 0.05$).

log pursuit times versus log standard length (Fig. 1):			
	Roach	Blue bream	Bleak
Linearity (F):	1.5**	1.9**	0.9**
Roach		3.6**	2.4**
Blue bream	4.4**		1.2 n.s.
Bleak	13.6**	10.5**	
Approach time versus standard length (Fig. 2):			
	Roach	Blue bream	Bleak
Linearity (F):	1.4**	2.0**	1.9**
Roach		7.8**	4.7**
Blue bream	3.3**		0.36 n.s.
Bleak	8.5**	10.5**	
log fixation time versus log standard length (Fig. 4):			
	Roach	Blue bream	Bleak
Linearity (F):	1.9**	1.9**	0.3**
Roach		0.02 n.s.	0.56 n.s.
Blue bream	2.2 n.s.		0.54 n.s.
Bleak	1.7 n.s.	0.39 n.s.	

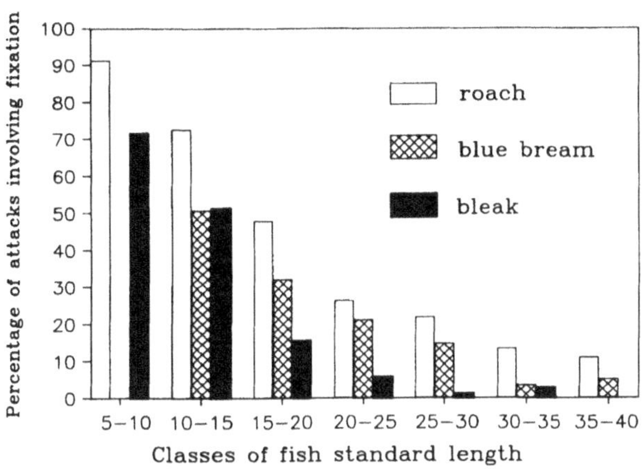

Fig. 3. Frequency attacks containing fixation periods within each standard length class of fish.

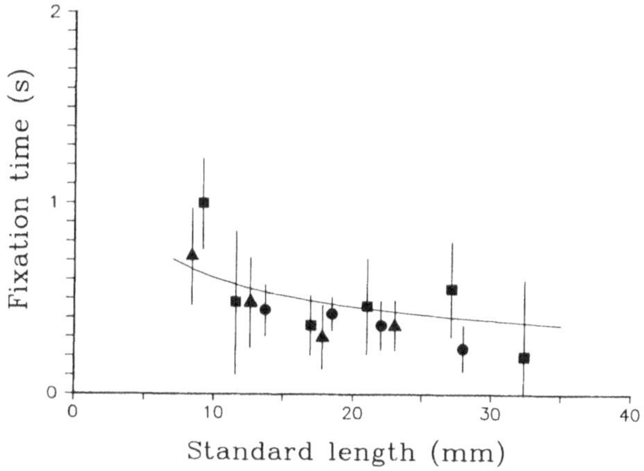

Fig. 4. Fixation time versus standard length of fish. Symbols represent medians, vertical bars indicate ± SE. Species indicated as in Figure 1. The curve derived from log-transformed data is given by: $y = 158 \times^{-0.42}$, $n = 515$.

tion time also decreased with fish size (Fig. 4). Prey size had no effect on fixation time. Comparisons within each size class did not reveal significant differences (Mann-Whitney U-test, $p > 0.05$). Therefore data obtained with the two prey sizes were pooled. The relationship between fixation time and fish standard length was described by a negative power function (Table 2).

Average swimming speed during approach was calculated from reactive distance and approach time. The correlation between approach speed and standard length was described by linear models (Fig. 5). For a given fish species the regression lines for the two prey size classes were parallel, i.e., slopes were not statistically different (analysis of covariance, Linder & Berchtold 1982, Table 3). However, the intercepts were different.

A more detailed analysis showed that swimming speed during approach was not constant. This is exemplified in Figure 6, which demonstrates swimming speed of large bleak (35–40 mm SL) during attack. Immediately after prey detection, swimming speed was highest, slowing down markedly during approach (regression analysis, F-test, $p < 0.05$).

Capture efficiency for *Ceriodaphnia reticulata* was near 100% in all three species (Fig. 7). Pairwise comparisons between size classes did not reveal significant changes in capture efficiency in roach (t-test of arc sin transformed data, $p > 0.05$). In blue bream the values of the first two size classes were significantly lower than the following ones. In bleak, capture efficiency of the smallest size class was significantly lower than that of other size classes. For *Eucyclops serrulatus* capture efficiency was initially low (10–20%) but increased slowly to 60% at the end of the study period. Pairwise comparisons between size classes revealed significant changes with fish size (t-tests of arc-sin-transformed data, $p < 0.05$)

Interspecific differences

Pursuit time, approach time, approach speed, and percentage of fixation periods differed among the three fish species. Total pursuit time was always lowest in bleak (Fig. 1). Analysis of covariance of log-transformed data showed that the intercepts of the three regression lines were statistically different (Table 2). Pursuit time of blue bream differed from that of roach, which showed the longest total pursuit time.

The change in approach time with fish length showed contrasting trends (Fig. 2). There was a slightly increasing trend in roach, whereas in blue bream and bleak the relationships decreased slightly. Although the slopes of the linear models for blue bream and bleak were not statistically differ-

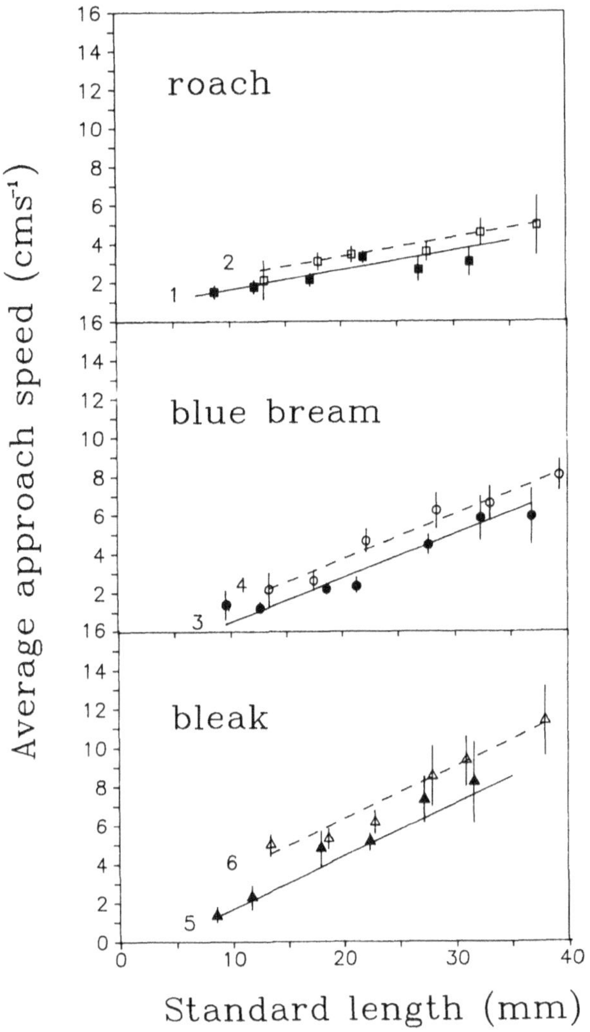

Fig. 5. Average approach speed versus standard length of fish for two size classes of *Ceriodaphnia retilulata* (solid symbols = 0.49 mm, open symbols = 0.65 mm). Symbols represent mean swimming speeds within each size category, vertical bars indicate ± 2SE. Regression equations: roach 1 y = 0.57 + 0.09 ×, n = 338; 2 y = 1.34 + 0.09 ×, n = 222; blue bream 3 y = −1.70 + 0.23 ×, n = 322; 4 y = −0.86 + 0.23 ×, n = 273; bleak 5 y = −1.06 + 0.27 ×, n = 282; 6 y = 0.88 + 0.27 ×, n = 241.

ent (Table 2), the intercepts were. The slope and intercept for roach were distinct from those of the other species. Differences in approach speed were statistically significant but slight.

Interspecific differences were also apparent with regard to fixation. Within each length class, bleak had fewer attacks containing fixations than roach (X^2 test, $p < 0.05$). Blue bream was intermediate and not always statistically different from roach (groups 4, 5, 7) or bleak (groups 2, 6).

No interspecific differences were found with regard to fixation time (Fig. 4). Linear models for the log-transformed data did not differ significantly in slope or in intercept (Table 2). Therefore data were pooled to calculate one regression line.

Average approach speed differed according to fish species: within each prey size class, bleak were faster than roach or blue bream (Fig. 5). Roach showed lowest approach speed with both prey size classes and blue bream was intermediate. Analysis of covariance revealed significant differences in slopes and/or intercepts between the corresponding regression lines (Table 3).

Although capture efficiency varied with fish size no statistical differences were found between fish species for either prey type within each fish length class (t-test of arc-sin-transformed data, $p < 0.05$).

Discussion

Handling time is a major component of foraging models used to explain niche separation in fish (Werner & Hall 1974). According to Eggers (1977)

Table 3. F-statistics for analysis of variance (linearity) and of covariance of regressions relating average approach speed to standard length. Regression numbers refer to Figure 5. Pairs tested: 1–2, 3–4, 5–6 (intraspecific, small prey versus large prey); 1–3, 1–5, 3–5 (interspecific, small prey); 2–4, 2–6, 4–6 (interspecific, large prey). Intercepts are compared for those pairs in which slopes are not significantly different. Values above the diagonal refer to slopes, below the diagonal to intercepts. Values marked: ** = significant ($p < 0.05$); n.s. = not significant ($p > 0.05$).

		Roach		Blue bream		Bleak	
		1	2	3	4	5	6
Linearity (F):		2.1**	0.68**	2.1**	0.69**	2.0**	1.88**
Roach	1		0.45 n.s.	6.9**		8.63**	
	2	4.13**			4.38**		2.4**
Blue	3				0.3 n.s.	3.04**	
bream	4			5.15**			1.0 n.s.
Bleak	5						0.8 n.s.
	6				7.56**	4.08**	

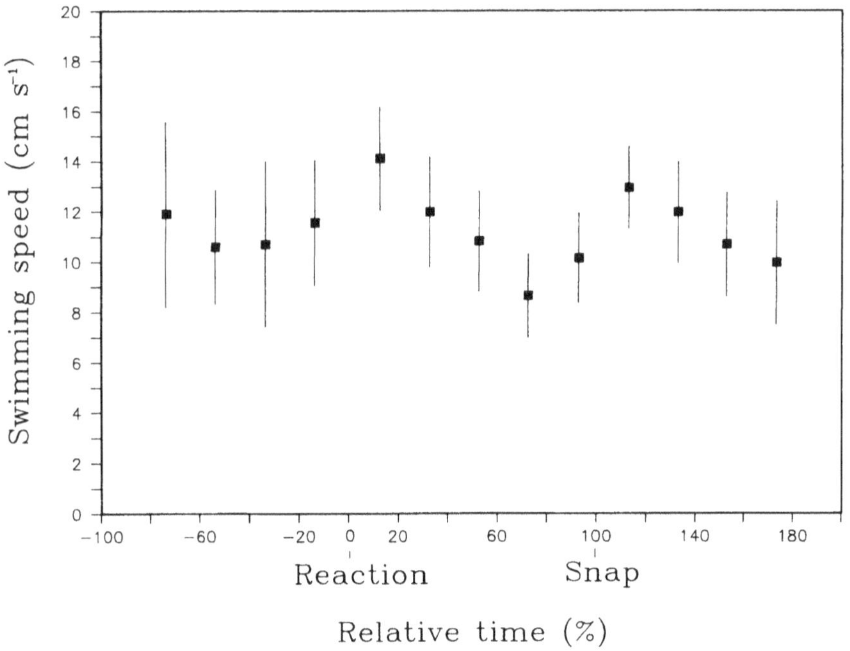

Fig. 6. Swimming speed of bleak (35–40 mm standard length) before, during, and after prey attack. Symbols represent means, vertical bars indicate ± 2SE.

it is defined as the time the predator is engaged in nonsearching activities for each pursuit initiated. This involves time spent in pursuing, capturing, and swallowing the prey. In the literature the term handling time is often reduced to the period required for processing the food item inside the mouth cavity (e.g., Stein et al. 1984). This may suffice in cases when handling time is dominated by this period. However in fish larvae, pursuit time (including fixation time) is of the same order of magnitude as food processing inside the mouth. It has been reported that carp larvae, after sucking in an *Artemia* nauplius, chews it for about one second before the nauplius enters the intestine (Drost & Van den Boogaart 1986). This indicates that, in cyprinid larvae, pursuit time is as long as (or even longer than) the time used for food processing inside the mouth, and that pursuit time is therefore an important part of overall handling time.

The pursuit time and fixation time determined in the present study agree with information on other species. Braum (1964) reported that for coregonid larvae in more than 90% of all pursuits their duration is in the range of 0.5 to 3 s. The time used to fix on a planktonic prey item varies between 1 and 3 s in northern pike, *Esox lucius,* larvae (Drost 1987). The same range of 1 to 3 s for fixation has been described for herring, *Clupea harengus,* larvae by Rosenthal & Hempel (1970) and for white crappie, *Pomoxis annularis,* by Browman (1989).

In any size class average pursuit time is strongly influenced by the proportion of attacks containing a fixation period. Interspecific differences in percentage of fixation periods dominate differences in pursuit time, since fixation time is identical in the three species, and differences in approach time are minimal. Also the marked decrease in pursuit time with length can be attributed to the diminishing percentage of fixation periods. A similar reduction in percentage of fixations has been reported by Brown & Colgan (1985) during the development of foraging behaviour in two centrachid fish. In the centrachids, interspecific differences in the frequency of attacks containing fixations correlated with differences in body size and in developmental stages at hatching. No such correlation was found in the cyprinids which were equal in body size as well as developmental stages at hatching (Wan-

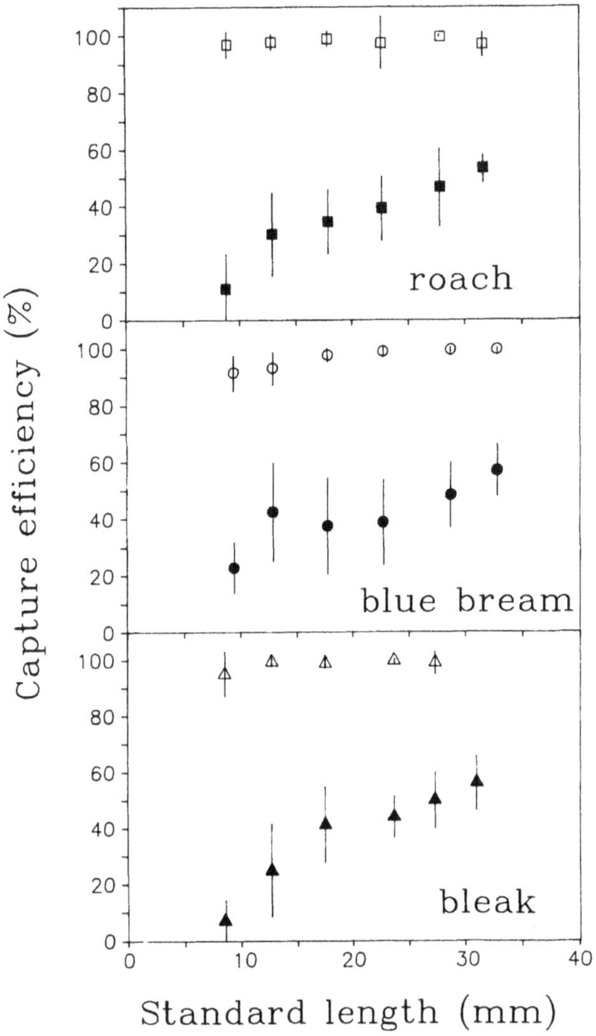

Fig. 7. Capture efficiency for *Ceriodaphnia reticulata* (open symbols) and for *Eucyclops serrulatus* (solid symbols) versus standard length of fish. Symbols represent mean values, vertical bars indicate ± 2SE.

zenböck unpublished data). Lower percentages of fixation periods in bleak correlate with the generally higher swimming speed during prey search (Wanzenböck & Schiemer 1989) as well as with the higher average swimming speed during approach of this species.

Approach speed was not found to have an important effect on approach time. Although the average speed during approach differed between species, this was compensated by differences in reactive distance. For example, bleak showing highest approach speed had the longest reactive distance (Wanzenböck & Schiemer 1989) and therefore needed the same amount of time for approach as the roach with its shorter reactive distance and with lower speed.

Approach speed was found to influence capture success because it determined the time available to the prey for detecting the approaching predator and for initiating escape movements (Nyberg 1971). However, in the present study average approach speed was not indicative of capture success because swimming speed was reduced during approach (to zero when fixation occured), perhaps for better aiming. The duration of fixation might influence capture success because it determines the time the fish stay close to the prey in an area of enhanced probability of detection by the prey. It is also essential whether fixation (approach speed reduced to zero) takes place or not. This is evident from the increasing capture efficiency for cyclopoid prey, accompanied by the decrease in frequency of fixation periods. This correspondence ought to be less important for cladoceran prey, which do not exhibit as distinct escape movements as copepods (Drenner et al. 1978). Capture success is furthermore governed by the speed of the actual snap and the accuracy of aiming (Drost 1987), and by the degree of the suction flow generated during mouth opening (Drenner et al. 1978, Winfield et al. 1983, Drost & Van den Boogaart 1986, Drost et al. 1988a, b).

Differences in capture efficiency could be one mechanism leading to partitioning of the planktonic food resource among cyprinid larvae and juveniles. However, the observed differences in prey attack behaviour do not lead to interspecific differences in capture efficiency. This similarity in capture efficiency towards the major planktonic crustacean types indicates that any difference in attack behaviour which might occur does not support resource partitioning during early development.

Nevertheless, the increase in capture efficiency for cyclopoids during development can help to explain the increasing preference of young of the year cyprinids for cyclopoids in the field (Keckeis et al. 1988).

Acknowledgements

For discussion and helpful critisism I am very much indebted to F. Schiemer, W. Wieser, H. Keckeis, and the other colleagues of the Fish Ecology Group Vienna. H.I. Browman, W. Hoogenboezem, J.W.M. Osse, and L. Fuiman made many helpful suggestions on an earlier draft of this manuscript. Many thanks to S.E. Endel for her assistance during the practical work. The study was supported by the Austrian Fonds zur Förderung der wissenschaftlichen Forschung, Project No. S-35.

References cited

Balon, E.K. 1985. Early life histories of fishes: new developmental, ecological and evolutionary perspectives. Developments in Env. Biol. Fish. 5, Dr W. Junk Publishers, Dordrecht. 280 pp.

Blaxter, J.H.S. & M. Staines. 1971. Food searching potential in marine fish larvae. pp. 467–485. *In:* D.J. Crisp (ed.) Fourth European Marine Biology Symposion, Cambridge University Press, Cambridge.

Braum, E. 1964. Experimentelle Untersuchungen zur ersten Nahrungsaufnahme und Biologie an Jungfischen von Blaufelchen (*Coregonus wartmanni* Bloch), Weißfelchen (*Coregonus fera* Jurine) und Hechten (*Esox lucius* L.). Arch. Hydrobiol. 28: 183–244.

Browman, H.I. 1989. Behavioural ecology of foraging in a zooplanktivorous fish, *Pomoxis annularis,* and a predaceous invertebrate, *Leptodora kindti:* ontogenetic and neuroethological perspectives. Ph.D. thesis, University of Kansas, Lawrence. 169 pp.

Brown, J.A. & P. Colgan. 1985. Interspecific differences in the ontogeny of feeding behaviour in two species of centrachid fish. Z. Tierpsychol. 70: 70–80.

Confer, J.L., G.L. Howick, M.H. Corzette, S.L. Kramer, S. Fitzgibbon & R. Landesberg. 1978. Visual predation by planktivores. Oikos 31: 27–37.

Cryer, M., G. Peirson & C.R. Townsend. 1986. Reciprocal interactions between roach, *Rutilus rutilus,* and zooplankton in a small lake: prey dynamics and fish growth and recruitment. Limnol Oceanogr. 31: 1022–1038.

Drenner, R.W., J.R. Strickler & J.W. O'Brien. 1978. Capture probability: the role of zooplankton escape in the selective feeding of planktivorous fish. J. Fish. Res. Board Can. 35: 1370–1373.

Drost, M.R. 1987. Relation between aiming and catch success in larval fishes. Can. J. Fish. Aquat. Sci. 44: 304–315.

Drost, M.R. & J.G.M. Van den Boogaart. 1986b. The energetics of feeding strikes in larval carp (*Cyprinus carpio*). J. Fish Biol. 29: 371–379.

Drost, M., J.W. Osse & M. Muller. 1988a: Prey capture by fish larvae, water flow patterns and the effect of excape movements of prey. Netherlands Journal of Zoology 38: 23–45.

Drost, M., M. Muller & J.W. Osse. 1988b: A quantitative hydrodynamical model of suction feeding in larval fishes: the role of frictional forces. Proc. R. Soc. Lond. B 234: 263–281.

Eggers, D.M. 1977. The nature of prey selection by planktivorous fish. Ecology 58: 46–59.

Houde, E.D. & R.C. Schekter. 1980. Feeding by marine fish larvae: developmental and functional responses. Env. Biol. Fish. 5: 315–334.

Hrbáček, J., M. Dvořaková & L. Prochazková. 1961. Demonstration of the effect of the fish stock on the species composition of zooplankton and the intensity of metabolism of the whole plankton association. Ver. int. Ver. Limnol. 14: 192–195.

Keckeis, H., J. Wanzenböck, S. Wurzian & F. Schiemer. 1988. Trophic ecology and bioenergetics of 0+ cyprinidae. pp. 142–163. *In:* Workshop on 'Fish, Fisheries and Natural Waters', University of Kesthely, Kesthely.

Lazzaro X. 1987. A review of planktivorous fish: their evolution, feeding behaviours, selectivities, and impacts. Hydrobiologia 146: 97–167.

Lelek, A. 1987. Threatened fishes of Europe. Vol. 9, Aula Verlag, Wiesbaden. 343 pp.

Linder, A. & W. Berchtold. 1982. Statistische Methoden II. UTB, Birkhäuser Verlag, Basel. 295 pp.

Mark, W., R. Hofer & W. Wieser. 1987. Diet spectra and resource partitioning in the larvae and juveniles of three species and six cohorts of cyprinids from a subalpine lake. Oecologia 71: 388–396.

Northcote, T.G. 1988. Fish in the structure and function of freshwater ecosystems: a 'top-down' view. Can. J. Fish. Aquat. Sci. 45: 361–379.

Nyberg, D.N. 1971. Prey capture in the largemouth bass. Amer. Midl. Nat. 86: 128–144.

O'Brien, W.J., N.A. Slade & G.L. Vinyard. 1976. Apparent size as the determinant of prey selection bluegill sunfish (*Lepomis macrochirus*). Ecology 57: 1304–1310.

Panova, G.L. 1966. On nutrition of *Abramis ballerus* L. larvae in the Rybinsk reservoir. pp. 100–107. *In:* B.S. Kusin & B.K. Stegmann (ed.) Fish Biology of Wolga Reservoirs, Acad. Sci. USSR, Moskaw. (In Russian).

Prejs, A. 1976. Fishes and their feeding habits. pp. 155–171. *In:* E. Pieczynska (ed.) Selected Problems of Lake Littoral Ecology, Wydawnietwo Uniwerzytetu Warszawskiego, Warsaw.

Rosenthal, H. & G. Hempel. 1970. Experimental studies in feeding and food requirements of herring larvae (*Clupea harengus* L.) pp. 344–364. *In:* J.H. Steel (ed.). Marine Food Chains, Oliver and Boyd, Edinburgh.

Schiemer, F. & T. Spindler. 1989. Endangered fish species of the Danube River in Austria. Regulated Rivers: Research & Management 4: 397–407.

Schiemer, F., H. Keckeis & J. Wanzenböck. 1989. Foraging in cyprinids during early development. Pol. Arch. Hydrobiol. 36: 467–474.

Stein, R.A., C.G. Goodman & E.A. Marschall. 1984. Using time and energetic measures of cost in estimating prey value for fish predators. Ecology 65: 702–715.

Wanzenböck, J. & F. Schiemer. 1989. Prey detection in cyprinids during early development. Can. J. Fish. Aquat. Sci. 46: 995–1001.

Werner, E.E. 1974. The fish size, prey size, handling time relation in several sunfishes and some implications. J. Fish. Res. Board Can. 31: 1531–1536.

Werner, E.E. & D.J. Hall. 1974. Optimal foraging and the size selection of prey by the bluegill sunfish (*Lepomis macrochirus*). Ecology 55: 1042–1052.

Whiteside, M.C. 1988. 0 + fish as major factors affecting abundance patterns of littoral zooplankton. Ver. int. Ver. Limnol. 23: 1710–1714.

Winfield, I.J., G. Peirson, M. Cryer & C.R. Townsend. 1983. The behavioural basis of prey selection by underyearling bream (*Abramis brama* L.) and roach (*Rutilus rutilus* L.). Freshwater Biol. 13: 139–149.

Environmental Biology of Fishes **33**: 33–45, 1992.

Food consumption and growth of larvae and juveniles of three cyprinid species at different food levels

Hubert Keckeis & Fritz Schiemer
Institute of Zoology, Department of Limnology, Althanstraße 14, A-1090 Vienna, Austria

Received 11.4. 1990   Accepted 24.9.1990

Key words: Cyprinid larvae, *Abramis ballerus, Alburnus alburnus, Rutilus rutilus,* Conversion efficiencies, Consumption rates, Energy partitioning, Growth, Oxygen consumption

Synopsis

The relationships between food availability, consumption and growth were analyzed from the onset of feeding to an age of 90 days in three cyprinid species. Fish were held at 20 ± 0.5° C and given two (three) constant rations of approximately 30, (40) or 100% dry body weight (dbw) ind^{-1} day^{-1}. Food consisted of living zooplankton, the size of which correlated with fish size. At high food densities consumption rates decreased rapidly with fish size in all three species. At reduced rations, fish consumed most of the food offered until they were larger than 10 mg dbw. In all species and at each feeding level daily rations consumed increased allometrically with body size. Respiration rate, expressed as routine metabolic rate differed insignificantly between the three species. At high ration levels, growth rates of small bleak, *Alburnus alburnus,* were distinctly lower than those of roach, *Rutilus rutilus,* and blue bream, *Abramis ballerus.* At low food supply all three species grew at similar rates. Assimilation efficiency at low food conditions was approximately twice that of the well-fed groups. If the caloric equivalents of oxygen consumption as measured in well-fed fish are applied to fish fed at low rations their energy budgets do not balance. This indicates the limitations of fish larvae in the partitioning of energy for growth or activity at such conditions.

Introduction

The preferential allocation of energy to growth is an essential feature of the early life history of fish. Rapid growth rates favor survival of young fish because mortality decreases with increasing size (Brett 1979, Calow 1985, Kiørboe 1988, McGurk 1986, Ware 1975b, Werner & Gilliam 1984). Development of morphological structures such as buccal gape allows the ingestion of larger prey (Dabrowski & Bardega 1984, Hartmann 1983). Differentiation of the digestive system enhances food uptake rates and conversion efficiencies (Govoni et al. 1986, Hofer & Uddin 1985, Pedersen & Hjelmeland 1988). Muscular development increases the scope for activity (El-Fiky & Wieser 1988, Lackner et al. 1988, Hinterleitner et al. 1987) which allows higher prey encounter rates (Wanzenböck & Schiemer 1989) and may be important for the pursuit of fast prey types, escape from predators, and habitat choice (e.g. resistance to water currents).

Growth patterns in the field are largely influenced by temperature and food availability. There is a significant relationship between growth rate and temperature (LeCren 1958, Elliott 1976, Houde 1990, Jobling 1983, Mooij & vanTongeren 1990, Wieser et al. 1988a) and also a positive relationship between food consumption and fish growth (Brett & Groves 1979, Condrey 1982, Elliott 1979, Paloheimo & Dickie 1966).

The present study was conducted to evaluate the relationship between food availability and growth throughout the critical period after the onset of exogenous feeding. The questions raised were: (a) how are differrent levels of food abundance expressed in the pattern of energy partitioning? (b) to which extent can low food availability be compensated by increased assimilation efficiency? These questions were addressed by analyzing consumption and growth of cyprinid larvae at two or three rations of live zooplankton. Experiments were conducted with three species: the bleak, *Alburnus alburnus* (L.), and the roach, *Rutilus rutilus* (L.), are abundant in Central Europe, whereas the blue bream, *Abramis ballerus* (L.) is common in the Danube River system. In contrast to the different habitat and food preferences of the adults (Schiemer & Spindler 1989), larval and juvenile stages may occur together at high densities in the inshore zone, feeding on the same types of zooplankton (Mark et al. 1987, Panova 1966).

Material and methods

Mass cultures: Fish were raised from laboratory fertilized eggs from adults of the Danube (Stopfenreuth, Austria). Mass cultures were kept in circular 60 l tanks, at a temperature of 20 ± 0.5° C with constant flow, and a light period of 16 h. Fish were fed *Artemia* sp. nauplii and living zooplankton (copepod nauplii, *Ceriodaphnia reticulata* and *Daphnia magna*).

Zooplankton cultures: Pure cultures of *C. reticulata,* held in 300 l tanks, provided standardized food. They were fed *Chlorella* sp., *Scenedesmus* sp., *Ankistrodesmus* sp. and yeast (see Hauser 1989). Since production of the zooplankton cultures fluctuated, food was supplemented with *D. magna* from a small pond in later phases of the experiments. Zooplankton dry weight was measured to the nearest 0.1 μg using batches of 3 to 10 individuals dried for 24 h at 60° C. Estimated mean individual weights and length-weight regression parameters are given in Table 1a.

Feeding experiments: Daily measurements of consumption and growth of young roach and bleak began 12 days after hatching. The blue bream experiments began when the fish were 18 days old. For each species, two batches of five individuals were held in black 5 l tanks with constant flow at 20° C. Each tank contained a perspex frame with a bottom net of 1000 μm mesh size to separate the fish and the zooplankton.

Food was provided by counting subsamples from 2 l samples of the zooplankton culture. Food item size increased with fish size: fish larvae were fed *A. salina* once a day, followed by different size classes

Table 1. (a) Length-weight regressions for *Ceriodaphnia reticulata* and *Daphnia magna* ($\log_{10} W = \log_{10} a + b \log_{10} X$, where W is the dry weight in μg, X is the length from the top of the head to the base of the spine, a and b are constants).

C. reticulata:	a ± 95% CL = 1.179 ± 0.23; b ± 95% CL = 3.31 ± 0.3 (n = 40)
D. magna:	a ± 95% CL = 0.858 ± 0.16; b ± 95% CL = 2.04 ± 0.3 (n = 30)

(b) Mean length and weight (± standard deviation) of different size classes of zooplankton used in the feeding experiments:

	A. salina		*C. reticulata*		*D. magna*	
Mesh size (μm)	Length (mm)	Weight (μg)	Length (mm)	Weight (μg)	Length (mm)	Weight (μg)
–	0.40 ± 0.1	2.2 ± 0.3	–	–	–	–
200–300	–	–	0.50 ± 0.05	1.52 ± 0.4	–	–
300–500	–	–	0.65 ± 0.08	3.62 ± 0.8	0.76 ± 0.06	4.12 ± 1.5
300–600	–	–	0.74 ± 0.05	5.57 ± 1.1	1.01 ± 0.08	7.37 ± 1.8
600–800	–	–	0.93 ± 0.07	11.87 ± 2.1	1.30 ± 1.02	12.34 ± 2.1

Table 2. Length-weight regressions ($\log_{10} W = \log_{10} a + b \log_{10} X$, W = dry weight im mg, X = total length in mm, a and b are constants, CL = 95% confidence limits) of fish from mass cultures (above), and of specifically fed fish (below) during the feeding experiments.

Mass cultures:

Species	Range (mm)	$\log_{10}$ a ± CL	b ± CL	r^2	n
A. alburnus	10–39	– 3.176 ± 0.155	3.119 ± 0.087	0.95	169
A. ballerus	10–52	– 3.904 ± 0.098	3.904 ± 0.064	0.99	57
R. rutilus	10–45	– 3.786 ± 0.130	3.636 ± 0.069	0.98	161

Feeding experiments:

Species	Food level	Range (mm)	$\log_{10}$ a ± CL	b ± CL	r^2	n
A. alburnus	high	8–37	– 3.406 ± 0.173	3.301 ± 0.138	0.99	15
	low	8–18	– 3.272 ± 0.188	3.131 ± 0.257	0.97	15
A. ballerus	high	13–49	– 3.870 ± 0.097	3.619 ± 0.093	0.99	10
	low	13–36	– 3.899 ± 0.099	3.646 ± 0.145	0.99	10
R. rutilus	high	8–45	– 3.481 ± 0.208	3.445 ± 0.140	0.99	25
	medium	8–30	– 3.295 ± 0.192	3.230- ± 0.175	0.98	25
	low	8–20	– 3.203 ± 0.208	3.139 ± 0.239	0.96	25

of *C. reticulata* and *D. magna* (Table 1b). Mean individual daily consumption rates (C) were estimated by counting the zooplankton in the experimental tanks after 24 h feeding period, thus

$$C = \frac{\text{(amount of food offered)} - \text{(remaining amount after 24 h)}}{\text{number of fish per tank.}}$$

Growth during the feeding experiments was determined by measuring total length from photographs taken every third day and estimating body mass using a length-weight relationship calculated from mass cultures (Table 2). Experimental animals were photographed after transfer with the perspex frame into a shallow water basin (depth = 1 cm) equipped with a 10 mm scale. No significant differences between this method and direct measurements were detected (t-test, $p \geq 0.05$, $n = 10$).

Rations of approximately 30, 40 and 100% ($dbw \cdot day^{-1}$) were provided once a day, based on the calculated weights of the fish. The actual values of the rations provided and consumed were calculated after termination of the feeding experiments using length-weight relationships obtained for each specific feeding group (see Table 2). Mean individual consumption rates were determined for groups of five fish.

Growth curves were calculated from the length-

Table 3. Relationship between age and weight of all three species and food levels: $\log_{10} W = a + b.X + c.X^2$; where W = dry weight in mg, X = age in days, a, b and c are constants, SE = error.

Species	Food level	Range (days)	a ± SE	b ± SE	c	r^2	n
A. alburnus	high	12–91	– 0.7396 ± 0.032	0.0339 ± 0.001	– 0.0000621	0.99	28
	low	12–91	– 0.7110 ± 0.040	0.0188 ± 0.002	– 0.0000120	0.99	28
A. ballerus	high	18–91	– 0.7507 ± 0.058	0.0582 ± 0.002	– 0.0002788	0.99	25
	low	18–91	– 0.0093 ± 0.063	0.0098 ± 0.003	0.0000663	0.99	25
R. rutilus	high	12–91	– 0.6270 ± 0.017	0.0502 ± 0.002	– 0.0002021	0.99	25
	medium	12–91	– 0.7079 ± 0.027	0.0333 ± 0.001	– 0.0000843	0.99	27
	low	12–91	– 0.7196 ± 0.035	0.0248 ± 0.001	– 0.0000639	0.99	28

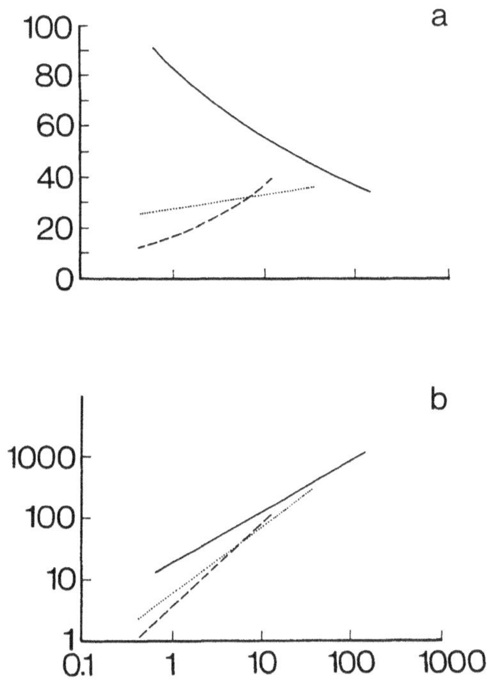

Fig. 1. Relationship, for *R. rutilus*, between consumption rates, expressed as (a) consumed rations (% dbw·day^{-1}·ind^{-1}) and (b) in J·day^{-1}·ind^{-1}, and body weight at three food levels: —— high ration; . . . medium ration; – – low ration.

weight relationship of the specific feeding groups. A second-order polynomial function was used to describe the age-weight relationship. Thus $Y = a + b.X + c.X^2$; where Y is $\log_{10}$ dry weight in mg, X age in days, and a, b and c are constants. The first derivative of the polynomial function was used to calculate daily growth increments: $Y = b + 2.c.X$ (Table 3).

Respirometry: A closed system respirometer with a polarographic oxygen sensor (POS) was used (Forstner 1983, Keckeis & Schiemer 1990). After two hours of feeding early in the morning the fish were put into the respirometers. Each experiment lasted 24 h, with an 18 h light period, at 20 ± 0.5° C. Measurements of oxygen consumption began between 1100 and 1200 h and continued at hourly intervals for 24 h. The number of fish in the chambers ranged from 1 to 200 individuals, depending on their average weight.

Table 4. Relationship between amount of food offered, consumption and body weight ($\log_{10} F = \log_{10} a + b.\log_{10} W$; $\log_{10} C = \log_{10} a + b.\log_{10} W$; F = food offered in J·day^{-1}·ind^{-1}, C = consumption, W = body dry weight in mg, a and b are constants, SE = standard error) at high and low food availability for all three species. Regression equation of pooled data for roach and bleak: $\log_{10} C = 1.1924 + 0.8133 \log_{10} W$. Regression equation of pooled data for all three species: $\log_{10} C = 1.1544 + 0.8393 \log_{10} W$.

Food offered: Species	Food level	Range (mg)	$\log_{10}$ a ± SE	b ± SE	r^2	n
A. alburnus	high	0.4– 65	1.65 ± 0.03	0.81 ± 0.03	0.92	79
	low	0.4– 8	0.78 ± 0.02	1.09 ± 0.03	0.92	79
A. ballerus	high	1.6–170	1.32 ± 0.04	0.91 ± 0.03	0.94	73
	low	1.6– 27	0.35 ± 0.03	1.50 ± 0.04	0.95	73
R. rutilus	high	0.4–147	1.38 ± 0.03	0.94 ± 0.03	0.95	79
	medium	0.4– 38	0.80 ± 0.03	1.29 ± 0.04	0.98	79
	low	0.4– 12	0.49 ± 0.02	1.48 ± 0.04	0.96	79
Consumption: Species	**Food level**	**Range (mg)**	**$\log_{10}$ a ± SE**	**b ± SE**	**r^2**	**n**
A. alburnus	high	0.4– 65	1.263 ± 0.11	0.81 ± 0.02	0.95	79
	low	0.4– 8	0.705 ± 0.12	1.04 ± 0.03	0.92	79
A. ballerus	high	1.6–170	1.030 ± 0.11	0.91 ± 0.02	0.96	73
	low	1.6– 27	0.255 ± 0.12	1.49 ± 0.04	0.95	73
R. rutilus	high	0.4–147	1.120 ± 0.19	0.82 ± 0.03	0.91	79
	medium	0.4– 38	0.786 ± 0.08	1.07 ± 0.02	0.98	79
	low	0.4– 12	0.469 ± 0.14	1.34 ± 0.04	0.94	79

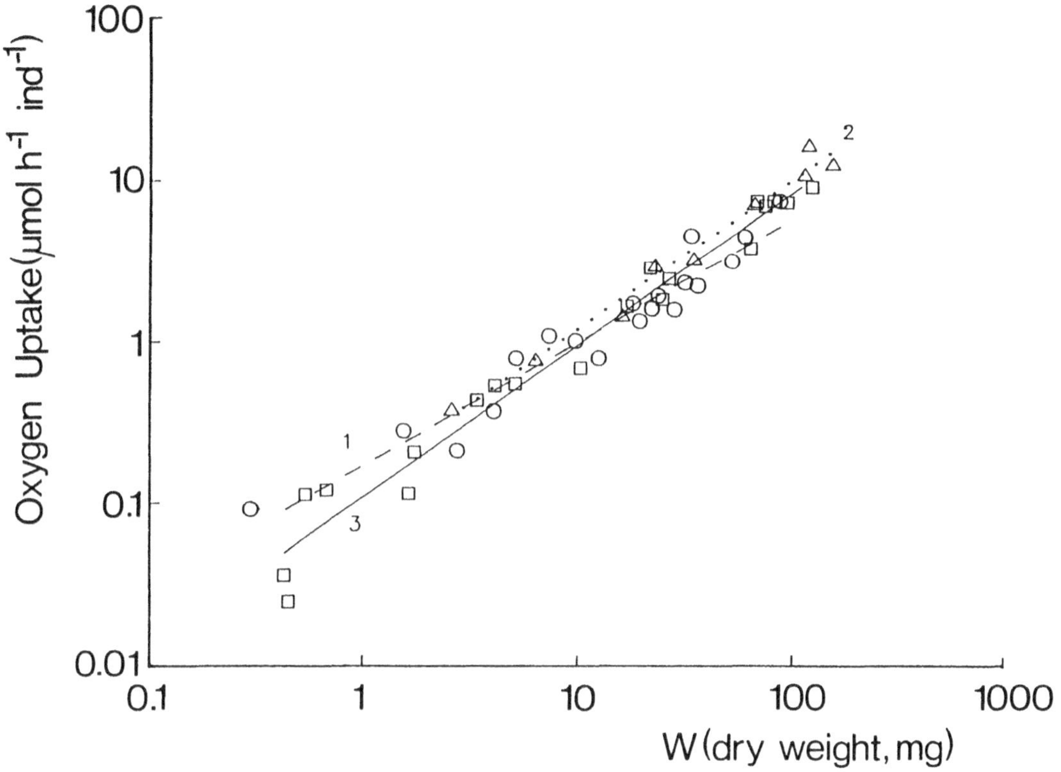

Fig. 2. Relationship between oxygen uptake (μmol·h^{-1}·ind^{-1}) and body weight for three cyprinid species. Data points represent mean values of 24 h experimental periods. ○ = bleak, △ = blue bream, □ = roach; 1, 2, 3 are regression lines for – – = bleak, . . . = blue bream and —— = roach, respectively. Regression statistics are given in Table 5.

Conversion factors

The energy value of *A. salina* was taken as 24.69 J·mg^{-1} while for *C. reticulata* and *D. magna* it was 22.18 J·mg^{-1} (Paffenhöfer 1967). Conversion of fish weights into energy units was based on their proximate composition (Rahman unpublished) using conversion factors of 23.63 J·mg^{-1} for protein, 17.15 J·mg^{-1} for carbohydrate and 39.54 J·mg^{-1} for fat (Kleiber 1961, Winberg 1971). One mg dry weight of roach or blue bream was equivalent to 26.61 J, and 1 mg dry weight of bleak was equivalent to 26.16 J. An oxycaloric coefficient of 14.3 J·$(gO_2)^{-1}$ was used (Winberg 1971, Niimi & Beamish 1973, Elliott & Davison 1975). Active metabolic rates were calculated by multiplying the routine metabolic rate (RMR) by a factor of 2.5 to correct for food searching activity and FIT (food induced thermogenesis, formerly SDA; Beamish 1974, Brett 1962, Kiørboe et al. 1987, Warren & Davis

Table 5. Relationship between respiration (RMR, routine metabolic rate) and body weight ($\log_{10} R = \log_{10} a + b.\log_{10} W$, R = oxygen consumption in μmol·h^{-1}·ind^{-1}, W = body dry weight in mg, a and b are constants, SE = standard error). Regression equation of pooled data of all three species: $\log_{10} R = -0.9546 + 0.92 \log_{10} W$.

Species	Range (mg)	$\log_{10}$ a ± SE	b ± SE	r^2	n
A. alburnus	0.3– 86	– 0.9645 ± 0.27	0.87 ± 0.02	0.89	394
A. ballerus	2.6–152	– 0.8658 ± 0.16	0.92 ± 0.02	0.91	220
R. rutilus	0.4–120	– 0.9534 ± 0.19	0.93 ± 0.02	0.94	418

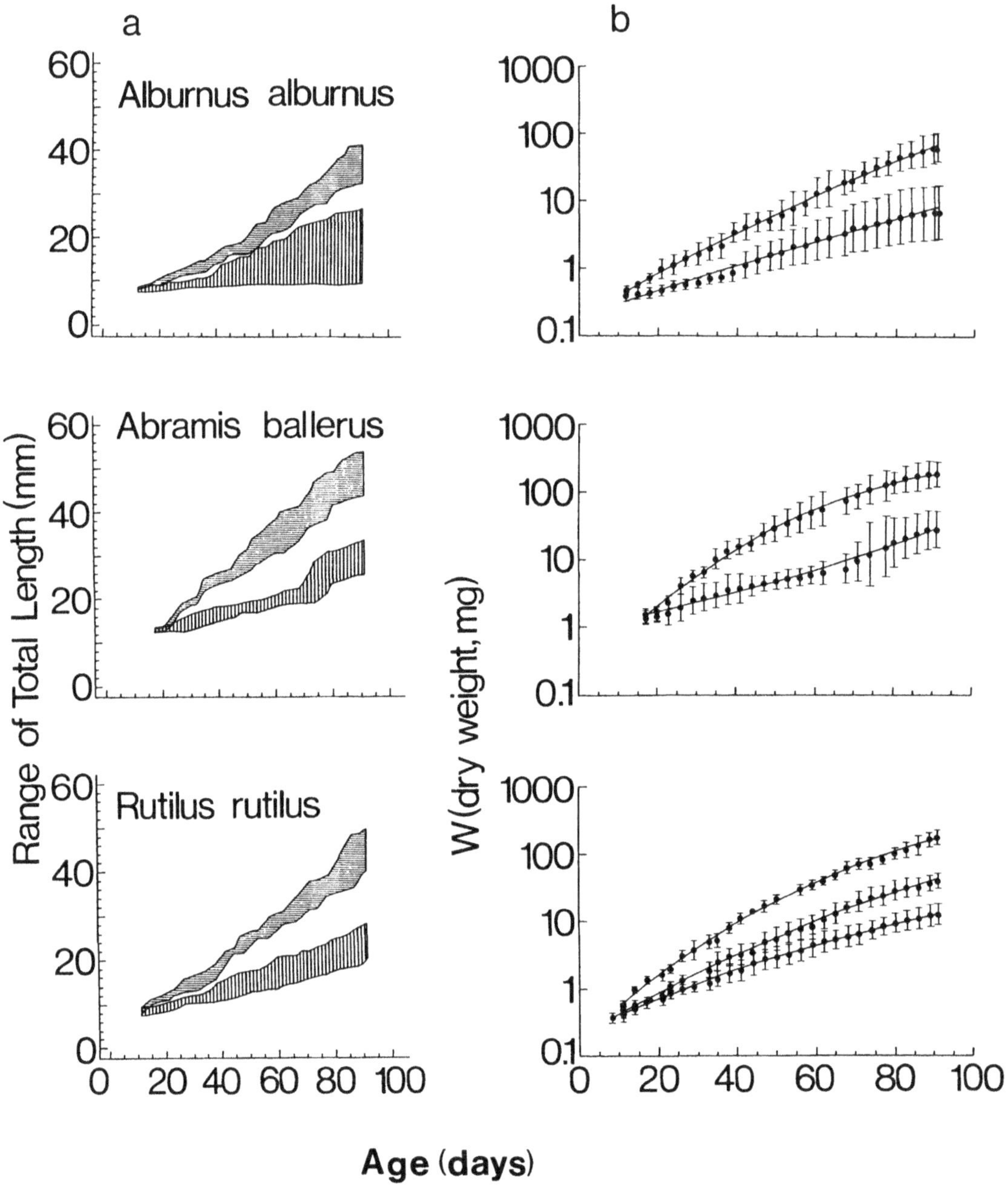

Fig. 3. (a) Maximal and minimal length of three cyprinid species with increasing age at two feeding regimes and (b) growth curves of the three species in terms of dry weight with arithmetic means (± 95% C.L.). □ = high ration (100% dbw·day^{-1}·ind^{-1}), □ = low ration (30% dbw·day^{-1}·ind^{-1}). Regression statistics are given in Table 3.

1967, Wieser & Medgyesy 1990, Ware 1975a, 1978).

Results

Consumption: The relationship between energy consumption rate (J·day^{-1}·fish^{-1}) and body weight (dry weight, mg) is described by the allometric function $\log_{10}C = \log_{10} a + b \log_{10}X$, for all three species at all food levels (Fig. 1, Table 4).

At high rations, the larval stages of all three species consumed up to 100% of their own weight per day. Food consumption rates (% dbw·day^{-1}·fish^{-1}) declined continuously to 50 and to 30% as fish weight increased to 10 and 100 mg, respectively. At restricted supply, the fish consumed almost all food offered.

Covariance analysis indicated that the regression lines of energy consumption for bleak and roach at high food availability were not significantly different ($p \geq 0.05$), whereas the regression line for the blue bream had a lower intercept and a higher slope compared with the other two species.

Respiration: Routine metabolic rates of all three species were measured using fish from the mass culture weighing from 0.3 to 150 mg dry weight (Fig. 2, Table 5). The relationship between weight and oxygen uptake is given by: $Y = a.W^b$, where Y is oxygen uptake in μmol·h^{-1}·ind^{-1} W is dry weight in mg, and a and b are constants. Covariance analysis indicated no significant differences between the three regression lines ($p \geq 0.05$; Table 5).

Growth: Plots of minimal and maximal total length (mm), converted weight and fitted growth function against fish age are presented in Figure 3. Interspecific comparison (ANOVA) indicated no significant differences in length or weight at the beginning of the experiments ($p > 0.05$; Table 6).

After 90 days at the high ration level bleak were significantly smaller than the other two species (ANOVA, $p < 0.05$). At reduced food availability, body length after 90 days differed insignificantly among the species (ANOVA, $p > 0.05$). Intraspecific variability was high, particularly in the bleak, indicating distinct differences in individual food uptake rates and growth patterns within each group of fish (Table 6).

Specific growth rate at high ration decreased with body size in all species (Table 7). Initial rates were highest in blue bream (12%·day^{-1}), followed by roach (11%·day^{-1}) and bleak (8%·day^{-1}). At a body size of about 60 mg bleak reached the same growth rate as the other two species. At lower

Table 6. Comparison of measured total lengths (L_t, mm) and dry weights (W, mg) of all three species at beginning and end of the feeding experiments (SD = standard deviation, CV = coefficient of variation = 100.SD.Mean^{-1}).

Mean lengths and weights at the beginning of the experiments:	Age (days)	$L_t \pm$ SD	W $\pm$ SD	n	CV(%)
A. alburnus	12	8.4 ± 0.3	0.34 ± 0.06	10	18
R. rutilus	12	8.3 ± 0.6	0.40 ± 0.07	10	17
A. ballerus	17	13.0 ± 0.5	1.50 ± 0.20	10	13

Mean lengths and weights at the end of the experiments:

High food level:	Age (days)	$L_t \pm$ SD	W $\pm$ SD	n	CV(%)
A. alburnus	91	37.0 ± 3.3	61.6 ± 18.0	5	29
A. ballerus	91	48.8 ± 3.7	178.0 ± 46.7	5	26
R. rutilus	91	44.8 ± 3.7	170.3 ± 52.3	5	31
Medium food level:					
R. rutilus	91	30.5 ± 3.4	38.1 ± 14.3	5	37

Low food level:	Age (days)	$L_t \pm$ SD	W $\pm$ SD	n	CV(%)
A. alburnus	91	17.8 ± 7.8	7.3 ± 8.2	5	112
A. ballerus	91	28.0 ± 2.9	28.0 ± 10.6	5	38
R. rutilus	91	23.8 ± 3.1	14.5 ± 7.4	5	51

rations the growth rates ranged between 3 to 5%·day^{-1} in all three species over the whole size range studied. The differences in growth rate between the three species observed at the high ration disappeared at reduced rations (Mann-Whitney U-Test, $p > 0.05$; Fig. 4).

Energy balance: Energy balance differed with size and ration as exemplified for roach in Figure 5. In small size classes (<2 mg dbw) both production and metabolic cost expressed as a percentage of energy intake decreased with increasing ration, reflecting a concomitant decline of assimilation efficiency (Fig. 5, Table 7). Below 5 mg dbw and at low rations, the sum of production and the hypothetical active feeding metabolic costs exceeded the measured amount of energy consumed in roach and in blue bream. This indicates that in the insufficiently fed small fish costs of feeding must have been less than assumed on the basis of the metabolic rates measured in well-fed animals (see Wieser & Medgyesy 1990).

Table 7. Components of energy budget of the three species (in Joule·day^{-1}·ind^{-1}) in different weight classes and at different food levels. C = consumption, P = production, RMR = routine metabolic rate, AMR = calculated active, feeding metabolic rate. Assimilation efficiency [AE = $(AMR + P).C^{-1}.100$], production efficiency [$K_1 = P \cdot C^{-1} \cdot 100$], and specific growth rate in % dbw·day^{-1} are also given. (Calculated AE values > 100% are set in brackets).

	Alburnus alburnus					*Abramis ballerus*					*Rutilus rutilus*				
Budget component	Dry weight (mg)					Dry weight (mg)					Dry weight (mg)				
	0.6	2	8	50	100	0.6	2	8	50	100	0.6	2	8	50	100
High food level:															
C	12.1	32.1	98.7	435.7	–	–	20.1	71.1	376.8	707.9	8.7	23.3	72.5	326.0	575.4
P	1.2	3.8	13.8	72.0	–	–	6.1	20.3	82.4	111.6	1.9	5.2	17.6	74.6	114.2
RMR	0.8	2.2	7.3	35.9	–	–	2.8	10.1	54.7	103.6	0.8	2.3	8.5	46.5	88.6
AMR	1.9	5.5	18.2	89.8	–	–	7.1	25.3	136.8	258.9	1.9	5.8	21.2	116.3	221.6
Efficiency															
AE	26	29	32	37	–	–	66	64	58	52	44	47	54	59	59
K1	10	12	14	17	–	–	30	29	22	16	22	22	24	23	20
Growth rate	7.6	7.1	6.5	5.5	–	–	11.4	9.5	6.0	4.2	11.1	10.0	8.4	5.7	4.2
Medium food level:															
C											3.7	13.1	55.4	–	–
P											1.2	3.5	11.5	–	–
RMR											0.8	2.3	8.5	–	–
AMR											1.9	5.8	21.2	–	–
Efficiency															
AE											84	71	59	–	–
K1											32	27	21	–	–
Growth rate											7.1	6.4	5.6	–	–
Low food level:															
C	3.1	10.5	44.1	–	–	–	5.2	39.9	–	–	1.9	7.4	45	–	–
P	0.7	2.2	7.8	–	–	–	1.7	9.3	–	–	0.9	2.9	10.5	–	–
RMR	0.8	2.2	7.3	–	–	–	2.8	10.1	–	–	0.8	2.3	8.5	–	–
AMR	1.9	5.5	18.2	–	–	–	7.1	25.3	–	–	1.9	5.8	21.2	–	–
Efficiency															
AE	87	76	59	–	–	–	(169)	87	–	–	(147)	(118)	70	–	–
K1	23	21	18	–	–	–	33	23	–	–	47	39	23	–	–
Growth rate	4.3	4.1	4.0	–	–	–	3.1	4.2	–	–	5.4	4.6	3.5	–	–

Discussion

Size-specific consumption rates at high food availability were similar in the three species studied. In Figure 6 an overall regression line for the three species is compared with literature data of consumption rates of fish larvae and juveniles measured under excess food supply. The comparison includes:

(1) three subtropical marine species at 26 to 28° C; 10 day feeding experiments using natural zooplankton as food (Houde & Schekter 1983);

(2) *Cyprinus carpio* fed natural zooplankton at high densities (> 100 ind·l^{-1}) at temperatures ranging from 18 to 21° C (Filatov 1972);

(3) *Alburnus alburnus* from a small pond in which temperature ranged from 19 to 22° C; food consisted of cyclopoids at densities > 70 ind·l^{-1} (Ivlev 1960);

(4) present study;

(5) a mixed culture of *Rutilus rutilus, Alburnus alburnus* and *Abramis brama* at 20° C fed high amount of living zooplankton (Marmulla 1985);

(6) *Coregonus lavaretus* (Coregonidae) at optimal temperatures of approximately 16° C and maximal rations using a mixture of wild zooplankton (Troschel 1988);

(7) *Coregonus lavaretus* fed *Daphnia pulicaria* at 16° C (Rösch 1985);

(8) sockeye salmon, *Oncorhynchus nerka,* at

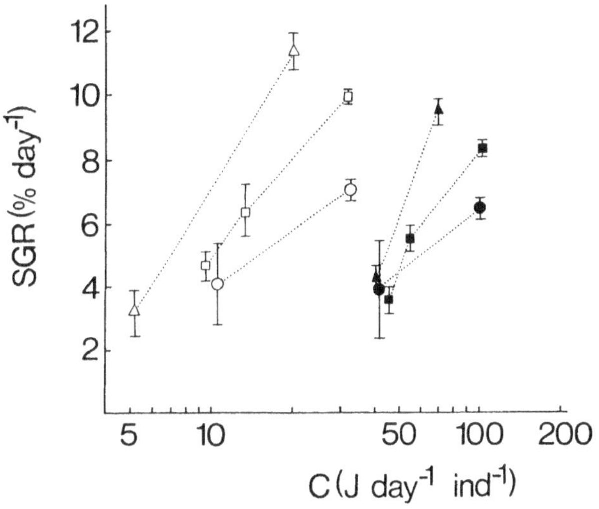

Fig. 4. Relationship between consumption rates and specific growth rates of two size classes (left 0.6 mg dbw, open symbols; right 8 mg dbw, full symbols) of bleak (○), blue bream (△) and roach (□) at different food levels.

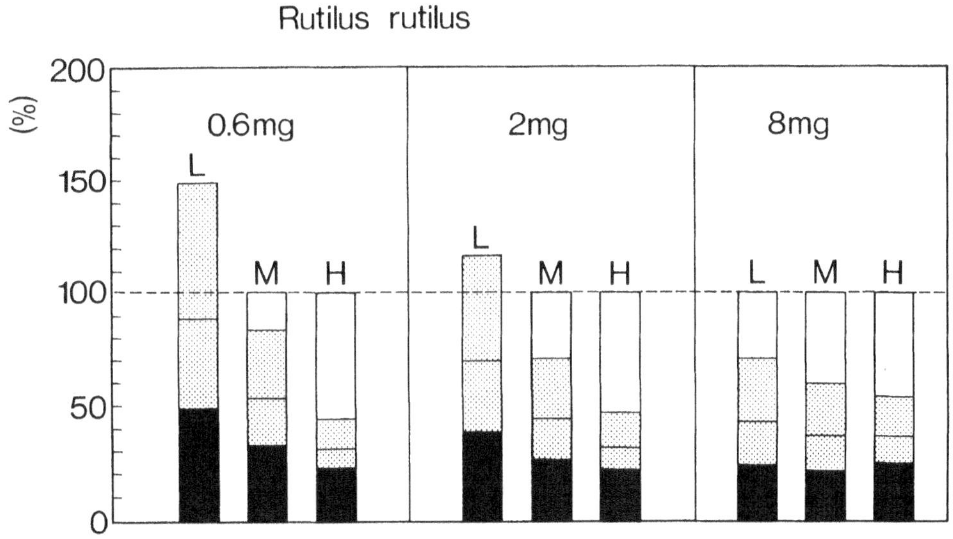

Fig. 5. Production and respiration rates of different weight classes of roach expressed as a percentage of measured consumption rates. ■ = production, ▨ = metabolism. Routine metabolic rate and active feeding metabolic rate are separated by a line. L = 30% dbw·day^{-1}·ind^{-1}, M = 40% dbw·day^{-1}·ind^{-1}, H = 100% dbw·day^{-1}·ind^{-1}. Numbers reflect body dry weight. Data from Table 7, for statistics see text.

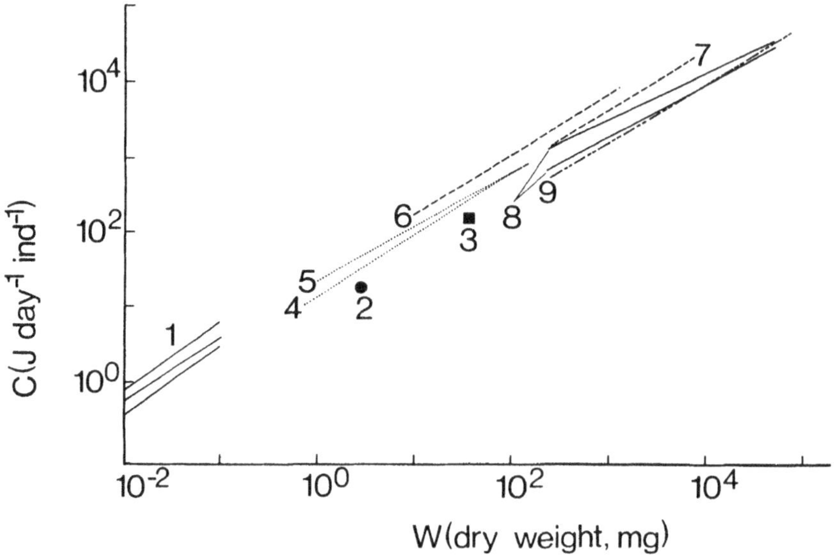

Fig. 6. Relationship between consumption and body weight of several fish species based on data in the literature. Values were transformed using specifications of the authors. For explanation see text.

15° C, fed artificial diets. The upper curve indicates consumption rates at triple, the lower curve at single feeding per day (Brett & Groves 1979);

(9) brown trout, *Salmo trutta,* at the optimal temperature of 13° C. Food consisted of freshwater invertebrates. Smallest size classes not included (Elliott 1975, 1979).

Two main conclusions can be drawn from this comparison:

(1) a relatively narrow range of the intercept, despite the wide range of temperature and size of species compared, and

(2) a weight exponent of the order of 0.7–0.8. Peters (1983) found a weight-exponent of 0.82 for food consumption of poikilotherms in general. Thus maximal consumption rate expressed as a function of ration shows a decreasing trend with increasing body size (Fig. 7).

Initial maximum production rates were significantly lower in bleak, which correlated with a distinctly lower assimilation efficiency at high ration. This relationship has been discussed by Keckeis & Schiemer (1990).

Lower food availability reduced growth rates of all three species to a similar degree, independent of body weight. Brett & Shelbourn (1975) obtained similar results with sockeye salmon fed different rations of artificial diet.

From Figure 4 it is evident that at low rations production is the same in all species, whereas at high ration bleak is at a disadvantage. This together with higher prey detection capacities (Wanzenböck & Schiemer 1989), makes the bleak well adapted to low nutritional environments.

The respiration rates measured are in accordance with those recently published by Wieser & Forstner 1986, Wieser et al. 1988b, and Kaufmann (1990) for roach during the early larval and juvenile period. They are also similar to those published by Kamler (1976) and Winberg (1961) for carp, *Cyprinus carpio,* and freshwater fishes in general, respectively. The slope of the allometric relationship between respiration and body weight is higher than that of the relationship between food consumption and body weight at high food availability. This indicates that metabolic expenditure increases faster with body size than energy uptake rates. This may explain the decrease of growth rate with increasing size (see Brett & Groves 1979).

The costs of metabolic processes greatly affect calculated conversion efficiencies. Based on data for the active feeding metabolic rate (AMR), as

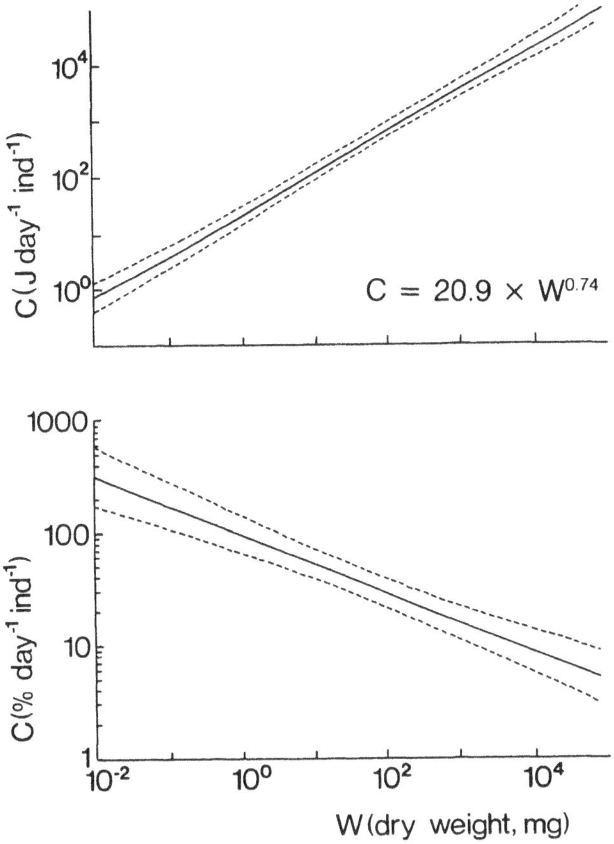

Fig. 7. Regression line of maximal consumption rates of several fish species determined at physiologically optimal or environmental temperature. C = consumption ($J \cdot day^{-1} \cdot ind^{-1}$); W = log dry weight (mg).

measured by Wieser & Medgyesy (1990), fish larvae at low food abundance are energy limited. Metabolic costs for food searching and for synthesis of body substances are well balanced at high rations, whereas consumption rates at reduced rations would not cover AMR in the small size classes of roach and blue bream. This means that the actual respiratory costs at low ration levels must have been well below the respiration rates measured at high rations. Above a dry body weight of 5 mg metabolic expenditure is well covered even at reduced rations. Consequently, the flexibility of small fish in the use of energy either for growth or for activity appears to be limited at low food levels. Our data suggest that under such conditions the energy gained is used preferentially for biosynthesis rather than for activity. Although production rates are lower than in the well fed groups, and low food intake is partially compensated by higher assimilation efficiency, the energy available does not sustain high active metabolic rates. This is in accordance with recent results on energy partitioning between growth and activity in cyprinid larvae by Wieser et al. (1988b). Therefore low food availability not only decreases growth but also reduces the scope for searching activity. These two constraints may explain the high mortality of small fish in nature.

Acknowledgements

This study was supported by the 'Fonds zur Förderung der wissenschaftlichen Forschung in Österreich' project No. S-35. We thank I. Zweimüller, JA. Gunatilaka and J. Wanzenböck for their helpful discussion and criticism and I. Winfield, W. Wieser and R. Tanasichuk for revision of an earlier version of this manuscript.

References cited

Beamish, F.W.H. 1974. Apparent specific action of largemouth bass, *Micropterus salmoides*. J. Fish. Res. Board Can. 31: 1763–1769.

Brett, J.R. 1962. Some considerations in the study of respiratory metabolism in fish, particularly salmon. J. Fish. Res. Board Can. 19: 1025–1037.

Brett, J.R. 1979. Environmental factors and growth. pp. 599–675. *In:* W.S. Hoar, D.J. Randall & J.R. Brett (ed.) Fish Physiology, Vol 8, Academic Press, New York.

Brett, J.R. & T.D.D. Groves. 1979. Physiological energetics. pp. 279–352. *In:* W.S. Hoar, D.J. Randall & J.R. Brett (ed.) Fish Physiology, Vol. 8, Academic Press, New York.

Brett, J.R. & J.E. Shelbourn. 1975. Growth rate of young sockeye salmon, *Oncorhynchus nerka*, in relation to fish size and ration level. J. Fish. Res. Board Can. 32: 2103–2110.

Calow, P. 1985. Adaptive aspects of energy allocation. pp. 13–31. *In:* P. Tytler & P. Calow (ed.) Fish Energetics, New Perspectives, Croom Helm, London.

Condrey, R.E. 1982. Ingestion-limited growth of aquatic animals: the case for Blackman kinetics. Can. J. Fish. Aquat. Sci. 39: 1585–1595.

Dabrowski, K. & R. Bardega. 1984. Mouth size and predicted food size preferences of larvae of three cyprinid species. Aquaculture 40: 41–46.

El-Fiky, N. & W. Wieser. 1988. Life styles and patterns of development of gills and muscles in larval cyprinids (Cyprinidae, Teleostei). J. Fish Biol. 33: 135–145.

Elliott, J.M. 1975. The growth rate of brown trout (*Salmo trutta* L.) fed on maximum rations. J. Anim. Ecol. 44: 805–821.

Elliott, J.M. 1976. The energetics of feeding, metabolism and growth of brown trout (*Salmo trutta* L.) in relation to body weight, water temperature and ration size. J. Anim. Ecol. 45: 923–948.

Elliott, J.M. 1979. Energetics of freshwater teleosts. Symp. Zool. Soc. Lond. 44: 29–61.

Elliott, J.M. & W. Davison. 1975. Energy equivalents of oxygen consumption in animal energetics. Oecologia 19: 195–201.

Filatov, V.I. 1972. Effectiveness of the utilization of natural foods by carp (*Cyprinus carpio* L.) larvae. J. Ichthyol. 12: 812–818.

Forstner, H. 1983. An automated multiple-chamber intermittent-flow respirometer. pp. 111–126. *In:* G. Gnaiger & H. Forstner (ed.) Polarographic Oxygen Sensors, Springer Verlag, Berlin.

Govoni, J.J., G.W. Boehlert & Y. Watanabe. 1986. The physiology of digestion in fish larvae. Env. Biol. Fish. 16: 59–77.

Hartmann, J. 1983. Two feeding strategies of young fishes. Arch. Hydrobiol. 96: 496–509.

Hauser, L. 1989. Experimentelle Untersuchungen über die Nahrungsselektivität von Rotauge (*Rutilus rutilus* L.) und Laube (*Alburnus alburnus* L.) in den ersten Lebensmonaten. Diplomarbeit, Universität Wien, Vienna. 82 pp.

Hinterleitner, L., U. Platzer & W. Wieser. 1987. Development of the activities of oxidative, glycolytic and muscle enzymes during early larval life in three families of freshwater fish. J. Fish Biol. 30: 315–326.

Hofer, R. & N. Uddin. 1985. Digestive processes during the development of the roach, *Rutilus rutilus* L. J. Fish Biol. 26: 683–689.

Houde, E.D. 1990. Comparative growth, mortality, and energetics of marine fish larvae: temperature and implied latitudinal effects. U.S. Fishery Bulletin 87: 471–495.

Houde, E.D. & R.C. Schekter. 1983. Oxygen uptake and comparative energetics among eggs and larvae of three subtropical marine fishes. Marine Biology 72: 283–293.

Ivlev, V.S. 1960. On the utilization of food by planktophage fishes. Bulletin of Mathematical Biophysics 22: 371–389.

Jobling, M. 1983. Influence of body weight and temperature on growth rates of arctic charr, *Salvelinus alpinus* (L.). J. Fish Biol. 22: 471–475.

Kamler, E. 1976. Variability of respiration and body composition during early developmental stages of carp. Pol. Arch. Hydrobiol. 23: 431–485.

Kaufmann, R. 1990. Respiratory cost of swimming in larval and juvenile cyprinids. I. Power-performance relations during early life. J. exp. Biol. 150: 343–366.

Keckeis, H. & F. Schiemer. 1990. Consumption, growth and respiration of bleak, *Alburnus alburnus* L. and roach, *Rutilus rutilus* L. during early ontogeny. J. Fish Biol. 36: 841–851.

Kiørboe, T. 1988. Growth in fish larvae: are they particularly efficient? Early Life History Symposium, Bergen. (in press).

Kiørboe, T., P. Munk & K. Richardson. 1987. Respiration and growth of larval herring *Clupea harengus:* relation between specific dynamic action and growth efficiency. Mar. Ecol. Prog. Ser. 40: 1–10.

Kleiber, M. 1961. The fire of life. Wiley, New York. 454 pp.

Lackner, R., W. Wieser, M. Huber & J. Dalla Via. 1988. Responses of intermediary metabolism to acute handling stress in untrained and trained *Leuciscus cephalus* (Cyprinidae, Teleostei). J. exp. Biol. 140: 393–404.

LeCren, E.D. 1958. Observations on the growth of perch (*Perca fluviatilus* L.) over twenty-two years with special reference to the effects of temperature and changes in population density. J. Anim. Ecol. 27: 287–334.

Mark, W., R. Hofer & W. Wieser. 1987. Diet spectra and resource partitioning in the larvae and juveniles of three species and six cohorts of cyprinids from a subalpine lake. Oecologia 71: 388–396.

Marmulla, G. 1985. Bestimmung der maximalen Tagesration von Jungfischen bei Überschussfütterung mit lebendem Zooplankton. Diplomarbeit, Universität Freiburg, Freiburg. 109 pp.

McGurk, M.D. 1986. Natural mortality of marine pelagic fish eggs and larvae: role of spatial patchiness. Mar. Ecol. Prog. Ser. 34: 227–242.

Mooij, W.M. & O.F.R. van Tongeren. 1990. Growth of 0^+ roach (*Rutilus rutilus*) in relation to temperature and size in a shallow eutrophic lake: comparison of field and laboratory observations. J. Fish. Res. Board Can. 47: 960–967.

Niimi, A.J. & F.W.H. Beamish. 1973. Bioenergetics and growth of largemouth bass (*Micropterus salmoides*) in relation to body weight and temperature. Can. J. Zool. 52: 447–456.

Paffenhöfer, G.A. 1967. Caloric content of larvae of the brine shrimp *Artemia salina.* Helgoländer Wiss. Meeresunters. 16: 130–135.

Paloheimo, J.E. & L.M. Dickie. 1966. Food and growth of fishes. III. Relations among food, body size, and growth efficiency. J. Fish. Res. Board Can. 23: 1209–1248.

Panova, G.L. 1966. On nutrition of *Abramis ballerus* L. larvae in the Rybinsk reservoir. pp. 100–107. *In:* B.S. Kusin & B.K. Stegman (ed.) Fish Biology of Wolga Reservoirs, Acad. Sci. USSR, Moscow. (In Russian).

Pedersen, B.H. & K. Hjelmeland. 1988. Fate of trypsin and assimilation efficiency in larval herring (*Clupea harengus*) following digestion of copepods. Marine Biology 97: 467–476.

Peters, R.H. 1983. The ecological implications of body size. Cambridge University Press, Cambridge. 329 pp.

Rösch, R. 1985. Nahrungsaufnahme und Nahrungsausnutzung von Sandfelchen (*Coregonus lavaretus* L.) bei Fütterung mit Lebendplankton (*Daphnia pulicaria*). Dissertation, Universität Freiburg, Freiburg. 124 pp.

Schiemer, F. & T. Spindler. 1989. Endangered fish species of the Danube river in Austria. Regulated Rivers: Research & Management 4: 397–407.

Troschel, H.J. 1988. Bestimmung der maximalen Tagesration

von Blaufelchen *Coregonus lavaretus* L. bei Fütterung mit lebendem Zooplankton in Überschuss. Diplomarbeit, Universität Freiburg, Freiburg. 90 pp.

Wanzenböck, J. & F. Schiemer. 1989. Prey detection in cyprinids during early development. Can. J. Fish. Aquat. Sci. 46: 995–1001.

Ware, D.M. 1975a. Growth, metabolism, and optimal swimming speed of a pelagic fish. J. Fish. Res. Board Can. 32: 33–41.

Ware, D.M. 1975b. Relation between egg size, growth and natural mortality of larval fish. J. Fish. Res. Board Can. 32: 2503–2512.

Ware, D.M. 1978. Bioenergetics of pelagic fish: theoretical change in swimming speed and ration with body size. J. Fish. Res. Board Can. 35: 220–228.

Warren, C.E. & G.E. Davis. 1967. Laboratory studies on the feeding bioenergetics and growth of fishes. pp. 175–214. *In:* S.D. Gerking (ed.) The Biological Basis of Freshwater Fish Production, Blackwell, Oxford.

Werner, E.E. & J.F. Gilliam. 1984. The ontogenetic niche and species interactions in size-structured populations. Ann. Rev. Ecol. Syst. 15: 393–425.

Wieser, W. & H. Forstner. 1986. Effects of temperature and size on the relative scope for activity in larval cyprinids. J. Comp. Physiol. B 156: 791–796.

Wieser, W., H. Forstner, F. Schiemer & W. Mark. 1988a. Growth rates and growth efficiencies in larvae and juveniles of *Rutilus rutilus* and other cyprinid species: effects of temperature and food in the laboratory and in the field. Can. J. Fish. Aquat. Sci. 45: 943–950.

Wieser, W., H. Forstner, N. Medgyesy & S. Hinterleitner. 1988b. To switch or not to switch: partitioning of energy between growth and activity in larval cyprinids (Cyprinidae: Teleostei). Functional Ecology 2: 499–507.

Wieser, W. & N. Medgyesy. 1990. Aeorobic maximum for growth in the larvae and juveniles of a cyprinid fish, *Rutilus rutilus* (L.): implications for energy budgeting in small poikilotherms. Functional Ecology 4: 233–242.

Winberg, G.G. 1961. New information on metabolic rate in fishes. Fish. Res. Board Can. Translation Series 362: 00–00.

Winberg, G.G. 1971. Methods for the estimation of production of aquatic animals. Translated from the Russian by A. Duncan. Academic Press, London. 175 pp.

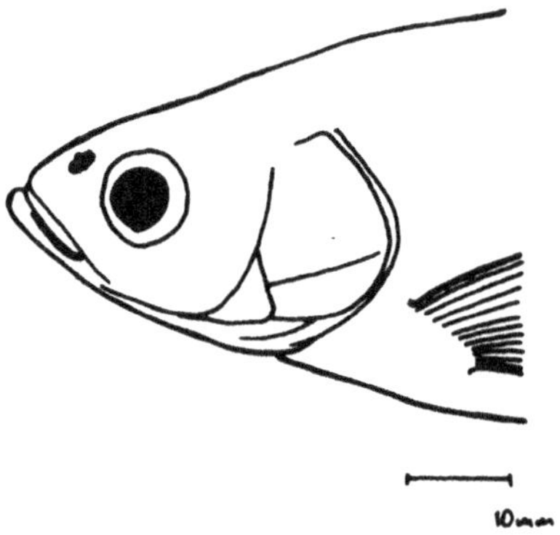
10mm

Environmental Biology of Fishes **33**: 47–52, 1992.

Food searching decisions in four cyprinid species

Franz Uiblein
Institut für Limnologie, Abteilung Mondsee, A-5310 Gaisberg 116, Austria

Received 19.1.1990 Accepted 14.8.1990

Keywords: Vimba elongata, Abramis brama, Chalcalburnus chalcoides, Rutilus rutilus, Interspecific differences, Sampling, Search rate, Decision rules, Resource use

Synopsis

Four species of cyprinids (vimba, bream, Danubian bleak and roach) which had been incubated and reared under identical conditions differed qualitatively as well as quantitatively in the use of prey search strategies during comparative experiments. While bream and vimba prefer to use sampling techniques under conditions of reduced visibility of benthic prey, roach favor almost exclusively directed search methods. Vimba and Danubian bleak, in contrast to bream and roach, rapidly increase their total search activity in response to successive presentations of novel prey. These results have several ecological implications which are confirmed by field data. It is concluded that food searching cyprinids follow decision rules which are species specific both in the amount of sampling and in the responsiveness to rapidly changing conditions.

Introduction

Temperate lakes often provide fluctuating conditions which require predators to cope with increased uncertainty of prey occurrence. Fish may respond to such conditions flexibly by choosing behavioral options which guarantee foraging success (Dill 1983). Recent investigations have tried to tackle the problem of determining which decision rules different species of fish and other predators may adopt in order to achieve foraging goals in situations with restricted predictability of prey occurrence (O'Brien et al. 1989, Roberts & Ilersich 1989, Uiblein 1991). The decisions involved in choosing between different behavioral options generally make use of both external and internal information (Staddon 1983, Real 1987, Stephens 1989). Because the availability of information is subjected to various constraints, different species (or even individuals) generate different strategies (Mitchell 1989). This may eventually lead to the partitioning patterns closely related fishes show in the field (Schmitt & Coyer 1982, Delbeek & Williams 1987, NcNeely 1987, Shpigel & Fishelson 1989).

Experimentally manipulated changes in prey novelty and visibility revealed that vimba, *Vimba elongata* Val., choose between directed and undirected search strategies based on external prey occurrence information and internal expectations of long term benefits (Uiblein 1991).

The fish used in these experiments originated from Mondsee, a mesotrophic temperate lake which offers rather fluctuating nutritional conditions (Herzig 1985), requiring predators to adopt flexible search strategies. Field data confirm this presumption by showing that vimba utilizes the resources of Mondsee with great flexibility (Uiblein & Winkler 1987).

Apart from vimba several other cyprinid species live in Mondsee which may use different strategies to adapt to environmental fluctuations. One major approach to reveal interspecific differences in resource utilization is to specify the decision rules animals may follow during foraging. The present study tests the hypothesis that four species of cyprinids which originate from Mondsee adopt species specific search decisions when faced with experimentally manipulated fluctuations in prey novelty and visibility.

Material and methods

Five specimens of each of four cyprinid species (vimba, *Vimba elongata* Val.; bream, *Abramis brama* L.; Danubian bleak, *Chalcalburnus chalcoides mento* Agassiz; roach, *Rutilus rutilus* L.) were tested individually. All fish were in year class 2 + (SL between 67 and 85 mm) and originated from populations of Mondsee (47° 52′ N, 13° 23′ W). Separated by species, they were reared in the laboratory under identical conditions and fed only with zooplankton. For the experiments the fish were kept in glass tanks (20 × 30 cm bottom) which contained 8 l of filtered water from Mondsee. Temperature was held constant at 19 ± 1° C and light was provided at 600 ± 50 Lux, alternating with total darkness in 12 h intervals. Experimental sessions were carried out during the two hour interval from 13:00 to 15:00, with each session starting 24 h after the start of the previous session. The experimental tanks were surrounded by white styropor-walls leaving a small window for observations.

Two experimental series separately tested the fish's response to changes in prey novelty and visibility. In the first experimental series 10 red, living chironomid larvae (*Chironomus* sp., 5–6 mm length) were offered as novel benthic prey during two successive sessions. After these initial sessions, several additional sessions followed until all fish had clearly increased their capture rate compared to the first session. In the second series, the chironomids (now familiar to the fish) were at first presented on fine sediment (0.3–0.6 mm), and in a successive session, on coarse sediment (1.0–1.5 mm). Both types of sediments formed a 5 mm cover on the glass bottom. During preparation of the experimental conditions the fish were enclosed behind a blind for about 5 min. The experiments started when the blind was lifted and the fish gained free access to the prey. During and between all sessions ten red, chironomid-like plastic dummies were in the tanks.

Snaps directed at prey items or dummies ('directed snaps') as well as snappings at inconspicuous particles ('undirected snaps') were recorded for a 20 sec period beginning with the first snap directed at a prey item. About half an hour before the start of the first experimental series two control observations recorded possible spontaneous snapping activity.

For statistical comparisons among sessions with G-tests the data from all individual fish within each session were pooled (see Sokal & Rohlf 1981). Pooling of the data appeared to be justified as no individual differences occurred within the sessions (G-tests, $p > 0.05$).

Results

In the absence of potential prey, all four species displayed a low spontaneous rate of between 0.03 and 0.28 snaps per 20 sec. Only vimba and bream showed a few snaps directed at the dummies. Therefore the percentage of undirected snaps was high for all species (between 81% and 100%).

After the first capture of novel prey (during the first session of the first experimental series) all species had markedly higher rates of total snaps than they exhibited during the earlier control trials when no potential food was provided. Considering the period after the first successful capture, the rates of total snaps were similar (between 1.3 and 2.8 per 20 sec; Fig. 1) and did not vary significantly among species (Kruskal-Wallis test, $p > 0.05$). Prey capture rate was low (from 0.2 to 1.2 per 20 sec) with no recognizable difference among species (Kruskal-Wallis test, $p > 0.05$). The ratio of undirected snaps ranged from 9.0% (roach) to 42.9% (vimba). No significant differences in the percentage of undirected snaps could be detected (G-test, $p > 0.05$).

The rate of total snaps significantly increased from the first to the second session in vimba (7.8

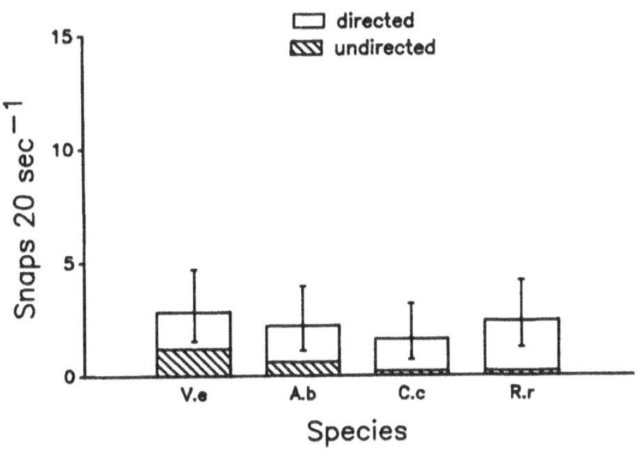

Fig. 1. Number of snaps per 20 sec following the first presentation of novel prey in 4 species of cyprinids (95% confidence limits). V.e = vimba, A.b = bream, C.c = Danubian bleak, R.r = roach.

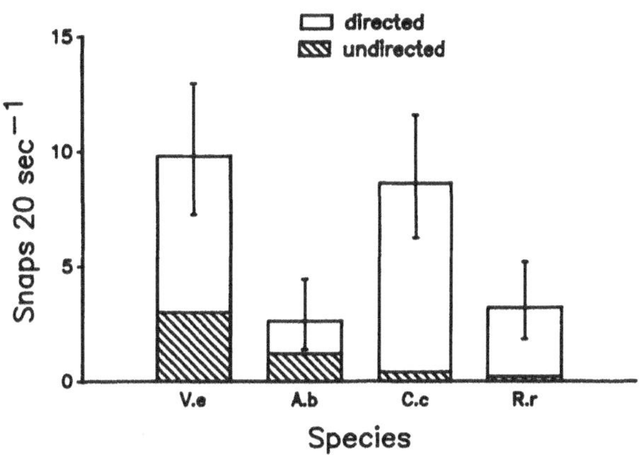

Fig. 2. Number of snaps per 20 sec following the second presentation of novel prey in 4 species of cyprinids (95% confidence limits). For abbreviations see Figure 1.

per 20 sec, Wilcoxon test, $p < 0.05$; Fig. 2) and Danubian bleak (8.6 per 20 sec, $p < 0.05$). Bream (2.6 per 20 sec) and roach (3.2 per 20 sec) showed no significant increase in the rate of total snaps from the first to the second session. As a consequence, the rate of total snaps in the second session differed significantly among species (Kruskal-Wallis test, $p < 0.025$). Pairwise comparisons showed that bream and roach both had significantly lower rates with respect to vimba and Danubian bleak (Mann-Whitney tests, $p < 0.05$).

Prey capture rate also changed from the first to second session, so that by the second session the capture rate differed among the total set of species (Kruskal-Wallis test, $p < 0.005$). Paired comparisons showed that the capture rate of bream (1.0 per 20 sec) was significantly lower than in vimba (5.4 per 20 sec; Mann-Whitney test, $p < 0.05$) and Danubian bleak (6.0 per 20 sec, $p < 0.05$).

The percentage of undirected snaps showed no significant changes from the first to the second session in all four species (G-tests, $p > 0.05$). By the second session it ranged from 4.7% (Danubian bleak) to 46.2% (bream) and was different within the total set of species (G-test, $P < 0.0005$). This analysis showed also that this result is due to differences between vimba and Danubian bleak, as well as between bream and Danubian bleak.

In the first session of the second experimental series the chironomids were presented on a layer of fine sediment. The rate of total snaps ranged between 5.8 and 8.0 per 20 sec (Fig. 3) and there were no significant differences among species (Kruskal-Wallis test, $p > 0.05$). Prey capture rate ranged from 3.2 to 6.2 captures per 20 sec which again did not significantly differ among the species (Kruskal-Wallis test, $p > 0.05$). The percentage of undirected snaps varied between 0.0% (roach) and 10.3% (bream). There were no significant differences among the four species (G-test, $p > 0.05$).

The rate of total snaps showed no significant changes from the first to the second session (with prey on coarse sediment) in all four species (Wilcoxon tests, $p > 0.05$). By the second session, the rate of total snaps ranged from 4.6 to 7.2 per 20 sec (Fig. 4), with no significant differences among the four species (Kruskal-Wallis test, $p > 0.05$). Prey capture rate ranged from 1.2 (bream) to 5.6 per 20 sec (roach) and varied significantly among species ($p < 0.005$). Pairwise comparisons revealed a significantly lower capture rate between bream and all other species (Mann-Whitney tests, $p < 0.05$), whereas vimba captured prey at a significantly lower rate than roach ($p < 0.025$).

The percentage of undirected snaps differed significantly between the first and the second session in vimba (G-test, $p < 0.0001$), bream ($p < 0.025$), and Danubian bleak ($p < 0.025$), but not in roach ($p > 0.05$). By the second session the percentage of

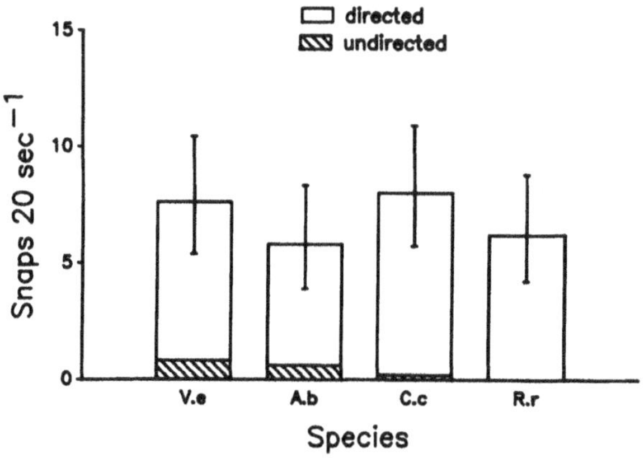

Fig. 3. Number of snaps per 20 sec following the presentation of familiar prey on fine sediment in 4 species of cyprinids (with 95% confidence limits). For abbreviations see Figure 1.

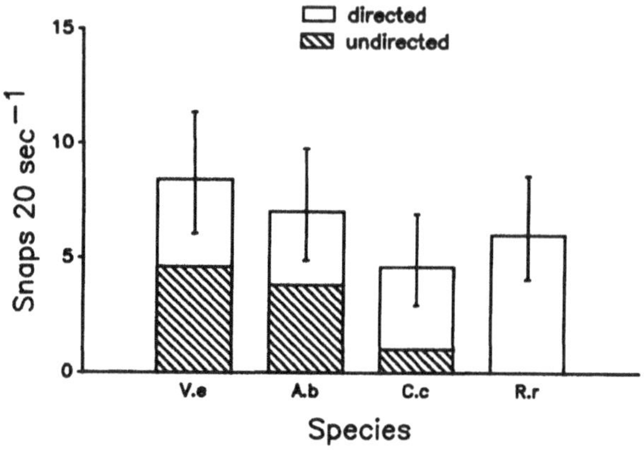

Fig. 4. Number of snaps per 20 sec following the presentation of familiar prey on coarse sediment in 4 species of cyprinids (with 95% confidence limits). For abbreviations see Figure 1.

undirected snaps varied between 0.0% (roach) and 54.8% (vimba). There were clear significant differences within all four species (G-test, $p < 0.0001$). This analysis also showed that this result is due to differences between roach and all other species.

Discussion

Both qualitative and quantitative differences were found in snapping activity among the four cyprinid species in relation to experimental manipulations of prey occurrence. The qualitative aspect concerns the choice between directed and undirected search methods. The frequency of undirected snaps indicates to which extent the fish decided to use search strategies beyond visually directed prey location. This choice of undirected search methods can be interpreted as sampling used to explore the environment for the potential occurrence of unpredictable resources (Krebs et al. 1978, Shettleworth et al. 1988).

Without immediate access to prey all fish performed a low number of total snaps which almost exclusively were undirected. When repeatedly exposed to novel prey on glass bottom each species continued to show a certain frequency of undirected snaps. Intraspecific differences in the percentage of undirected snaps occurred in the second series when prey were offered on different backgrounds. According to earlier results (see Uiblein 1991), the present data suggest that the four tested species choose undirected sampling methods based upon external information on prey visibility and internal expectations of long-term benefits gained from sampling.

Interspecific qualitative differences occurred during both experimental series. Vimba and bream showed a higher percentage of undirected snaps than Danubian bleak in the second experiment with novel prey. In the experiment with coarse background vimba, bream, and Danubian bleak performed a higher sampling frequency than roach. During all experimental sessions roach showed an extremely low frequency of undirected snaps. These data indicate that sampling may be of different value to the four species. Vimba and bream, especially under conditions of low prey visibility, adopt undirected snaps to explore the environment. In contrast, roach almost exclusively persist in using visually mediated, directed locating techniques. Danubian bleak preferentially used a directed search strategy with increased sampling frequency only under conditions of reduced visibility caused by coarse background.

These results suggest that with respect to different conditions of prey visibility, the expectations of sampling benefits may vary significantly among species. Roach persisted in the use of visual location techniques even in situations of reduced visibility of the prey, while all other species increased their sampling activity. Furthermore, in the experiments with coarse background roach achieved a much higher capture success compared to bream and vimba. Thus roach appears to have the ability to discern prey signals above backgrounds which provide bad visibility, where other species prefer switching to random sampling. In the field, these interspecific differences may lead to distinct patterns of resource utilization. While the decisions adopted by vimba and bream predispose them to search for food of rather unpredictable occurrence such as sediment-dwelling prey, Danubian bleak and especially roach may decide to locate prey by search techniques directed to immediately accessible individual prey items.

Some field investigations support these ideas on species specific tendencies of resource utilization. Danubian bleak in the Mondsee feeds mainly on planktonic prey by directed, selective feeding methods (Orellana 1985), whereas vimba from the same lake utilizes other food resources including sediment dwelling prey by adopting both directed and sampling methods (Uiblein & Winkler 1987). Field studies from other areas report that bream preferentially feed on prey hidden under sediment cover (Brabrand 1984, Lammens et al. 1987) or in clouds of plankton (Lammens et al. 1987) which points to the use of undirected search methods. Roach, apart from quite inefficiently foraging upon sediment dwelling prey (Lammens et al. 1987), exhibit mainly selective predation on large food particles (Lammens et al. 1987).

While all fish display a low rate of spontaneous snapping activity, this rate obviously increases when novel prey is encountered for the first time. Vimba and Danubian bleak further increased their rate of total snaps, including prey capture rate, in the second session of the first series. Bream and roach both needed seven consecutive sessions to achieve a similar increase (Uiblein 1988). These results provide evidence that vimba and Danubian bleak have the ability of maximizing their behavioral response within comparatively few encounters with novel prey. Hence these two species should be well equipped to live in habitats where the time to accurately examine single prey items is limited, for instance in fast flowing rivers (see Scott 1987). In the field, fish of the genus *Vimba* and *Chalcalburnus* often are described as typically rheophilous species (Busniţa 1961, Holčík & Bastl 1976). Even in Mondsee the two species demonstrate the ability of utilizing the discharging rivers at least during the spawning season (Uiblein et al. 1987).

While vimba and Danubian bleak seem to be well suited for opportunistically expanding their habitats, roach and especially bream may be less equipped to fast-changing conditions. This line of reasoning is confirmed by field data on bream which characterize this species as a conservative forager preferentially living in environments with moderate change (Löffler 1984).

All the fish tested experienced the same treatment during the experiments. They had been reared from the eggs under identical conditions and were of the same age and size. Hence the qualitative and quantitative results in snapping behavior indicate differences of probably phylogenetic origin in the fish's search decisions. When searching for food, cyprinids obviously follow decision rules which are species specific with regard to the amount of sampling and to the responsiveness to rapidly changing conditions.

Acknowledgements

I thank Hans Winkler for discussions and much useful advice. Further stimulating discussions were provided by Raymond Colley, Dan Danielopol, Walter Geiger, Erich Kainz, Eddy Lammens and Manfred Pöckl. J.P. Ball, Eddy Lammens and Wolfgang Wieser helped to improve the manuscript. This study is part of project S-35 supported by the 'Fonds zur Förderung der wissenschaftlichen Forschung' of Austria.

References cited

Brabrand, A. 1984. Microhabitat segregation between bream (*Abramis brama* (L.)) and white bream (*Blicca bjoerkna* (L.))

in a mesotrophic lake, SE Norway. Pol. Arch. Hydrobiol. 31: 99–108.

Busniţa, T. 1961. Die Wandlungen der Fischfauna der unteren Donau während der letzten hundert Jahre. Verh. Int. Ver. Limnol. 14: 381–385.

Delbeek, J.C. & D.D. Williams. 1987. Food resource partitioning between sympatric populations of brackishwater sticklebacks. J. Anim. Ecol. 56: 949–967.

Dill, L.M. 1983. Adaptive flexibility in the foraging behavior of fish. Can. J. Zool. 65: 803–811.

Herzig, A. 1985. Fischnährtier-Almanach für den Mondsee. Österr. Fischerei 38: 97–108.

Holčík, J. & I. Bastl. 1976. Ecological effects of waterlevel fluctuations upon fish populations in the Danube river floodplain. Acta Sc. Nat. Brno 10: 1–46.

Krebs, J.R., A. Kacelnik & P. Taylor. 1978. Test of optimal sampling by foraging great tits. Nature 275: 27–31.

Lammens, E.H.R.R., J. Geursen & P.J. McGillavry. 1987. Diet shifts, feeding efficiency and coexistence of bream *Abramis brama,* roach *Rutilus rutilus* and white bream *Blicca björkna* in hypertrophic lakes. pp. 153–162. *In:* S. Kullander & B. Ferholm (ed.) Proceedings of the V. Congress of European Ichthyologists, Stockholm.

Löffler, H. 1984. Zur Ökologie des Brachsen (*Abramis brama* L.) im Bodensee. Schweiz. Z. Hydrobiol. 46: 147–162.

NcNeely, D.L. 1987. Niche relations within an Ozark stream cyprinid assemblage. Env. Biol. Fish. 18: 195–208.

Mitchell, W.A. 1989. Informational constraints on optimally foraging hummingbirds. Oikos 55: 145–154.

O'Brien, W.J., B.I. Evans & H.I. Browman. 1989. Flexible search tactics and efficient foraging in saltatory searching animals. Oecologia 80: 100–110.

Orellana, C.P. 1985. Nahrungserwerb und Biologie der Seelaube *Chalcalburnus chalcoides mento* (Agassiz) im Mondsee. Diplomarbeit, Universität Salzburg, Salzburg. 69 pp.

Real, L. 1987. Objective benefit versus subjective perception in the theory of risk-sensitive foraging. Amer. Nat. 130: 399–411.

Roberts, W.A. & T.J. Ilersich. 1989. Foraging on the radial maze: the role of travel time, food accessibility, and the predictability of food location. J. Exp. Psych. (Anim. Beh. Proc.) 15: 274–285.

Schmitt, R.J. & J.A. Coyer. 1982. The foraging ecology of sympatric marine fish in the genus *Embiotica* (Embioticidae): importance of foraging behavior in prey size selection. Oecologia 55: 369–378.

Scott, A. 1987. Prey selection of juvenile cyprinids from running water. Freshw. Biol. 17: 129–142.

Shettleworth, S.J., J.R. Krebs, D.W. Stephens & J. Gibbon. 1988. Tracking a fluctuating environment: a study of sampling. Anim. Behav. 36: 87–105.

Shpigel, M. & L. Fishelson. 1989. Food habits and prey selection of three species of groupers from the genus *Cephalopholis* (Serranidae: Teleostei). Env. Biol. Fish. 24: 67–73.

Sokal, R.R. & F.J. Rohlf. 1981. Biometry, 2nd. edition. W.H. Freeman & Company, San Francisco. 859 pp.

Staddon, J.E.R. 1983. Adaptive behavior and learning. Cambridge University Press, Cambridge. 555 pp.

Stephens, D.W. 1989. Variance and the value of information. Amer. Nat. 134: 128–133.

Uiblein, F. 1988. Experimente zur Öko-Ethologie des Nahrungserwerbs einheimischer Karpfenfische (Cyprinidae). Doctoral Dissertation, Universität Wien, Wien. 95 pp.

Uiblein, F. 1991. Expectancy controlled sampling decisions in *Vimba elongata.* Env. Biol. Fish. (in press).

Uiblein, F., E. Kainz & H.P. Gollmann. 1987. Wanderungen, Wachstum und Fortpflanzung der Mondsee-Zährte (*Vimba elongata* Val., Cyprinidae). Österr. Fischerei 40: 282–295.

Uiblein, F. & H. Winkler. 1987. Der Nahrungserwerb der Mondsee-Zährte. Sitzber. Österr. Akad. Wiss., math. – naturwiss. Klasse 196: 311–322.

Environmental Biology of Fishes **33**: 53–62, 1992.

Functional responses of five cyprinid species to planktonic prey

Hans Winkler & Carmen P. Orellana
Institut für Limnologie der Österreichischen Akademie der Wissenschaften, Gaisberg 116, A-5310 Mondsee, Austria
Present address: Konrad Lorenz-Institut für Vergleichende Verhaltensforschung der Österreichischen Akademie der Wissenschaften, Savoyenstraße 1A, A-1160 Vienna, Austria

Received 30.4.1990 Accepted 1.10.1990

Key words: Cladocera, Copepoda, Foraging behavior, *Chalcalburnus chalcoides, Vimba vimba, Abramis brama, Rutilus rutilus, Scardinius erythrophthalmus*

Synopsis

The functional responses of five species of cyprinids (*Chalcalburnus chalcoides, Vimba vimba, Abramis brama, Rutilus rutilus,* and *Scardinius erythrophthalmus*) feeding on four planktonic prey types were measured in the laboratory. Although no alternative prey types were present, the response curves were sigmoid in most cases, because attack rates were not independent of prey density. The findings are explained as being the overt expression of the fishes' foraging tactics. The chief way of maximizing food uptake, according to our interpretation, is accelerating attack rates with increasing prey density. The ability of prey to escape or relative prey size may interfere with this strategy. *C. chalcoides,* the only obligatory planktivore among the species studied, attacks at higher rates and responds most markedly to changes in prey density.

Introduction

The functional response of a predator is the relationship between prey density and the number of prey items it ingests per unit time, and is an important component of foraging models and predator-prey models. Holling (1965) suggested three simple forms of functional responses. In a type 1 response, the 'filter feeder response', feeding rate increases linearly to a maximum as prey density increases. In type 2, the 'invertebrate response', the feeding rate increases asymptotically at a continuously decreasing rate. The disc-equation is the most commonly applied representation of this response (Holling 1959, Rashevsky 1959). Type 3, the 'vertebrate response', is described by a sigmoid function. Abrams (1982), however, has shown that for some predators this traditional classification may not be adequate.

Some of the more important assumptions of the type-2 functional response model (see also Stephens & Krebs 1986 for a discussion) are that (1) searching and handling are mutually exclusive activities, (2) encounter with prey is sequential and a Poisson process, (3) net energy gain, handling time, and attack rate are independent of prey density. The wide occurrence of sigmoid functional responses in vertebrates requires some additional explanation (Krebs 1973, Curio 1976, McNair 1980). In most cases they are discussed in the context of encounters with multiple prey types; the mechanisms responsible are thought to be learning, search image formation, and prey patchiness (Holling 1965, Krebs 1973, Curio 1976, Real 1977,

McNair 1980, 1981). Holling (1965) suggested that learning changes the threshold at which the predator attacks the prey. Real (1977) assumed that it takes several encounters with a specific prey type before a predator becomes maximally efficient at feeding on that prey. The authors cited so far imply that the change in encounter rate necessary in sigmoid functional responses are due to a change either in the readiness to attack or in prey detection. In particular, McNair (1980) showed that search image, prey preference formation or training effects on prey capture probability generate sigmoid responses, whereas training in handling time does not.

From a model developed by Marten (1973) a sigmoid response may be expected if a change in food density causes a predator to shift its emphasis on different foods or different search modes. Regardless of the mechanisms, whenever the encounter rate is not independent of prey density, deviations from the conventional type 2 response are to be expected. An obvious way to control encounter rate for a mobile predator, such as a fish, is to alter the speed of locomotion. Although Ivlev (1944) found that speed of locomotion may increase at high particle densities, this was later dismissed as an important factor for the quantitative analysis of feeding responses by Rashevsky (1959). Other investigations also suggested that swimming velocity is not a continuous function of prey density, but changes abruptly and then stays constant over a wide range of prey densities (Durbin & Durbin 1975). While learning models assume that encounter rate is a continuous function of prey density without sudden changes (e.g. Real 1977, where a simple power function is used), locomotor responses may be more erratic.

In this paper, we demonstrate that predators encountering a single type of prey, randomly distributed in space, exhibit a rich variety of functional responses. Thus the description and analysis of the functional responses of five species of cyprinids to various prey types may serve as a basis for the understanding of species-specific solutions to the problem of feeding on zooplankton. The results also have some bearing on current foraging models in general.

Methods

Laboratory experiments

The fish species used for the experiments were the cyprinids *Chalcalburnus chalcoides mento* (A.) (Danube bleak), *Vimba vimba* (L.) (zährte), *Rutilus rutilus* (L.) (roach), *Abramis brama* (L.) (bream), *Scardinius erythrophtalmus* (L.) (rudd). All fish (total body length 77–92 mm) were incubated and reared in the laboratory (see also Herzig & Winkler 1986 for details). The stock was held in separate, large tanks according to species, fed with fresh zooplankton when available and with frozen zooplankton otherwise. Zooplankton for feeding and for experimental purposes was caught from lake Mondsee (Upper Austria). As a result, the fish had sufficient experience with the prey offered in the experiments.

The experiments were set up in a room maintained at 20° C, unless stated otherwise. Fish in the stock tanks were kept at 20° C from May to November and at 10° C during the remainder of the year.

All experiments were carried out in glass aquaria (42 × 29 × 26 cm) containing 20 l of water. Each was illuminated by one 300 W tungsten halogen lamp suspended 110 cm above the water surface and connected to a rheostat for regulation of the light intensity. An intensity of 1200 lux at the water surface was fixed for most of the experiments.

Three sides of the aquaria were covered with pale green paper to avoid disturbances from the outside, while the fourth was kept free to allow observation.

Prey types used were the cladocerans *Daphnia hyalina* (L.) (1.8 mm), *Leptodora kindti* (F.) (5 mm) and *Bythotrephes longimanus* (L.) (3 mm) and the copepod *Cyclops* spp. (1.4 mm) (all were ovigerous individuals). All prey animals were collected from the pelagic zone of lake Mondsee, separated and counted in the laboratory using a wide pipette and kept in large petri dishes until used for the experiments. Damaged specimens were discarded and replaced by healthy ones.

An individual fish was introduced into an observation aquarium and not fed for 24 h prior to each trial to standardize hunger. New fish were used in

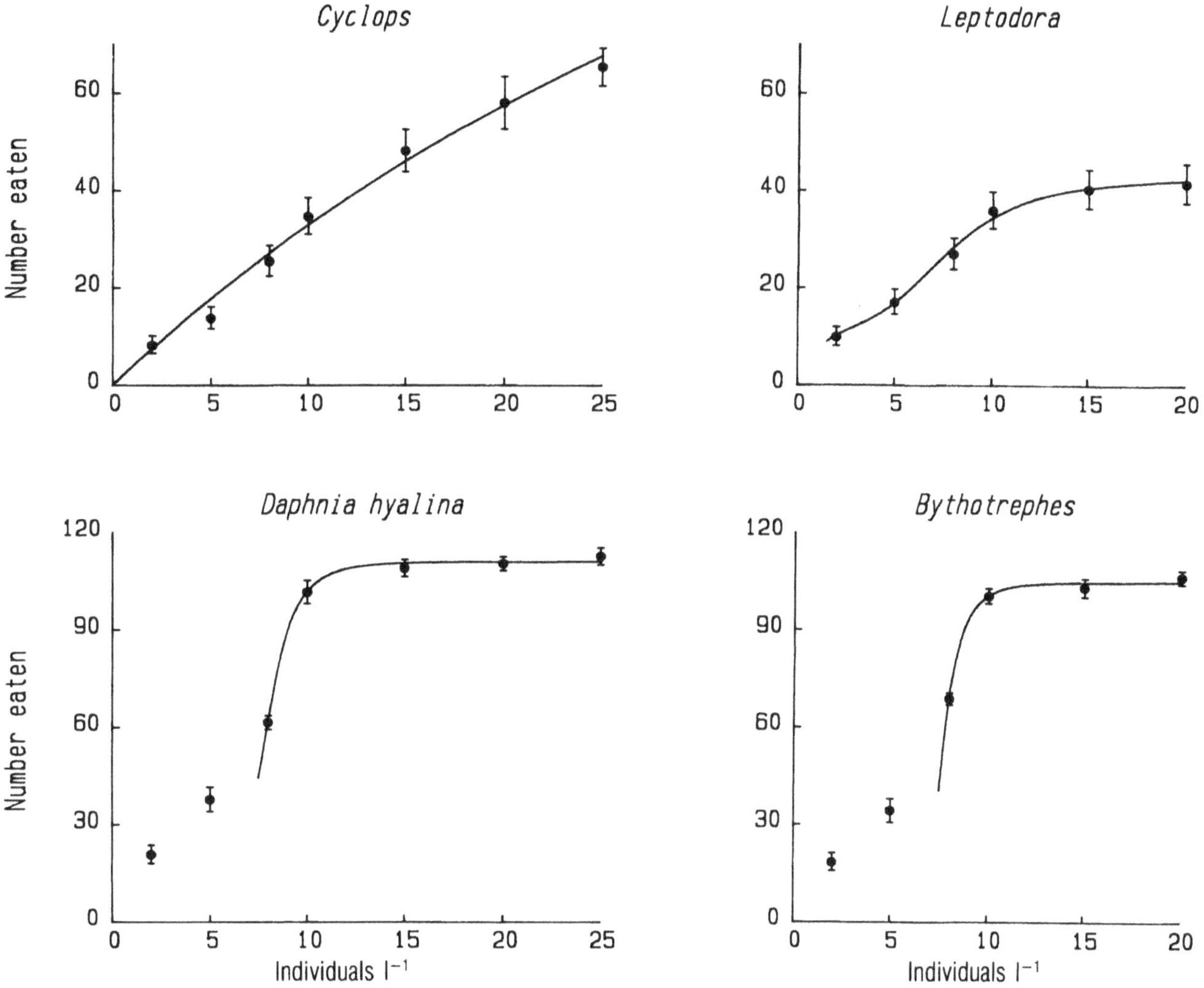

Fig. 1. The functional response of *Chalcalburnus chalcoides* with *Cyclops, Leptodora kindti, Daphnia hyalina* and *Bythotrephes longimanus* as prey. Means of the number of prey items eaten within 3 minutes in 10 experiments at each prey density are shown, together with the corresponding 95% confidence limits. For the parameters of the curve see Table 1.

each experiment. Prey were then introduced into the aquarium. The fish usually started feeding between 2 and 5 min after prey introduction, during which time no aggregation of the prey took place. Experiments in which the fish did not start feeding before 15 min were discarded. Visual recording began after the fish made its first attack and was continued for a period of 3 min. Fish were exposed to a range of initial densities of 2, 5, 8, 10, 15, and 20 prey l^{-1}. Prey eaten by the fish were not replaced during the experiment. Each prey density was replicated ten times and all experiments were carried out in a random order with regard to prey type and density. As the Danube bleak is the only strict zooplanktivore, all prey species were tested with this fish. Since the experiments with the Danube bleak showed that *B. longimanus* was the preferred prey species, all the other species of fish were tested with this prey species. The response of Danube bleak to *D. hyalina* was very similar, but due to preparation difficulties (air bubbles in the carapax) this prey species was not used with other fish species. Bream and zährte, which are common in habitats also used by the Danube bleak, were tested with *B. longimanus, L. kindti* and *Cyclops*. Roach of the size used had severe problems in handling *L. kindti* and were therefore not tested with this prey species. Rudd is not a potential competitor for the Danube bleak and therefore was only tested with *B. longimanus* for the sake of comparison.

Data analysis

In order to fit appropriate functional response equations to our data, a general extension of the classical type 2 response model (Holling 1959, 1966) was developed. Its parameters are the attack rate per unit search time per unit prey density and the handling time, which by definition includes the time spent identifying, pursuing, subduing, and eating prey. Both are assumed to be constant in the classical model. Since a generalization suggested by Real (1977, 1979) was inadequate for many of our data, we used a function derived from the conventional one by expanding the constant (a) commonly interpreted as attack or encounter rate, into a polynomial function of prey density. The equation used was

$$y = f(a,x)/(1 + h\, f(a,x)), \quad (1)$$

where y represents the number of prey items taken up per unit time and h handling time. The function

$$f(a,x) = x(a_0 + a_1x + \ldots a_4x^4) \quad (2)$$

expresses the relationship between attack rate and prey density x, the a_i representing the constants for the polynomial.

Because the experimental results are essentially counts, the fitting criterion was to minimize

$$E(f(a,h,x_i) - y_i)^2/f(a,h,x_i),\ i = 1, \ldots n, \quad (3)$$

where x_i and y_i are the values for x and y observed in the experiments. This is equivalent to minimizing the associated Chi-Square and corresponds to a maximum likelihood estimate. For the evaluation of the model parameters one has to take into account the fact that the data are based on the first 3 min of foraging time. Although the functions were fitted using all the original data, graphical representations of the results show data means at the initial food densities only. The confidence limits of the means were constructed under the assumption that the counts are samples from a Poisson-distributed variate. Because in most cases variances were lower than expected from this distribution, the limits represent conservative estimates.

The fitting process began with the simple type 2 model and expanded until a Chi-Square of a cumulative probability of less than 5% was reached or until the addition did not result in a change of Chi-Square greater than 3.841. In the most extreme cases (responses of *Chalcalburnus*), although a good fit to the data points could be achieved, the functions thus generated were unsatisfactory because they showed local minima between adjacent experimental densities (see Fig. 1). Alternatives,

Table 1. Functional responses of *Chalcalburnus chalcoides*. The columns give the parameters estimated according to the general functional response equation discussed in the text. Jack-knife estimates of the 95% confidence limits are presented below each parameter.

Prey	h	a_0	a_1	a_2	a_3	a_4
Cyclops spec.	0.00436	3.834				
	0.0024–0.063	3.33–4.34				
D. hyalina	0.00901	29.24	– 23.47	11.43	– 2.163	0.1347
	0.0089–0.0091	29.2 –29.3	– 23.5 —23.4	11.4 –11.5	– 2.18—2.14	0.132–0.137
B. longimanus	0.00953	13.41	– 19.17	14.90	– 3.403	0.2328
	0.0094–0.0097	13.2 –13.6	– 19.7—18.6	14.8 –15.0	– 3.46—3.34	0.227–0.239
L. kindti	0.02316	10.79	– 2.707	0.3302		
	0.0222–0.0241	7.58–14.0	– 3.87-1.54	0.224–0.436		

such as piecewise linear and quadratic functions could not amend this problem which will be discussed in more detail later.

The problem of assigning at least crude confidence limits to the parameter estimates was solved using the jack-knife estimate (Miller 1964, Efron 1982). The 95% confidence limits given in the results were computed with the appropriate t-value and the jack-knife estimate of the variance.

Other models involving density dependence of the handling time did not provide an adequate fit for our data.

Theoretically, one could adjust the data for food depletion during one experimental run. However, analyses showed this did not change the principal shape of the fitted curves.

Results

Chalcalburnus chalcoides

Danube bleak were highly efficient at catching planktonic prey at all densities (Table 1). All prey animals were caught with 100% efficiency except of *Cyclops*, which were caught with a success rate ranging from 76.6% to 83.6% under different prey densities. *Cyclops* were also the prey species to which the fish reacted with an almost linear response (Fig. 1). Although there are indications of a more sigmoid response at densities between 2 and 10 individuals l^{-1}, statistical analysis did not allow expansion of the basic disc equation.

Responses were quite different with regard to the other prey types to which Danube bleak reacted in a strongly sigmoid fashion which cannot be

Table 2. Functional responses of *Vimba vimba*. See Table 1 for further explanations.

Prey	h	a_0	a_1
Cyclops	0.02023	2.747	
	0.0173–0.0231	2.45–3.04	
Bythotrephes	0.01521	0.00002	2.004
	0.0147–0.0157	−0.0001–0.0001	1.83 –2.18
Leptodora	0.03212	0.8031	0.7326
	0.0289–0.0353	−0.80–2.41	0.239–1.226

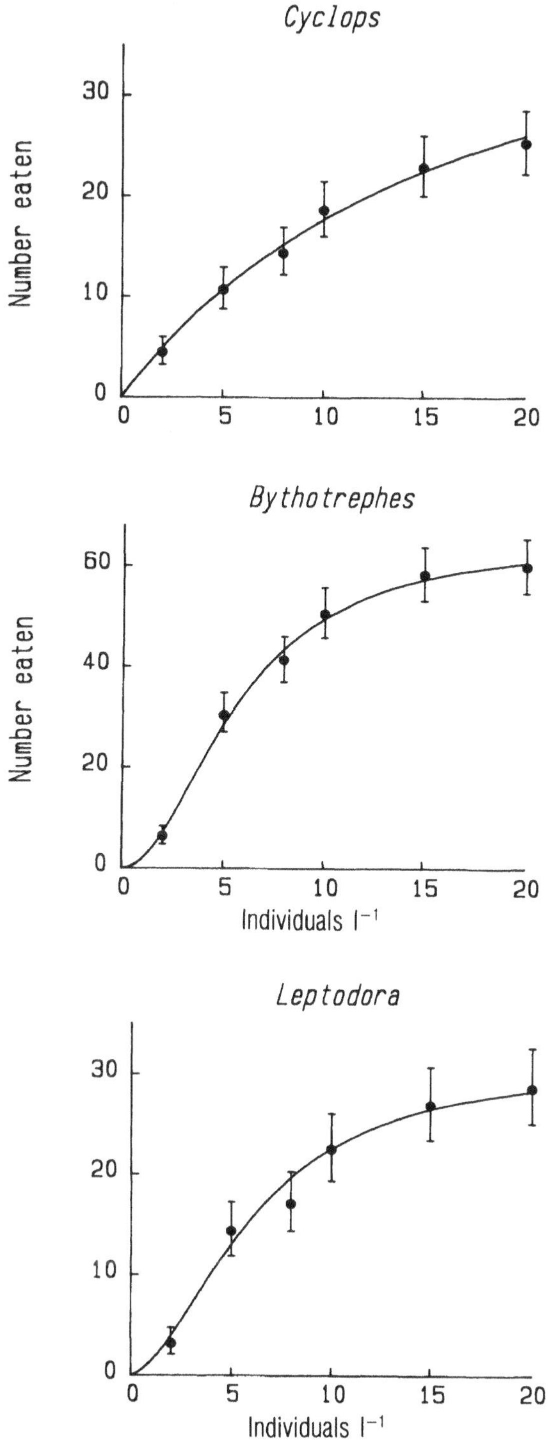

Fig. 2. The functional response of *Vimba vimba* with *Cyclops*, *Leptodora kindti* and *Bythotrephes longimanus*. Means of the number of prey items eaten within 3 minutes in 8 experiments at each prey density are shown, together with the corresponding 95% confidence limits. For the parameters of the curve see Table 2.

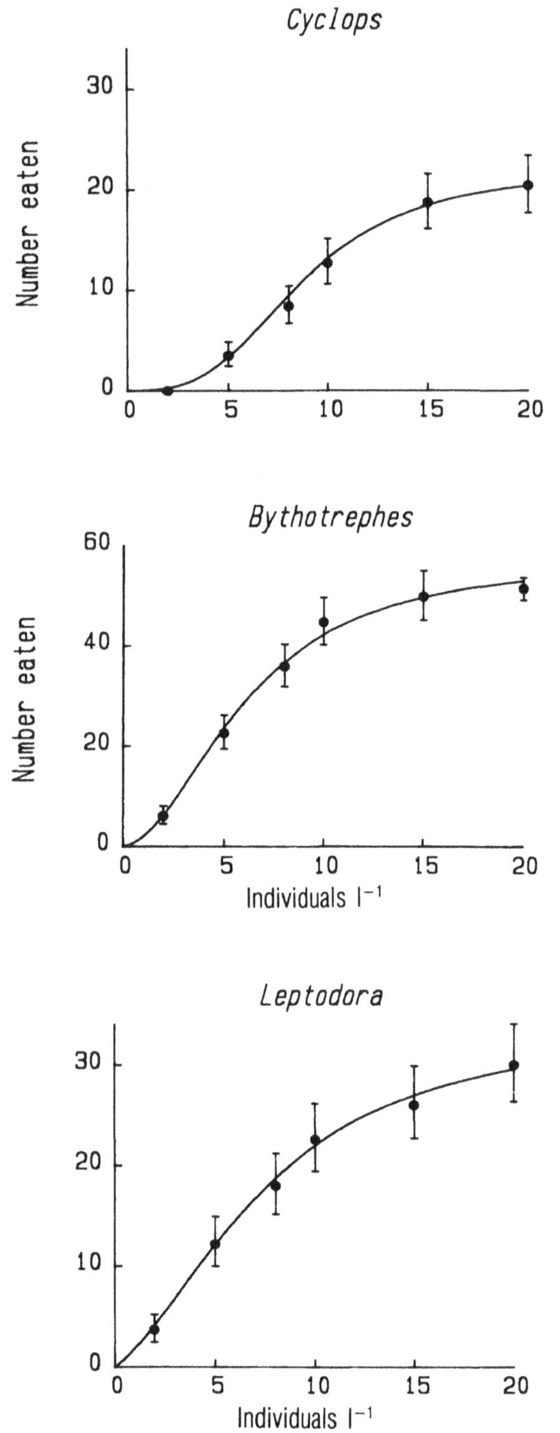

Fig. 3. The functional response of *Abramis brama* with *Cyclops, Leptodora kindti* and *Bythotrephes longimanus*. Means of the number of prey items eaten within 3 minutes in 10 experiments at each prey density are shown, together with the corresponding 95% confidence limits. For the parameters see Table 3.

adequately described by the type of equations used. For *Daphnia* and *Bythotrephes,* it was necessary to include terms of order three and four into the attack equation (2) while for *Leptodora,* expansion of the attack equation up to the order of two sufficed. In all cases a sharp increase in the attack rates occurred at a density above 5 individuals per liter.

When Danube bleak fed on *Leptodora* and *Cyclops* it was possible to measure the time needed to manipulate one prey item. The activity measured was essentially mastication which occupies most of the total handling time. The prey densities used for these measurements were in both cases 20 individuals l^{-1}, resulting in an average duration of mastication of 1.95 sec for *Leptodora* and 0.16 sec for *Cyclops*. The theoretical handling times derived from the fitted equations are 4.17 sec and 0.78 sec, respectively. The theoretical values are higher because, of course, they also include prey recognition and attack time which were not measured in the experiments. The theoretical handling times for *Bythotrephes* and *Daphnia* were respectively 1.72 and 1.62 sec.

Vimba vimba

At the highest prey density used with this fish species, namely 20 individuals l^{-1}, around 60 *Bythotrephes* were caught in 180 sec, which is close to twice the number of *Cyclops* or *Leptodora.* Only the results for *Bythotrephes* and *Leptodora* (Fig. 2) are directly comparable because in both cases the fish increased their attack rates linearly with prey density (Table 2). The low total handling time with *Bythotrephes* derived from the corresponding equation is directly associated with the high feeding rates at prey densities of 15 l^{-1} and higher. For *Bythotrephes* the computed value for the handling time is 2.74 sec, whereas for *Leptodora*, in accordance with the lower feeding rate, it is 5.78 sec. Like the previous species, *Vimba* reacted to *Cyclops*- with a constant, density-independent attack rate with a theoretical handling time of 3.64 sec. The functional response curve does not come close to its asymptote at the experimental food densities used.

Abramis brama

The responses of bream (Table 3) towards *Bythotrephes* and *Leptodora* (Fig. 3) are similar to those of *Vimba*. At prey densities of 20 individuals l^{-1}, roughly 50 *Bythotrephes* and 30 *Leptodora* were caught. With both prey types, there was a linear increase in attack rate with prey density. The derived handling time was 3.10 sec for the former prey species and 5.26 sec for the latter. Compared with that of the previous fish species, the response of bream to *Cyclops* is remarkably strong: the attack rates increased quadratically with prey density, resulting in a clearly sigmoid response curve. Nevertheless, maximum feeding rates did not reach the values for the two cladoceran prey species due to the relatively long total handling time of 8.15 sec for *Cyclops*.

Rutilus rutilus

Like bream, roach were also able to respond to copepod prey in a sigmoid fashion, while their response to *Bythotrephes* (Fig. 4) was even stronger. In the first case, attack rates increased linearly with prey density, while in the second, the attack function was quadratic (Table 4). Total handling times derived from the equations were 4.82 sec for the copepod and 3.59 sec for the cladoceran.

Scardinius erythrophthalmus

Rudd were tested with *Bythotrephes* only (Table 5). Their response to this prey was similar to that of Danube bleak or zährte on copepods. Feeding rate increased almost linearly with prey density (Fig. 5). Within the range of prey densities used, the response curve did not level off significantly. Due to the low constant attack rate, the maximum feeding rate attained at the highest prey density remained rather low at around 40 individuals eaten in 180 sec. The theoretical handling time of 1.71 sec would allow for much higher feeding rates.

Discussion

Rich diversity of responses of different cyprinid species emerged. This is particularly remarkable because prey were not clumped in space and no alternative food was present during the experiments. Visual observations did not give any indications that the fish had to decide between feeding and other activities. Thus, the main reasons for sigmoid functional responses commonly discussed in the literature (e.g. Curio 1976) were excluded from these experiments. It is therefore necessary to consider alternative possibilities for this rather unexpected variation. Basically, the particular shape of the response curve could be associated (1) with some trait of the prey, (2) with the behavior of the fish, or (3) could be a result of complex interactions between both.

If prey type is the only variable affecting the feeding response, one would predict that each prey species has a specific function regardless of which fish species acts as the predator. The shape of the empirical response function applied here is influenced mainly by the number of terms into which the attack function is expanded. The shape of the at-

Table 3. Functional responses of *Abramis brama*. See Table 1 for further explanations.

Prey	h	a_0	a_1	a_2
Cyclops	0.04528	0.0000	– 0.00914	0.03334
	0.0421–0.0484	– 0.0003–0.0003	– 0.122–0.103	0.016–0.050
Bythotrephes	0.01721	0.4642	1.502	
	0.0165–0.0179	– 0.682 –1.610	1.19 –1.81	
Leptodora	0.02924	1.373	0.4774	
	0.0267–0.0317	0.312 –2.44	0.243–0.712	

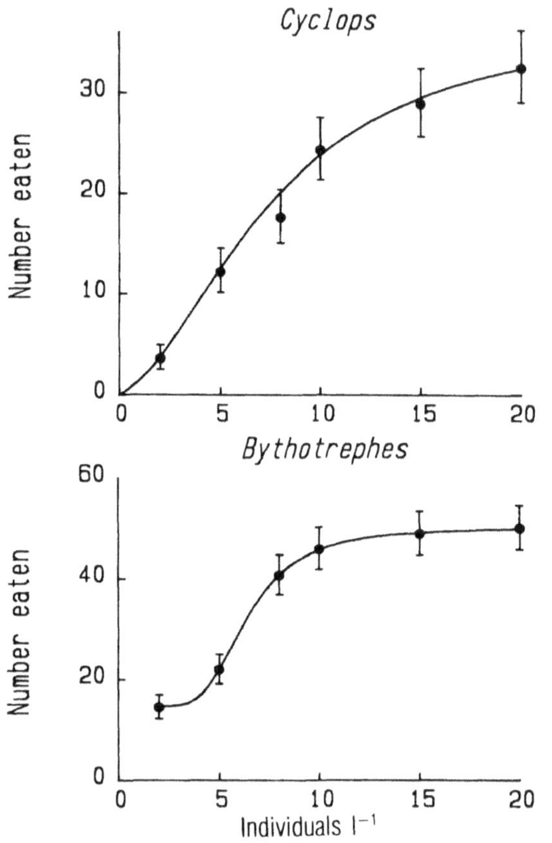

Fig. 4. The functional response of *Rutilus rutilus* with *Cyclops* and *Bythotrephes longimanus*. Means of the number of prey items eaten within 3 minutes in 10 experiments at each prey density are shown, together with the corresponding 95% confidence limits. For the parameters of the curve see Table 4.

tack function, and its relation to prey density, may thus serve as a first guide for the interpretation.

In two cases *Cyclops* induced constant attack rates, in one a linear, and in another one a parabolic attack function. *Leptodora* was never hunted at a constant attack rate, with attack rates increasing linearly with prey density by two fish species, and with a non-linearly increasing attack rate by one species. Finally, the functional response to *Bythotrephes* was characterized by a constant attack rate in one case, a linear increase in two, and a non-linear relationship in two further instances. *Daphnia hyalina,* which was used only in the series of experiments with the Danube bleak, elicited a non-linear response, rather like *Bythotrephes.*

The corresponding account for the different fish species shows that the Danube bleak responded with a constant attack rate in the presence of copepods only. Any other prey response was distinctly non-linear; that is, at initial densities above 5 prey individuals per liter, attack rates increased with prey density. Roach showed a linear increase in attack rate when presented with copepods, and a parabolic one with *Bythotrephes.* The attacks of zährte upon copepods were characterized by a constant attack rate and those on *Bythotrephes* and *Leptodora* by rates increasing linearly with prey density. Bream behaved like zährte with the two large predatory cladocerans. However, the response of this species to copepods is best described by a quadratic attack function. Rudd was the only species which reacted to *Bythotrephes* with a constant attack rate.

Thus it is reasonable to conclude that characteristics of both prey and predator interact to determine the shape of the feeding response curves. In particular the responses can be viewed as the outcome of the fishes' attempts to accelerate attack rates, and the opposing effects of prey traits which act to constrain the predator more or less efficiently in this enterprise.

Copepods are very efficient at escaping compared to cladocerans. In the laboratory, all fish species exhibited more efficient foraging (in terms of numbers eaten) while feeding on cladocerans

Table 4. Functional responses of *Rutilus rutilus*. See Table 1 for further explanations.

Prey	h	a_0	a_1	a_2
Cyclops	0.02677	0.9890	0.5579	
	0.0247–0.0288	0.181–1.80	0.361–0.754	
Bythotrephes	0.01994	23.75	−9.139	1.191
	0.0196–0.0202	22.0 −25.4	−9.70 —8.57	1.09–1.28

than on copepods. This observation conforms with findings from a variety of zooplanktivorous fishes (Drenner et al. 1978, Furnass 1979, Drenner & McComas 1980, McComas & Drenner 1982, Winfield et al. 1983). In light of the interpretation given above there are two possible ways in which traits of the prey may interfere with the predator's attempt to maximize its feeding rate. First, it could be that at low densities of *Cyclops* in combination with its efficient escape reactions, the fish 'shut off' the attack response because the possible gain of energy is too low (Townsend & Risebrow 1982, Winkler & Moreno 1984), or use a different attack behavior for evasive prey (e.g. Wright & O'Brien 1984). The second mechanism may involve the intermittent movement pattern of copepods which is such that only a small proportion of the prey stock is in motion at any given time. In the study of Wright & O'Brien (1984) only 3.2% of the *Diaptomus* individuals were active simultaneously. The operational density for copepods may therefore be much lower than the actual density. The differences between the movement patterns of *Leptodora, Bythotrephes* and *Daphnia* do not seem to have similar consequences.

As observed by Lazzaro (1987), most studies dealing with fish feeding on zooplankton are not concerned with obligate planktivores. Even though many such fishes are dependent on zooplankton during some period of their life, they may be correctly classified as facultative planktivores. Feeding techniques of these two types may differ considerably (e.g. Janssen 1982). While all species investigated in this study feed on zooplankton when young, the Danube bleak is the only true planktivore among them. Its diet consists mainly of zooplankton and neuston, although sometimes even larger prey, such as small fish, are taken (Herzig, Orellana & Winkler unpublished). The present ex-

Table 5. Functional response of *Scardinius erythrophthalmus*. See Table 1 for further explanations.

Prey	h	a_0
Bythotrephes	0.00950	3.144
	0.0073–0.0117	2.83–3.46

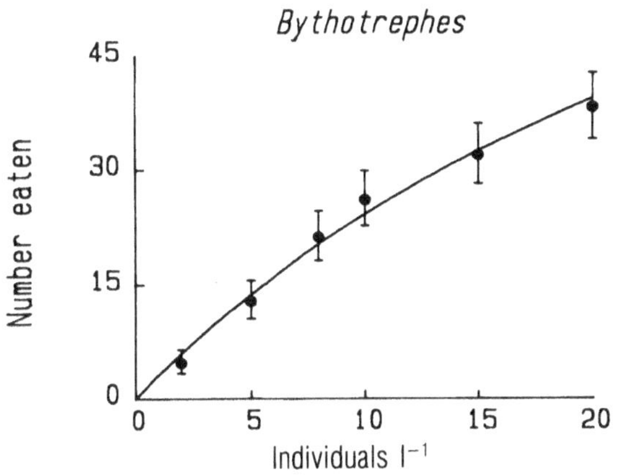

Fig. 5. The functional response of *Scardinius erythrophthalmus* with *Bythotrephes longimanus*. Means of the number of prey items eaten within 3 minutes in 8 experiments at each prey density are shown, together with the corresponding 95% confidence limits. For the parameters of the curve see Table 5.

periments have revealed that this species exhibits high attack rates which strongly increase with prey density. In these respects, the Danube bleak has no match among the cyprinid species studied. Other measurements, to be presented elsewhere (Winkler unpublished), clearly show that the main mechanism for this attack behavior is the locomotor response of the species. Swimming speed when feeding is higher than in any of the other species studied and since high encounter rates tend to reduce the time spent for examining prey, one would predict that either the fish fails to discriminate among different prey types or it has to reduce its attack rate to facilitate prey choice (Gendron & Staddon 1983). Bluegills, *Lepomis gibbosa* (L.), for instance, lower their feeding rates in turbid water, such that their prey selectivity remains the same as in clear water (Gardner 1981).

At the sizes of the fish studied in the laboratory, problems of handling larger prey, especially *Leptodora,* were of some importance. Older Danube bleak (Orellana 1985) strongly favour both species of predatory cladocerans. However for larger fish, handling prey is no longer a problem. If gape-limitation (Schmitt & Holbrook 1984) becomes less important as the main correlate of prey choice, locomotion is left as the prime candidate for explaining species differences in natural communities. Consequently, morphological and physiological traits associated with locomotion, and therefore attack behavior, ought to be powerful variables in the explanation of feeding niches and species diversity.

Acknowledgements

We are grateful to the staff of the Limnological Institute of the Austrian Academy of Sciences for their encouragement, discussion, and technical assistance; we should especially like to mention A. Herzig, K. Mayrhofer, and R. Niederreiter. This work was carried out in the course of a research program financed with a grant (S35-2) from the Fonds zur Förderung der Wissenschaftlichen Forschung.

References cited

Abrams, P.A. 1982. Functional responses of optimal foragers. Amer. Nat. 120: 382–390.

Curio, E. 1976. The ethology of predation. Springer-Verlag, Berlin. 250 pp.

Drenner, R.W. & S.R. McComas. 1980. The roles of zooplankter escape ability and fish size selectivity in the selective feeding and impact of planktivorous fish. pp. 587–593. *In:* W.C. Kerfoot (ed.) Evolution and Ecology of Zooplankton Communities, University Press of New England, Hannover.

Drenner, R.W., J.R. Strickler & W.J.O'Brien. 1978. Capture probability: the role of zooplankter escape in the selective feeding of planktivorous fish. J. Fish. Res. Board Can. 35: 1370–1373.

Durbin, A.G. & E.G. Durbin. 1975. Grazing rates of the Atlantic menhaden *Brevoortia tyrannus* as a function of particle size and concentration. Mar. Biol. 33: 265–277.

Efron, B. 1982. The jackknife, the bootstrap and other resampling plans. Soc. Ind. Appl. Math., Philadelphia. 92 pp.

Furnass, T.I. 1979.Laboratory experiments on prey selection by perch (*Perca fluviatilis*). Freshwat. Biol. 9: 33–43.

Gardner, M.B. 1981. Effects of turbidity on feeding rates and selectivity of bluegills. Trans. Amer. Fish Soc. 110: 446–450.

Gendron, R.P. & J.E.R. Staddon. 1983. Searching for cryptic prey: the effect of search rate. Amer. Nat. 121: 172–186.

Herzig, A. & H. Winkler. 1986. The influence of temperature on the embryonic development of three cyprinid fishes, *Abramis brama, Chalcalburnus chalcoides mento* and *Vimba vimba*. J. Fish Biol. 28: 171–181.

Holling, C.S. 1959. Some characteristics of simple types of predation and parasitism. Can. Entomol. 91: 385–398.

Holling, C.S. 1965. The functional response of predators to prey density and its role in mimicry and population regulation. Mem. Entomol. Soc. Can. 45: 1–60.

Ivlev, V.S. 1944. The time of hunting and distance travelled by the predator in relation to the population density of the prey. Zoologicheskij Zhurnal 23: 139–144. (In Russian).

Janssen, J. 1982. Comparison of searching behavior for zooplankton in an obligate planktivore, blueback herring (*Alosa aestivalis*) and a facultative planktivore, bluegill (*Lepomis macrochirus*). Can. J. Fish. Aquat. Sci. 39: 1649–1654.

Krebs, J.R. 1973. Behavioral aspects of predation. pp. 73–111. *In:* P.P.G. Bateson & P.H. Klopfer (ed.) Perspectives in Ethology, Plenum Press, New York.

Lazzaro, X. 1987. A review of planktivorous fishes: their evolution, feeding behaviours, selectivities, and impacts. Hydrobiologia 146: 97–167.

McComas, S.R. & R.W. Drenner. 1982. Species replacement in a reservoir fish community: silverside feeding mechanics and competition. Can. J. Fish. Aquat. Sci. 39: 815–821.

McNair, J.N. 1980. A stochastic foraging model with predator training effects: I. Functional response, switching, and run lengths. Theor. Pop. Biol. 17: 141–166.

McNair, J.N. 1981. A stochastic foraging model with predator training effects. II. Optimal diets. Theor. Pop. Biol. 19: 147–162.

Marten, G.G. 1973. An optimization equation for predation. Ecology 54: 92–101.

Miller, R.G. 1964. The jackknife – a review. Biometrika 61: 1–17.

Orellana, C.P. 1985. Nahrungserwerb und Biologie der Seelaube *Chalcalburnus chalcoides mento* (Agassiz) im Mondsee. M.Sc. Thesis, University of Salzburg, Salzburg. 69 pp.

Rashevsky, N. 1959. Some remarks on the mathematical theory of nutrition of fishes. Bull. Math. Bioph. 21: 161–183.

Real, L.A. 1977. The kinetics of functional response. Amer. Nat. 111: 289–300.

Real, L.A. 1979. Ecological determinants of functional response. Ecology 60: 481–485.

Schmitt, R.J. & S.J. Holbrook. 1984. Gape-limitation, foraging tactics and prey size selectivity of two microcarnivorous species of fish. Oecologia 63: 6–12.

Stephens, D.W. & J.R. Krebs. 1986. Foraging theory. Princeton University Press, Princeton. 247 pp.

Townsend, C.R. & A.J. Risebrow. 1982. The influence of light level on the functional response of a zooplanktonivorous fish. Oecologia 53: 293–295.

Winfield, I.J., G. Peirson, M. Cryer & C.R. Townsend. 1983. The behavioural basis of prey selection by underyearling bream (*Abramis brama* (L.)) and roach (*Rutilus rutilus* (L.)). Freshwat. Biol. 13: 139–149.

Winkler, H. & T. Moreno. 1984. Die Funktionelle Reaktion planktonfressender Fische. Sitzber. Österr. Akad. Wiss., Mathem. Naturw. Kl., Abt. I, 193: 23–31.

Wright, D.I. & W.J. O'Brien. 1984. The development and field test of a tactical model of the planktivorous feeding of white crappie (*Pomoxis annularis*). Ecol. Monogr. 54: 69–98.

Environmental Biology of Fishes **33**: 63–71, 1992.

The energetics of starvation and growth after refeeding in juveniles of three cyprinid species

Wolfgang Wieser, Gerhard Krumschnabel & Julius P. Ojwang-Okwor
Institut für Zoologie, Universität Innsbruck, Technikerstrasse 25, A-6020 Innsbruck, Austria

Received 2.11.1989 Accepted 7.5.1990

Key words: Leuciscus cephalus, Chalcalburnus chalcoides mento, Scardinius erythrophthalmus, Compensatory growth, Cost of growth, Oxygen consumption, Routine activity, Maintenance

Synopsis

Experiments were conducted to monitor changes in body mass and metabolic energy expenditure before, during, and after periods of starvation in juveniles of three species of cyprinids: *Leuciscus cephalus, Chalcalburnus chalcoides mento,* and *Scardinius erythrophthalmus.* During the starvation period all fish lost weight at about the same rate and the total amount of oxygen consumed during an experimental period of 20 h was about 40% lower in the starved than in the fed groups. Upon refeeding, both mass specific maintenance and routine rates of metabolism as well as relative growth rates increased rapidly, the peaks of these increases being directly proportional to the length of the starvation period. Maximum compensatory growth was observed after four weeks of starvation in *C. chalcoides* and *S. erythrophthalmus,* with relative growth rates reaching 30% d^{-1} during the first measuring interval after refeeding. The pattern of 'time-dependent compensatory growth' displayed by these fish is similar to the responses of a colonial hydroid in which the rate of catch-up growth increased with the amount of stress to which the animals had been exposed. The exact cost of compensatory growth cannot be calculated because oxygen consumption and growth were not measured simultaneously. However, on the basis of data and calculations reported by Wieser & Medgyesy (1990) it appears that compensatory growth, if fuelled by the metabolic power indicated by our measurements of oxygen consumption, would have to be about twice as efficient as normal growth in the related species *Rutilus rutilus.*

Introduction

Due to the spatial and temporal patchiness of food in the aquatic environment, fish experience periods of reduced food availability which may last several days. This holds particularly for larvae and juveniles which because of their selective mode of feeding, limited powers of prey detection, and elevated metabolic rates require high densities of prey (Wanzenböck & Schiemer 1989, Wieser & Medgyesy 1990). Blaxter's (1969, p. 218) observation 'that larvae in the sea were often very near the point of starvation' may also describe the situation in oligotrophic lakes. Although the effects of starvation in fish have been discussed extensively (Ivlev 1961, Blaxter 1969, 1988, Love 1970, Moon & Johnston 1980, Gadomski & Petersen 1988) many aspects are insufficiently understood. Two questions are of particular ecological interest: first, how do young fish adjust to periods of food shortage so as to maximize their chance of survival and second, how do they react to the presence of food so as to regain as quickly as possible the body mass lost during the preceding period of shortage?

Physiological and behavioral responses to food shortage are likely to include the reduction of all superfluous activity, both metabolic and locomotory (Beamish 1964, Blaxter & Ehrlich 1974), but during the initial phase of food shortage increased locomotory activity might be a prerequisite for obtaining the little food that is still available. Such a strategy has been described (Huse & Skiftesvik 1985, Blaxter 1988, Wieser et al. 1988b), and a trade-off must be sought between these two contradictory options.

As far as the reactions of the organism to refeeding are concerned, compensatory growth after starvation has been described in adult rainbow trout (Dobson & Holmes 1984), juvenile arctic charr (Miglavs & Jobling 1989), and in young coregonids (Gunkel 1979), but contradictory results have been obtained by Dabrowski et al. (1986) with coregonid larvae, and by Schwarz et al. (1985) with common carp. Certainly, the energetics of growth during and after periods of starvation has not been investigated with sufficient precision.

The experiments were designed to monitor changes in body mass and metabolic energy expenditure before, during, and after periods of starvation in juveniles of three species of cyprinids. Our aim was three-fold: first, to record changes in the patterns of oxygen consumption and growth in the same groups of animals; second, to contribute to an understanding of the energetics of growth after refeeding; and third, to compare the responses to starvation and refeeding of closely related species.

Material and methods

Experiments were conducted with rudd, *Scardinius erythrophthalmus* (L.), chub, *Leuciscus cephalus* (L.), and Danubian bleak, *Chalcalburnus chalcoides mento* (Agassiz). Eggs stripped from wild-caught fish were fertilized and left to develop attached to glass plates; later resultant larvae were fed *Artemia salina* nauplii ad libitum. According to our previous work with these and other species of cyprinids (Wieser et al. 1988a, b) the ration required to achieve ad libitum feeding decreased from about 200% fbw d^{-1} for the first feeding larvae to about 75% d^{-1} for juveniles weighing 150 mg. Beyond this size *Artemia* does not constitute an optimal diet for the cyprinids studied (Wieser & Medgyesy 1990). For our experiments we used juveniles weighing at the beginning between 50 and 100 mg, corresponding to an age from 35 to 55 days after fertilization. The fish were maintained in 20 l aquaria at a photoperiod of 16L : 8D. The water in the aquaria was replaced approximately every hour and *Artemia* were dispensed automatically into the aquaria as described before (Wieser et al. 1988b). Two groups, each consisting of about 600 fish, were set up for each species, one fed to satiation with *Artemia,* the other deprived of food. After 7, 14, 21, and 28 days of starvation 150 fish were transferred to other tanks and refed with *Artemia* ad libitum. Thus the following experimental regimes were established: (1) fed ad libitum (i.e. about 75% fbw d^{-1}) throughout (control); (2) starved throughout (starved); (3) starved one week and refed (SRF1); (4) starved two weeks and refed (SRF2); (5) starved three weeks and refed (SRF3); (6) starved four weeks and refed (SRF4). All tanks were well aerated and the water was maintained at 20° C. Even after four weeks of starvation the point of no return (Blaxter & Hempel 1963) was not reached in any of the species.

Growth

Every 2nd or 3rd day 100 fish were removed from each tank, lightly anaesthetized with MS 222, blotted and weighed in ten lots. After weighing, the fish were put into well-aerated water and after recovery returned to the experimental tank. Mortality was very low throughout the experiment. This procedure ensured that each weight determination was based on at least two-thirds of the number of fish exposed to each experimental regime, thus minimizing the problem of individual variability. Fish that had been fed were left in aerated vessels for 12 h before weighing.

Growth rates were calculated for each measuring interval by assuming exponential growth and determining the exponent r from $\ln M_2 - \ln M_1/\Delta t$, M_1 and M_2 representing the mean weights of the younger and the older group, respectively. Relative growth rate per day, g, is then given by $e^r - 1$.

Respirometry

Oxygen consumption of groups of fish, 5–15 according to weight, was measured at 20° C under continuous illumination, using the intermittent 3-chamber system (one chamber for each species) described by Wieser et al. (1988b). Each monitoring period lasted 20 h and consisted of 10 one-hour measurement periods. One sample of fish from each of the different regimes and species was taken on the following days after the beginning of the experiment:
control – 7, 15, 22; starved – 2, 9, 14, 21, 28; SRF1 – 10; SRF2 – 17, 24, 29; SRF3 – 23, 30. No measurement could be carried out on SRF4.

Thus, oxygen consumption of the starved/refed fish was measured for the first time 2–3 days after the resumption of feeding. The time schedule of the investigation did not permit more than one measurement of oxygen consumption per treatment and species. However, the metabolic responses of the three species investigated to starvation and refeeding were very similar (interaction species × treatment: $p = 0.127$, n.s.; ANOVA; see Table 1 for species-specific averages) so that each treatment is represented by three quasi-replicates.

Fish that had been feeding were left without food for about 12 h before the start of a period of oxygen consumption measurement.

Table 1. Relative growth rates (g in % d^{-1}) and routine rates of oxygen consumption ($\dot{V}O_2$ in μmol $g^{-1}h^{-1}$) in continuously fed (control), continuously starved, and starved + refed (SRF) groups of juvenile cyprinids belonging to three species. Refeeding began after 1, 2, 3, or 4 weeks of starvation. All experiments for the determination of routine oxygen consumption are shown, each represented by means and SD of all (10) hourly values taken during a measuring period of 20 h. The number of days listed in column 2 begins with the day on which the experiment (control, starved) or refeeding (SRF) was started. The days indicated are those on which oxygen consumption was measured. The relative growth rates indicated represent averages for the periods (difference between two consecutive dates) listed in column 2. n.d. = not determined.

Experiment	Days	*Leuciscus cephalus*		*Chalcalburnus chalcoides*		*Scardinius erythrophthalmus*	
		g	$\dot{V}O_2r$ $\bar{x}$ (SD)	g	$\dot{V}O_2r$ $\bar{x}$ (SD)	g	$\dot{V}O_2r$ $\bar{x}$ (SD)
control	7	4.7	21.4(3.7)	2.2	20.5(6.0)	3.0	16.4(5.1)
	15	3.4	20.3(3.0)	5.6	19.9(5.9)	2.0	16.5(4.3)
	22	3.8	17.7(3.9)	3.8	18.6(6.2)	4.3	n.d.
starved	2	– 1.0	12.0(1.4)	– 0.6	9.9(2.6)	– 0.3	11.6(3.7)
	9	– 1.2	11.4(2.5)	– 4.2	11.4(1.9)	– 1.6	10.5(1.1)
	14	– 1.6	11.0(1.3)	– 0.4	10.1(2.4)	– 6.1	9.0(1.0)
	21	– 4.3	9.1(1.1)	– 3.6	8.0(1.1)	– 2.4	n.d.
	28	– 1.7	10.8(2.4)	– 3.8	10.6(2.6)	– 3.0	10.6(2.8)
SRF1	3	0.6	13.9(4.7)	7.6	14.4(6.7)	5.4	12.6(4.0)
SRF2	3	0.7	16.2(5.4)	8.0	15.8(4.6)	9.6	14.4(3.7)
	10	6.3	19.6(5.8)	5.4	15.0(5.6)	3.0	15.6(5.3)
	15	6.0	16.9(3.3)	6.6	15.6(4.3)	6.6	13.8(2.3)
SRF3	2	10.5	21.7(5.6)	20.5	18.6(3.4)	14.6	18.3(4.7)
	9	5.9	16.8(3.1)	8.4	17.7(4.9)	6.8	14.3(2.9)
SRF4	2	11.2	n.d.	30.7	n.d.	30.0	n.d.
	9	9.0	n.d.	9.6	n.d.	9.2	n.d.
	18	4.9	n.d.	3.7	n.d.	4.8	n.d.

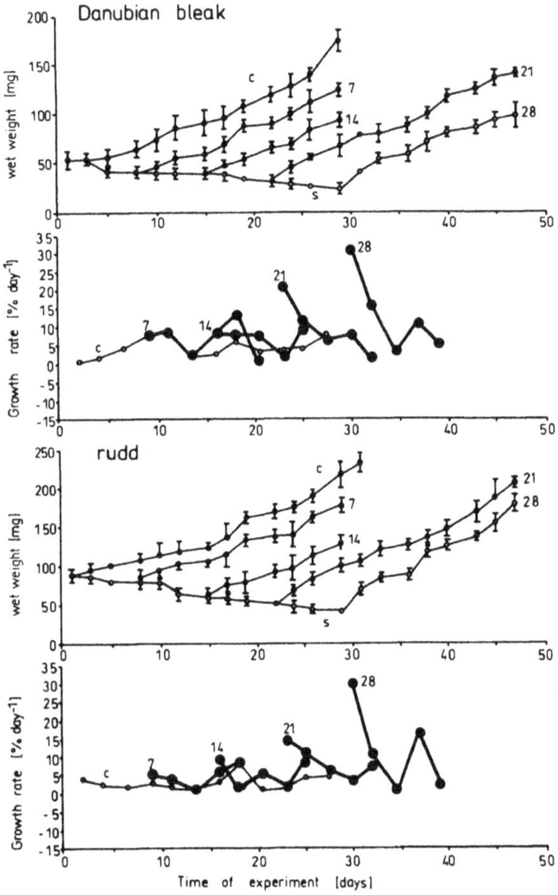

Fig. 1. Body weights (upper panels) and relative growth rates (lower panels) in juveniles of two species of cyprinids: Danubian bleak, *Chalcalburnus chalcoides mento,* and rudd, *Scardinius erythrophthalmus.* Curves are for continuously fed (c), continuously starved (s), and starved + refed fish. Numbers next to curves indicate duration of the starvation period (in days) before refeeding was initiated. For clarity, mean growth rates are shown for the first four measuring intervals only. Each weight determination is based on 100 specimens; fresh weights are represented by means ± SD. Experimental temperature 20° C.

Results

Growth

The three species of cyprinids responded to the experimental treatments with uniform patterns of growth. Average initial weight was 50–100 mg fresh body weight (fbw). Controls grew uniformly with average rates of from 2 to 5% d^{-1}. The starving groups lost weight at rates ranging from 0.3 to 6.1% d^{-1} (Table 1, Fig. 1).

Growth after refeeding showed characteristic patterns: the initial growth rate after refeeding increased in proportion to the length of the starvation period in all three species (Fig. 1, only two species shown). In *L. cephalus* the growth rate was 5% d^{-1} on refeeding after 7 days starvation and 11.2% d^{-1} after resumption of feeding following 28 days of starvation; in *C. chalcoides,* 7.6 and 30.7% d^{-1}, and in *S. erythrophthalmus,* 5.4 and 30.0% d^{-1}. Thus the three species clearly showed 'compensatory growth' after refeeding, most noticeably when they had been starved for at least two weeks. In *L. cephalus* growth rates were generally lower than in the other two species and the compensatory growth response, after one and two weeks of starvation, was delayed (Table 1). However, the duration of the compensatory response was short in all species. In most trials the relative growth rate returned to that of the controls after the second or third measuring interval (i.e. after 6–9 days). Only in the groups that had been starved longest did the specific growth rate show a second peak 9–12 days after refeeding (Fig. 1).

Respiration

A clear response of oxygen consumption to handling and general excitement was seen after the fish had been put into the respirometer. This effect took about 10 h to subside (Fig. 2), and the rate of oxygen consumption remained fairly constant thereafter. In accordance with previous experiments (Wieser et al. 1988b, Wieser & Medgyesy 1990) this stable level of respiratory rate may be called maintenance rate ($\dot{V}O_2m$) but is actually very close to the standard rate as usually defined (see Wieser & Medgyesy 1990 for a more extensive discussion of this point). The metabolic responses of the three species investigated to starvation and refeeding were three-fold: (1) the level of excitement during the first ten hours, and thus the range of $\dot{V}O_2m$-values during the whole measurement period (corresponding to the routine rate of oxygen consumption ($\dot{V}O_2r$) decreased with starvation and

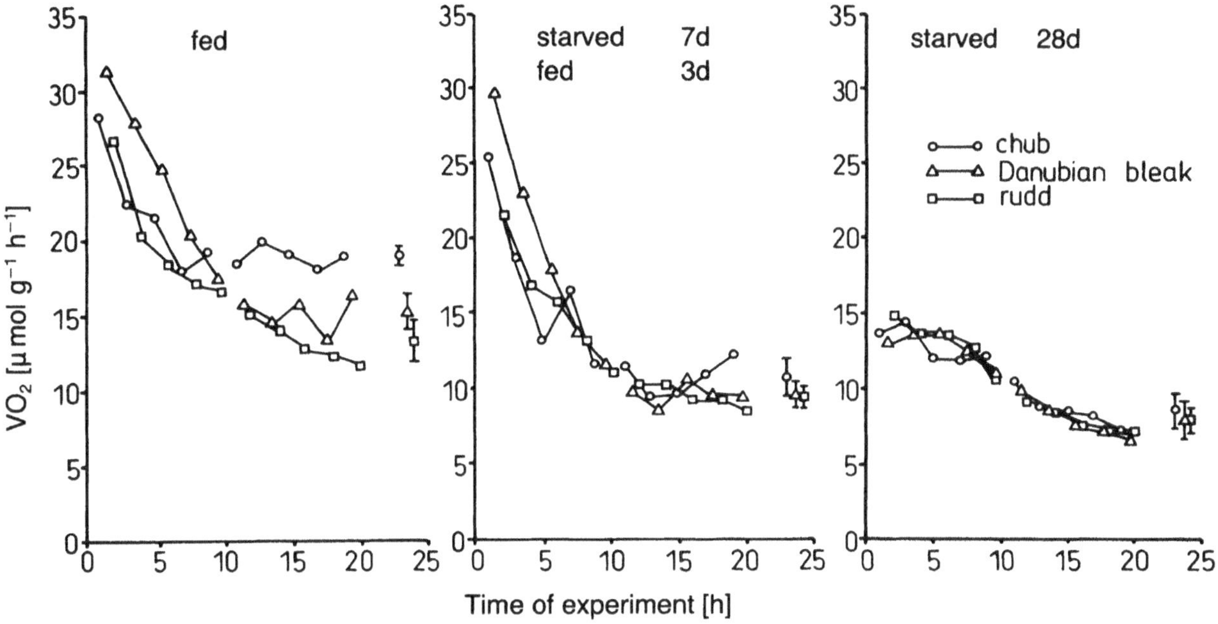

Fig. 2. Three experiments showing weight specific rates of oxygen consumption (based on fresh body weight) of fed (controls), starved, and starved + refed groups of young fish of three species of cyprinids. Each symbol represents an average based on 40 one-minute values extrapolated to 60 minutes (see Wieser & Medgyesy 1990). Measurements were taken every other hour for 20 hours. Means ± SD of all measurements during the last 10 h of an experiment are indicated to the right of each diagram. These rates are considered to be maintenance rates. Experimental temperature 20° C.

increased again upon refeeding; (2) $\dot{V}O_2m$ also decreased with starvation and increased upon refeeding; (3) within two to three days after refeeding both the routine and the maintenance rate of oxygen consumption increased in proportion to the length of the starvation period.

The first statement is illustrated by Figure 2 which gives an example of the metabolic responses to stress in the respirometer of fed, starved, and refed juvenile cyprinids. A more general test is provided by comparing the coefficients of variation ($SD/\bar{x} \cdot 100$) of all experiments on fed and refed fish

Table 2. Maintenance rates of oxygen consumption ($\mu mol\ g^{-1}\ h^{-1}$) of juvenile cyprinids (values of three species pooled). Metabolic level comprises the following groups: control (continuously fed); starved; starved one week and refed (SRF1); starved two weeks and refed (SRF2); starved three weeks and refed (SRF3). Duration of treatment (days), mean values ($\bar{x}$), standard deviations (SD), and number of independent experiments (n) are given.

Metabolic level	Days	$\bar{x}$ (SD)	n	Metabolic level	Days	$\bar{x}$	n
control	7	15.8(3.1)	14	SRF1	3	9.8(1.0)	12
	15	15.5(3.1)	12	SRF2	3	11.7(1.2)	12
	22	14.0(0.9)	10		10	12.8(2.3)	12
					15	13.5(2.4)	15
starved	2	8.9(1.3)	13				
	9	9.7(0.6)	12	SRF3	2	14.9(1.2)	12
	14	8.6(1.4)	12		9	14.0(1.9)	12
	21	6.2(1.1)	11				
	28	7.9(0.7)	12				

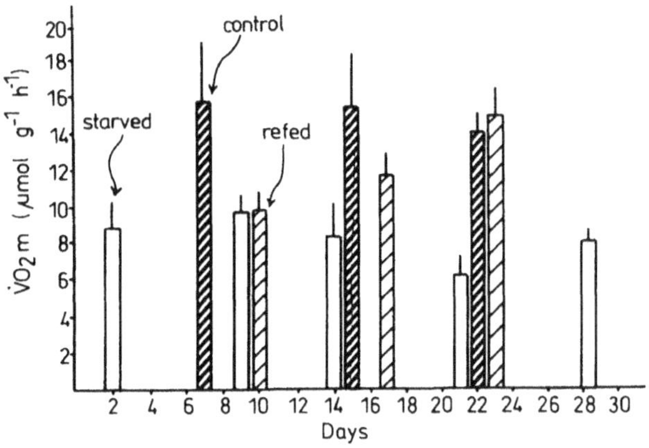

Fig. 3. Maintenance rates of oxygen consumption of starved (open bars), control (full bars) and refed (hatched bars) juvenile cyprinids (three species pooled). In refed fish measurement of oxygen consumption started 2–3 days after commencement of refeeding. Height of bars: $\bar{x}$; thin lines: SD (in μmol O_2 g^{-1} h^{-1}). Experimental temperature 20° C.

(27.0 ± 7.2; n = 27) with those on starved fish (17.7 ± 7.4; n = 15). The difference is significant ($p \leq 0.001$, t-test).

The validity of the second statement is shown by pooling the values of oxygen consumption during the second halves of all experiments, i.e. what here are called maintenance rates ($\dot{V}O_2m$) (Table 2). During the period of starvation $\dot{V}O_2m$ (y) decreased slightly with duration (x in days):

$$y = 9.5–0.08x; \; r = -0.64, \; p \leq 0.001, \; n = 60.$$

However, since the slope of this relationship is shallow (8% d^{-1}), the average rate of starvation metabolism of all three species may be approximated by ($\bar{x}$ ± SD) 8.3 ± 1.3 μmol g^{-1} h^{-1}. This value represents 55% of the average maintenance rate of all control fish and 62% of that of all fish refed for 9–10 days (Table 2).

The third statement is illustrated by Figure 3 which shows that average $\dot{V}O_2m$ remained constant or decreased slightly with time in the controls and in starving fish, whereas in refed fish it increased strikingly in proportion to the length of the starvation period. This is true also for most of the routine values of oxygen consumption. In Table 3 the first determinations of oxygen consumption after the beginning of refeeding in SRF1 and SRF3 are summarized, species specific values for $\dot{V}O_2r$, pooled values for $\dot{V}O_2m$. With exception of $\dot{V}O_2r$ in Danube bleak all differences are highly significant ($p \leq 0.01$; t-test), indicating that refed fish consumed about 50% more oxygen after three weeks than after one week of starvation. Since the same trend holds for the growth rates (Table 3) the increased amount of oxygen consumed is likely to represent the increased cost of growth after the fish had resumed feeding.

By plotting growth rate (% fbw d^{-1}) against routine metabolic power ($\dot{V}O_2r$) of all experiments (Fig. 4) a functional relationship between the two variables is suggested. Since oxygen consumption and growth were not measured simultaneously (as by Wieser & Medgyesy 1990) we refrain from calculating a hypothetical cost of growth. The two pairs of values corresponding to the highest growth rates (14.6% d^{-1} in rudd, 20.5% in Danube bleak) also corroborate the conclusion reached by Wieser & Medgyesy (1990) that in juvenile cyprinids an aerobic maximum is reached beyond which growth can be supported only by suppressing some mainte-

Table 3. Routine ($\dot{V}O_2r$) and maintenance ($\dot{V}O_2m$) rates of oxygen consumption (μmol g^{-1} h^{-1}: means (SD) and number of one h measurements (n)) and relative rates of growth (g in % d^{-1}), of fish starved for one (SRF1) or three (SRF3) weeks and refed for 2–3 days. Routine rates are listed separately for the three species investigated, whereas maintenance rates are pooled values. Levels of significance indicate whether SRF3 and SRF1 differ from each other (ns = non-significant; x = $p \leq 0.01$ xx = $p \leq 0.001$).

	Leuciscus cephalus	*Chalcalburnus chalcoides*	*Scardinius erythrophthalmus*
SRF1			
$\dot{V}O_2r$	13.8(4.7)	14.4(6.7)	12.6(4.0)
n	10	10	10
g	0.6	7.6	5.5
$\dot{V}O_2m$		9.8(1.0)	
n		12	
SRF3			
$\dot{V}O_2r$	21.7(5.6)^{x}	18.6(3.4)^{ns}	18.3(4.7)^{x}
n	10	10	10
g	10.5	20.5	14.6
$\dot{V}O_2m$		14.9(1.2)^{xx}	
n		12	

nance function or by increasing the efficiency of converting metabolic energy into body substance.

Discussion

Both maintenance and routine rates of metabolism decreased with starvation in all three species (see also Beamish 1964, Blaxter & Ehrlich 1974). Thus starving fish saved energy by reducing locomotory activity and perhaps also by reducing the cost of some maintenance functions. A comparison of total oxygen consumption in control and in starved fish yields an estimate of about 40% reduction of metabolic power in the latter. Weight loss during 28 days of starvation was greater than reported by Ivlev (1961) for a number of cyprinids: *L. cephalus* lost about 43%, *C. chalcoides* 56%, and *S. erythrophthalmus* 53%, of their initial wet weights, whereas the values given by Ivlev range between 30 and 40%. In agreement with published information we have to assume that the starving cyprinids experienced a decrease of stamina (Ivlev 1961), RNA/DNA ratio (Buckley 1981, Clemmesen 1987, Miglavs & Jobling 1989), energy reserves (Ehrlich 1974), protein turnover (Loughna & Goldspink 1984), and muscle enzymes (Moon & Johnston 1980, Moon 1983, Lowery et al. 1987). Yet, upon refeeding, mass specific maintenance and routine rates of metabolism as well as growth rates increased rapidly, the most surprising finding being that the peaks of these increases were directly proportional to the length of the starvation period (Fig. 3). While the metabolic rates did not increase beyond the levels of the control (continuously fed) fish, the maximum growth rates (up to 30% d^{-1}: Table 1) are higher than have ever been found in experiments with aquarium raised cyprinids and as high as the highest rates observed in nature (Wieser et al. 1988a).

Thus our experiments confirm the existence of compensatory growth in the juveniles of three cyprinid species. However, the compensatory response was of short duration only since after 10–18 days of refeeding the rates of growth had returned to the level of the controls (Table 1, Fig. 1).

The pattern of what here is called *time-dependent*

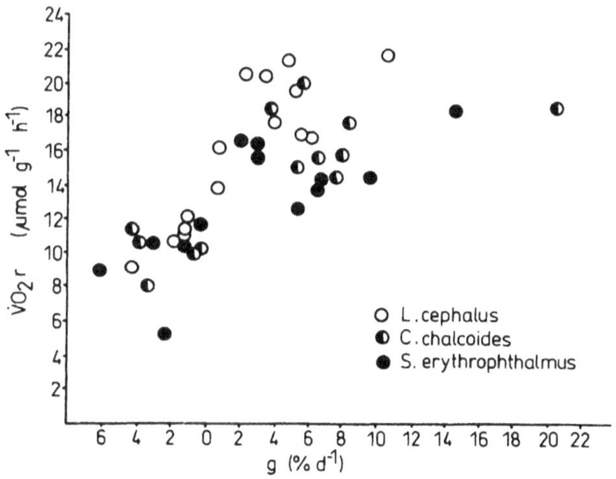

Fig. 4. Routine metabolic rate ($\dot{V}O_2$ r in μmol g^{-1} h^{-1}) plotted against relative growth rate (g in % d^{-1}) of juveniles of three species of cyprinids. Each symbol represents the average rate of oxygen consumption during a measurement period of 20 h, and the average growth rate during an interval ranging from 2 to 7 days (listed in Table 1), in fed, starved, and refed fish.

compensatory growth has recently been described also for another cyprinid, the minnow *Phoxinus phoxinus* by Russell & Wootton (personal communication). In a most stimulating paper, Fenchel (1989) illustrated several metabolic strategies by which protozoans respond to starvation and refeeding. The strategy employed depends on the life style and the ecology of the species, but in all species studied the length of the starvation period was *indirectly* proportional to the rate at which oxygen consumption increased when food became available again. In other words, the longer the cells remained without food the longer it took to put the metabolic machinery in gear again. The opposite is true for the juvenile cyprinids. The longer the fish remained without food, the faster was the metabolic response to the presentation of food, as if, with prolonged food shortage, an emergency machinery was brought into action to guarantee the most rapid and efficient exploitation of a new food source. This pattern resembles that described for the colonial hydroid *Campanularia flexuosa* by Stebbing (1981) and discussed by Sibly & Calow (1986). Colonies of this species showed catch-up growth after exposure to a range of copper concentrations. As in our fish, the rate of catch-up growth was of short duration only and increased with the amount of stress (the concentration of copper), to which the animals had been exposed.

We cannot calculate the exact cost of compensatory growth in the juvenile cyprinids investigated because oxygen consumption and growth were not measured simultaneously. However, on the basis of the data presented by Wieser & Medgyesy (1990) it can be said that a growth rate of 30% d^{-1} – the highest rate of compensatory growth measured – translates into a production figure of approximately 54 J $gfbw^{-1}$ h^{-1}. If the relationship derived in our previous paper for roach, *Rutilus rutilus*, is applied to the cyprinids studied in the present investigation, 55 μmol O_2 g^{-1} h^{-1} of metabolic power would be required to support such a level of somatic production. This is about twice as much as the maximum scope for activity (i.e. the difference between the highest and the lowest rates of oxygen consumption) measured in fed and starved fish (Fig. 2). Compensatory growth would thus have to be about twice as efficient as normal growth in *R. rutilus* if, under otherwise identical conditions, it were to be fuelled by the metabolic power available to the fish used in our experiments.

Acknowledgements

This investigation was carried out within the framework of the Austrian Cyprinid Project, supported by the 'Fonds zur Förderung der wissenschaftlichen Forschung in Österreich', project no. S-35/04. J.P.O.-O. was the recipient of a UNESCO-fellowship. We thank Wolfgang Mark, Niki Medgyesy, Rüdiger Kaufmann, and Manfred Krotthammer for help at various stages of this project.

References cited

Beamish, F.W.H. 1964. Influence of starvation on standard and routine oxygen consumption. Trans. Amer. Fish. Soc. 93: 103–107.

Blaxter, J.H.S. 1969. Development: eggs and larvae. pp. 177–252. *In:* W.S. Hoar & D.J. Randall (ed.) Fish Physiology, vol. 3, Academic Press, New York.

Blaxter, J.H.S. 1988. Pattern and variety in development. pp. 1–58. *In:* W.S. Hoar & D.J. Randall (ed.) Fish Physiology, vol. 11A, Academic Press, New York.

Blaxter, J.H.S. & K.F. Ehrlich. 1974. Change in behavior during starvation of herring and plaice larvae. pp. 575–588. *In:* J.H.S. Blaxter (ed.) The Early Life History of Fish, Springer Verlag, Heidelberg.

Blaxter, J.H.S. & G. Hempel. 1963. The influence of egg size on herring larvae (*Clupea harengus* L.). J. Cons. Int. Explor. Mer 28: 211–240.

Buckley, L.J. 1981. Biochemical changes during ontogenesis of cod (*Gadus morhua* L.) and winter flounder (*Pseudopleuronectes americanus*) larvae. Rapp. P.-V. Reun. Cons. Int. Explor. Mer 178: 547–552.

Clemmesen, C.M. 1987. Laboratory studies on RNA/DNA ratios of starved and fed herring (*Clupea harengus* and turbot (*Scophthalmus maximus*) larvae. J. Cons. Int. Explor. Mer 43: 122–128.

Dabrowski, K., F. Takashima & C. Strüssmann. 1986. Does recovery growth occur in larval fish? Bull. Jap. Soc. Sci. Fish. 52: 1869.

Dobson, S.H. & R.M. Holmes. 1984. Compensatory growth in the rainbow trout, *Salmo gairdneri* Richardson. J. Fish Biol. 25: 649–656.

Ehrlich, K.F. 1974. Chemical changes during growth and starvation of herring larvae. pp. 301–323. *In:* J.H.S. Blaxter (ed.) The Early Life History of Fish, Springer Verlag. Heidelberg.

Fenchel T. 1989. Adaptations to a feast and famine existence in Protozoa. pp. 290–295. *In:* W. Wieser & E. Gnaiger (ed.) Energy Transformations in Cells and Organisms, G. Thieme Verlag, Stuttgart.

Gadomski, D.M. & J.H. Petersen. 1988. Effects of food deprivation on the larvae of two flatfishes. Mar. Ecol. Progr. Ser. 44: 103–111.

Gunkel, G. 1979. Laboratory experiments on the cultivation of young whitefish, *Coregonus fera* J.: food intake, gross growth-efficiency and growth of the fry. Europ. Maricult. Soc. spec. Publ. 4: 211–242. Bredene.

Huse, I. & A.B. Skiftesvik. 1985. Qualitative and quantitative behaviour studies in starving and feeding turbot (*Scophthalmus maximus* L.) larvae. Maricult. Comm. Int. Counc. Explor. Sea, Copenhagen. Pap. F: 38 (mimeo).

Ivlev, V.S. 1961. Experimental ecology of the feeding of fishes. Yale University Press, New Haven. 302 pp.

Loughna, P.T. & G. Goldspink. 1984. The effects of starvation upon protein turnover in red and white myotomal muscle of rainbow trout, *Salmo gairdneri* Richardson. J. Fish Biol. 25: 223–230.

Love, M. 1970. The chemical biology of fishes. Academic Press, London. 547pp.

Lowery, M.S., S.J. Roberts & G.N. Somero. 1987. Effects of starvation on the activities and localization of glycolytic enzymes in the white muscle of the barred sand bass *Paralabrax nebulifer*. Physiol. Zool. 60: 538–549.

Miglavs, I. & M. Jobling. 1989. Effects of feeding regime on food consumption. Growth rate and tissue nucleic acids in

juvenile Arctic charr, *Salvelinus alpinus*, with particular respect to compensatory growth. J. Fish Biol. 34: 947–957.
Moon, T.W. 1983. Metabolic reserves and enzyme activities with food deprivation in immature American eels, *Anguilla rostrata* (LeSueur). Can. J. Zool. 61: 802–811.
Moon, T.W. & I.A. Johnston. 1980. Starvation and the activities of glycolytic and gluconeogenic enzymes in skeletal muscles and liver of the plaice, *Pleuronectes platessa*. J. comp. Physiol. 136: 31–38.
Schwarz. F.J., J. Plank & M. Kirchgessner. 1985. Effects of protein or energy restriction with subsequent realimentation on performance parameters of carp (*Cyprinus carpio* L.). Aquaculture 48: 23–33.
Sibly, R.M. & P. Calow. 1986. Physiological ecology of animals. An evolutionary approach. Blackwell Scientific Publications, Oxford. 179pp.
Stebbing, A.R.D. 1981. The kinetics of growth in a colonial hydroid. J. mar. biol. Ass. U.K. 61: 35–63.
Wanzenböck, J. & F. Schiemer. 1989. Prey detection in cyprinids during early development. Can. J. Fish. Aqu. Sci. 46: 995–1001.
Wieser, W. 1985. Developmental and metabolic constraints of the scope for activity in young rainbow trout (*Salmo gairdneri*). J. exp. Biol. 118: 133–142.
Wieser, W., H. Forstner, F. Schiemer & W. Mark 1988a. Growth rates and growth efficiencies in larvae and juveniles of *Rutilus rutilus* and other cyprinid species: effects of temperature and food in the laboratory and in the field. Can. J. Fish. Aqu. Sci. 45: 943–950.
Wieser, W., H. Forstner, N. Medgyesy & S. Hinterleitner. 1988b. To switch or not to switch: partitioning of energy between growth and activity in larval cyprinids. Functional Ecology 2: 499–507.
Wieser, W. & N. Medgyesy. 1990. Aerobic maximum for growth in the larvae and juveniles of a cyprinid fish, *Rutilus rutilus* (L.): implications for energy budgeting in small poikilotherms. Functional Ecology 4: 233–242.

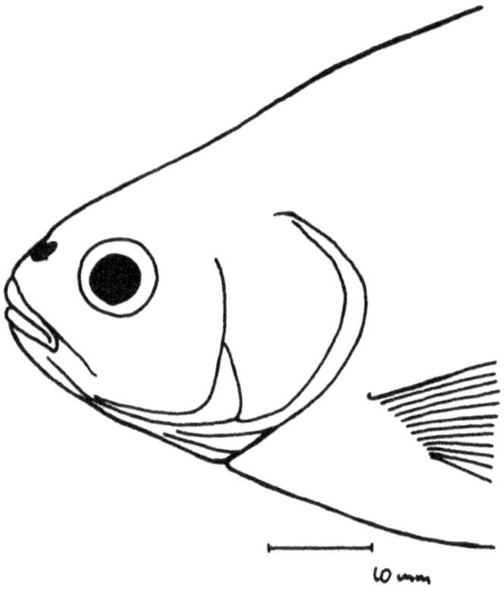
10 mm

Environmental Biology of Fishes **33**: 73–85, 1992.

Interactive effects of season and temperature on enzyme activities, tissue and whole animal respiration in roach, *Rutilus rutilus*

Franz Koch, Wolfgang Wieser & Harald Niederstätter
Institut für Zoologie, Universität Innsbruck, Technikerstrasse 25, A-6020 Innsbruck, Austria

Received 26.3.1990 Accepted 12.8.1990

Key words: Temperature compensation, Thermal acclimation, Oxygen consumption, Tissue respiration, Body mass relationship, Reproductive cycle, Cytochrome oxidase, Phosphofructokinase, Acetoacetyl-CoA thiolase, Na^+, K^+-ATPase, Amylase

Synopsis

This paper reviews investigations on the ecophysiology of a population of roach, *Rutilus rutilus,* from a subalpine oligotrophic lake in the Austrian Tirol. Metabolic responses to season and temperature were studied in whole animals, tissues and selected enzymes. The exponent of the relationship between body mass and three levels of the metabolic rate of acclimated fish was 0.82 ± 0.02, 0.60 ± 0.15, and 0.75 ± 0.01 at 4, 12, and 20° C respectively. Various combinations of long-term acclimation to constant or seasonally fluctuating temperatures and long-term (up to 14 days) monitoring of $\dot{V}O_2$ at the acclimation temperature led to the conclusion that the aerobic power of fish swimming in the routine mode does not show any sign of being temperature compensated. On the other hand, there are several indications that the energy expenditure of spontaneously swimming fish is adjusted to the seasonal pattern of environmental change and that these responses of metabolism and behaviour are controlled by both endogenous and exogenous factors. The rate of oxygen consumption of gill and muscle tissue 'brei' from fish caught during a seasonal cycle and measured at 15° C appears to follow closely the reproductive and gonadal cycle of the living fish. The same holds for the activities of phosphofructokinase, acetoacetyl-CoA thiolase, and cytochrome oxidase. On the other hand, the Na^+, K^+-ATPase of the kidney shows near perfect temperature compensation when fish acclimated to 5 and 25° C are compared, whereas an equally pronounced case of *inverse* temperature acclimation has been reported for the activity of digestive enzymes in the gut. Summarizing these data it is pointed out that the temperature relationship of a poikilothermic organism is the sum of often very diverse temperature relationships of specific metabolic and behavioural functions. In the case of the roach, strong effects of acclimation temperature on the molecular level, sometimes in the opposite direction, combine with seasonal effects on enzyme activities and tissue respiration. However, on the whole animal level the fish behave as strictly non-compensating poikilotherms, the reproductive cycle being the only detectable influence capable of modulating the basic temperature relationship of energy expenditure.

Introduction

As pointed out in a review some time ago (Wieser 1973), the interpretation of temperature relations of poikilothermic animals has already run through two phases and is now in a third phase. In the first phase poikilothermic animals were regarded as being at the mercy of the thermal environment, an interpretation best expressed by Krogh's so-called 'normal curve' (Krogh 1914). In the second phase, dating from about 1936 (Spärck 1936, Schlieper 1950, Precht et al. 1955, Bullock 1955, Prosser 1958), the emphasis was placed upon the *homoeostatic* properties of metabolism. For some time it was held that most poikilothermic animals are capable of compensating for the effects of low temperature on the rates of chemical reactions by a variety of biochemical strategies, so that in their natural environment poikilothermic animals may, in fact, be 'homoio-metabolic'. This simplistic view soon gave way to a more differentiating one. It was realized that homoeostasis is but one of several physiological principles controlling the relationships between an organism and its environment. As far as temperature is concerned the most important conclusion is that different body functions may be characterized by different temperature relationships and that a seemingly simple function like Krogh's normal curve may actually represent the sum of a number of elementary functions with diverging properties (Wieser 1973).

The experimental consequences of such a concept are that in order to be able to define the thermal niche of an animal, temperature effects have to be studied on all levels of organization, from molecular to systemic (Roberts 1966), and that a distinction has to be made between processes serving different functions, such as enzymatic reactions in different matabolic pathways (Hazel & Prosser 1970), or locomotor activity and maintenance metabolism in whole animals (Brett 1964, Forstner & Wieser 1990). Moreover, the effects of temperature on organisms cannot be isolated from the effects of other environmental factors, particularly season, or from the internal programmes that allow an organism to respond to one and the same stimulus in many different ways, all of which may be 'adaptive' in terms of the principle of inclusive fitness (Roberts 1961, 1966).

The major goal of this investigation was a better understanding of the temperature relationships of energy metabolism of a eurythermal species of fish, the roach *Rutilus rutilus* (L.), in its natural environment. This called for a combination of long-term laboratory studies, in which the animals were acclimated to specific temperature regimes, with acute measurements on animals collected from their natural habitat throughout the year. Energy metabolism was monitored by measuring oxygen consumption of whole animals and of tissues, thus permitting an estimate of the role of locomotor activity and other forms of integrative functions in determining the temperature relationship of total aerobic metabolism. To make such a comparison possible it was necessary to use respirometers allowing long-term measurements of more or less undisturbed, spontaneously swimming groups of fish. In addition to the studies on the whole organism and at tissue level, measurements of the activities of selected enzymes of energy metabolism were expected to provide insights into the relationship between molecular and systemic properties.

Material and methods

Animals

This study was conducted in 1978–1979 (whole animal respirometry) and in 1988–1989 (tissue respiration, enzyme activities) on a population of roach, *Rutilus rutilus*, the only cyprinid species occurring in a subalpine lake, the Seefelder See, near Innsbruck. For a description of the habitat and further work on the ecology of this population of roach, see Wieser et al. (1988) and Mark et al. (1989). The fish were caught by net, trap or hook and line, and transferred to the Department's aquarium. They were fed with Tetramin 1000 SDE plus trout pellets, and, in all long-term experiments, occasional rations of *Tubifex*, mealworms, and fresh grass.

Temperature regimes and maintenance

In some experiments groups of fish were fed ad libitum and kept in slowly circulating water at constant temperatures of 4, 12, or 20° C (each set temperature ± 1° C). The tanks used for maintenance and acclimation had a volume of 60 l. Temperature was controlled by the combined action of cooling and heating units. Circulating water was well aerated and filtered through a biological filter. Each tank was stocked with 6–8 fish which were used as a

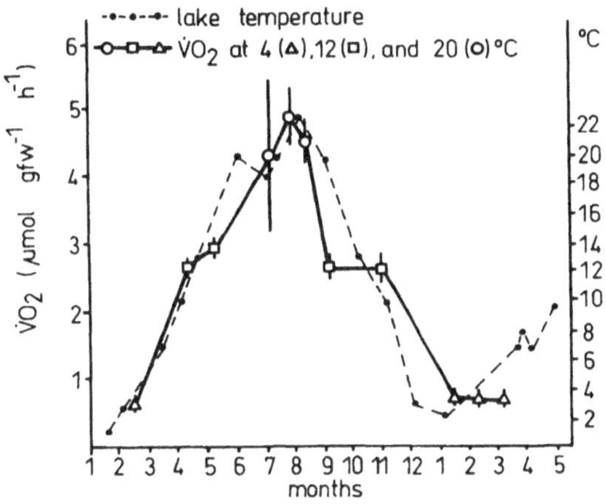

Fig. 1. Routine rate of oxygen consumption of groups of *Rutilus rutilus* caught at different months of the year, acclimated and measured at selected lake temperatures (4, 12, and 20° C). Monthly lake temperatures at depth where the fish had been caught are also shown. Means and SD of 168 one-hour measuring periods per experiment.

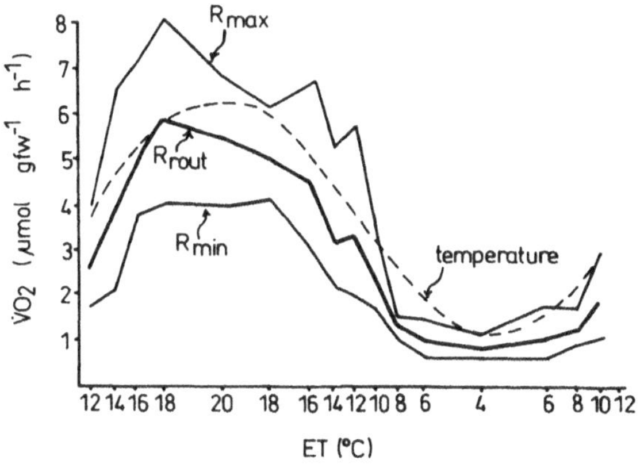

Fig. 2. Rate of oxygen consumption of groups of *Rutilus rutilus* acclimated to a temperature sinus with a period length of one year, simulating the seasonal cycle of temperatures in Seefelder See. Oxygen consumption of the fish was measured for three days at 2° C intervals (indicated on the abscissa). Three levels of metabolic rate are shown: R_{min} = average of daily minima; R_{max} = average of daily maxima; R_{rout} = average of all one-hour values.

group for the measurement of oxygen consumption. Damaged or diseased fish were removed and replaced by healthy specimens. From time to time the tanks were cleaned and disinfected with potassium permanganate and malachite green.

In another experiment the lake temperature was simulated by a temperature sinus with an amplitude from 4 to 20° C and a period length of one year, which translates into a rate of temperature change of approximately, 0.1° C · d^{-1}. Comparison of Figure 1 and 2 illustrates that this programme closely resembled the annual temperature regime of the lake. A large group of fish collected in February of 1978 was subjected to this temperature regime in a holding tank and treated as described above. At temperature intervals of 2° C a subgroup was removed, weighed and put into a respirometer where oxygen consumption was measured for 72 h at the temperature prevailing in the holding tank at the time of removal of the fish. During the experiment the fish were not fed. After termination of the measurement the fish were reweighed and returned to the holding tank. The photoperiod was kept constant at 12: 12 (light from 0800 to 2000) throughout all experimental and acclimatory treatments.

Whole animal respirometry

The respirometers consisted of large (30 or 60 l) tanks made of glass and closed with a tightly fitting lid, in which the six to eight fish measured at one time could swim freely. Two such measuring tanks were contained in one large outer wooden tank, and covered with plastic. The water in the large outer tank exchanged slowly and was maintained at constant temperature, saturated with air. The respirometers were of the intermittent type described by Forstner (1983) and used before (Wieser et al. 1988). Water from the sealed measuring tank was pumped by means of Eheim underwater pumps through an electrode chamber fitted with an YSI electrode and magnetic stirrer. After a measuring period of 1 h a valve opened automatically and water in the measuring tank was replaced by water flowing in from the outer tank; this flushing phase lasting for another 1 h. In this way 12 measuring periods per day were achieved. A servo unit controlled the alternation of measuring and flushing phases. A three way valve allowed calibration of the electrode with oxygen saturated water from the outer tank. The experiments lasted either three or 14 days during which time the fish were not fed.

Because of the large volume of the measuring tanks, bacterial metabolic activity was low and after an initial increase remained fairly constant (between 0.5 and 3% of fish respiration depending on temperature). Feeding of the fish was stopped 24 h before their transfer to the respirometer tanks. They were weighed before and after the experiment. It was assumed that weight loss was linear and the interpolated body weights were used for the calculation of weight specific oxygen consumption.

Levels of metabolic activity

The large volume of the respirometer chamber allowed the fish to swim about freely during the measuring period. As far as can be concluded from visual observation, swimming bahavior in the measuring chamber was the same as in the larger holding tanks. Based on the intermittent recording of oxygen consumption during experiments lasting either 3 or 14 days (with 12 one-hour measuring periods per day) the following levels of metabolic activity will be distinguished: *minimum* (R_{min}) = mean of all lowest daily values; *routine* (R_{rout}) = mean of all one-hour values; *maximum* (R_{max}) = mean of all highest daily values. As discussed in a previous paper (Forstner & Wieser 1990), the lowest oxygen consumption values of spontaneously swimming fish in long-term experiments are indistinguishable from the standard metabolic rate, defined as the rate of oxygen consumption at zero swimming velocity determined in a tunnel respirometer (Brett 1964).

Tissue respiration

Fish were caught in February, March, May, June, August, September, and November of 1988, and maintained in the laboratory for a few days at the appropriate lake temperature. For every experiment three fish of similar weight (10–20 g) were killed by a blow on the head and chilled immediately with ice. White muscles (a piece extending from the posterior edge of the dorsal fin to the tail region), all the gills and all of the red muscles were dissected, the latter carefully freed of connective tissue and white fibres, and minced with scissors until the tissue became pasty ('brei'). This operation (carried out on ice) took about 30 minutes, after which the tissue breis were transferred to Warburg vessels, two vessels for one fish, and suspended in 2 ml of a buffer of the following composition: 25 mM phosphate buffer (Sørensen), 60 mM sodium succinate, 100 mM sucrose, 15 mM glucose, 0.5 mM $CaCl_2$, pH = 7.4. The whole preparation was performed at 0–4° C and took about 120 minutes, including 20 minutes for thermal equilibration. Although live fish may use lipids and amino acids as major metabolic fuels, the combination of glucose, succinate, and a high concentration of sucrose (which stabilizes mitochondrial function) ensures high and stable rates of oxygen consumption of the more or less intact cells (see also Oikawa & Itazawa 1983). Oxygen was determined at 15° C with a Warburg manometer (shaking $80 \cdot min^{-1}$) with 10% NaOH as CO_2 absorbent. Consumption rates were virtually constant for about 90 minutes when the experiment was terminated. Results are expressed as $\mu mol\ O_2 \cdot h^{-1}$. (gfw of tissue brei)$^{-1}$.

Enzyme activities

From the material collected for the measurement of tissue respiration a number of fish were used for the determination of enzyme activities. Tissues were dissected as described above and subsequently homogenized with an Ultraturrax homogenizer in icecold extraction medium and constant dilution (0.015 g tissue per 1 ml medium). Treatment of the homogenates and measurement of the activities of phosphofructokinase (PFK; EC 2.7.1.11.), acetoacetyl-CoA thiolase (Thiolase; EC 2.3.1.9.), and cytochrome oxidase (COX; EC 1.9.3.1.) followed the procedure used previously (Hinterleitner et al. 1989). The three enzymes were chosen as representing the three major pathways of energy metabolism, glycolysis, β-oxidation of fatty acids, and electron transport chain, respectively.

Results

Whole animal respiration

Relation between body size, temperature and $\dot{V}O_2$
Roach ranging in size from 10 to 250 g were caught in the lake at approximately 4, 12 or 20° C, and acclimated to laboratory conditions at these temperatures for four weeks. Oxygen consumption of groups of similarly sized fish was measured at the acclimation temperature for 14 days. The relationship between different metabolic levels (see Methods) and body weight can be represented by a power function of the form $y = ax^b$, or $\log y = \log a + b \log x$, with y = rate of oxygen consumption (μmol · $fish^{-1}$ · h^{-1}), and x = fresh body weight in g. The results as set out in Table 1 indicate that the mass exponent b depends on temperature but hardly at all on the level of activity. Moreover, it appears that the average value of the exponent (mean of all experiments: b = 0.72) is somewhat lower than the values for other adult fish which are usually approximated by b = 0.8–0.85 (Winberg 1961, Brett & Groves 1979). Since our values are based on long-term measurements of well adjusted fish it is possible that they are closer to the situation prevailing in nature than other values based on short-term experiments, in which excitement and restlessness of the fish cannot be excluded. Other experiments of at least 24 h duration have also led to lower values of the mass exponent (Du Preez et al. 1986a, b).

Table 1. Mass exponents (b) of the power function $y = ax^b$ where y is rate of routine oxygen consumption (μmol $fish^{-1}$ h^{-1}), and x fresh body weight (g). n refers to number of experiments, each with 6–8 fish of similar weight, across a weight range from 10 to 250 g. AT = acclimation temperature, ET = experimental temperature.

Level of metabolism	AT = ET (° C)		
	4	12	20
minumum	0.84	0.59	0.74
routine	0.82	0.60	0.74
maximum	0.80	0.62	0.76
n	7	7	6

$\dot{V}O_2$ of fish acclimated to selected field temperatures
At different times of year, when the lake temperature was approximately 4, 12 or 20° C, fish were caught and transferred to the aquarium where they were fed and maintained at these temperatures for 10–14 days. After the acclimation phase the oxygen consumption of groups of fish, all in the weight range of 20 g ± 10%, was measured at the appropriate acclimation temperature for 14 days. If the metabolism of the fish had shown perfect temperature compensation, this procedure would have led to temperature independence of the rate of oxygen consumption of the experimental fish. However, as Figure 1 demonstrates, the opposite was true, the metabolic rate at routine activity following almost exactly the curve of field temperatures.

$\dot{V}O_2$ of fish acclimated to an annual temperature sinus
A temperature sinus with an amplitude between 4 and 20° C and a period of one year, roughly corresponding to the annual temperature cycle in the lake, was created in the laboratory. A large group of roach, caught in February and consisting of specimens weighing 20 g ± 10%, was subjected to this temperature regime. Although the photoperiod was kept constant at L : D = 12 : 12, the fish began to display typical spawning behaviour between the end of May and the beginning of June, when temperatures were 16–18° C. We also found a few eggs in the holding tanks, indicating that the temperature regime imposed on the fish induced at least part of their reproductive cycle. This population of roach spawns at about the same time in its original habitat (Koch & Wieser 1983, Mark et al. 1989). Oxygen consumption was measured for three days at 2° C intervals. As Figure 2 shows, the curves for three levels of metabolic intensity follow the temperature sinus fairly closely. Deviations of the curves for metabolic rate reflect the elevated energy turnover during the prespawning period in spring and the period of gonadal synthesis in late summer to autumn. However, correct interpreta-

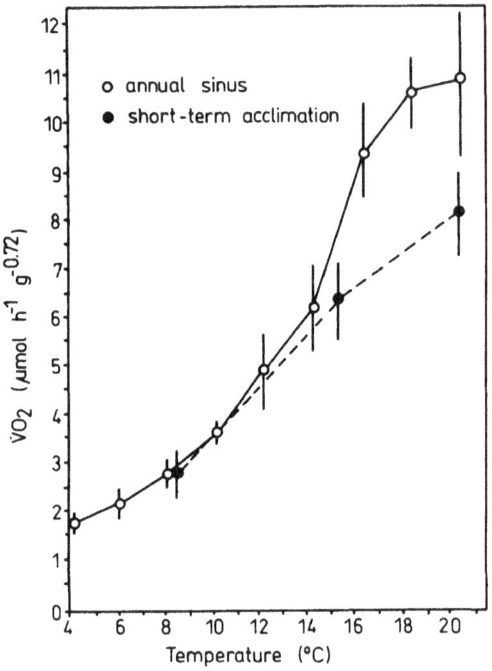

Fig. 3. Temperature relationship of mass-independent routine rate of oxygen consumption of two groups of *Rutilus rutilus* from Seefelder See. One group (open circles) was acclimated to an annual temperature sinus and measured at the acclimation temperature at 2° C intervals; the other group (solid circles) was acclimated to 8 or 20° C for two weeks and measured at 8, 15, and 20° C (after Forstner & Wieser 1990). For each temperature means and SD of 36 one-hour values are shown.

tion of these deviations requires the simultaneous monitoring of swimming activity and oxygen consumption, as described by Koch & Wieser (1983). When the minimum rate of oxygen consumption [corresponding to the standard rate as defined by Forstner & Wieser (1990)] is plotted against temperature rather than against time (Fig. 3) it emerges that the temperature range between 14 and 16° C is characterized by high Q_{10}-values (Table 2). This probably reflects the elevation of energy requirements during critical phases of the reproductive cycle, like the prespawning period of spring.

In Figure 3 and Table 2 the rate of oxygen consumption is represented by the mass independent coefficient (a) in the equation $y = ax^b$ i.e. by $a = yx^{-b}$, where y is $\dot{V}O_2$ in μmol fish^{-1} h^{-1}, x body mass in g, and $b = 0.72$ which is the average of all measurements with fish of different body weight carried out at 4, 12, or 20° C (Table 1). In this way it was possible to compare our results with those of another study with the same population but represented by juveniles weighing 1–2 g (Forstner & Wieser 1990). These fish had been acclimated to 8 or 20° C for two weeks and measured at 8, 15 and 20° C, but the acclimation temperature had had no effect on the metabolic rate. Figure 3 shows that the standard rates of the two groups of roach obey the same temperature relationship up to about 14° C. At higher temperatures standard oxygen consumption of the group acclimated to an annual temperature sinus increases and surpasses by far that of the short-term acclimated group. This demonstrates the value of long-term acclimation for estimating the annual energy requirements of poikilothermic animals in nature.

$\dot{V}O_2$ of fish acclimated to annual constant temperatures

Fish caught in February were maintained in the laboratory at 4, 12 or 20° C and a constant photoperiod of L : D = 12 : 12 for a whole year. Once a month oxygen consumption of groups of fish was

Table 2. Q_{10}-values and mass-independent oxygen consumption (in μmol g$^{-0.72}$ h^{-1}) of two groups of roach from Seefelder See. One group was acclimatized to a temperature sinus simulating the seasonal pattern of lake temperature, and measured for 3 days at 2° C intervals at the appropriate acclimation temperature. The values shown represent the routine activity of the fish, each based on 36 one-h measurements. The second group was measured at three temperatures after a few weeks of acclimation to 8 or 20° C. (after Forstner & Wieser 1990).

AT (°C)	Seasonal acclimation		Non-seasonal acclimation	
	$\dot{V}O_2$	Q_{10}	$\dot{V}O_2$	Q_{10}
4	1.7			
6	2.1	2.9		
8	2.7	3.5	2.7	
10	3.5	3.7		
12	4.8	4.8		
14	6.1	3.3		
15			6.3	3.4
16	9.3	8.2		
18	10.5	1.8		
20	10.8	1.1	8.1	1.7

measured at each acclimation temperature for three days. This experiment was designed to bring to light all changes of metabolic and locomotor activity due exclusively to endogenous processes. As Figure 4 demonstrates, very pronounced seasonal fluctuations of the routine rate of oxygen consumption occurred at the three acclimation (= experimental) temperatures. The fluctuation of R_{rout} reflects equivalent fluctuations of R_{max} and R_{min} which are not shown here. How well the fish expressed the endogenous circannual rhythm becomes apparent when one compares the values of experiments one year apart ($\dot{V}O_2$ in μmol gfw^{-1} h^{-1}; $\bar{x} \pm$ SD of 36 one-h measurements per experiment):

(°C)	April 1978	April 1979
20	7.9 ± 1.6	7.4 ± 1.0
12	4.7 ± 0.7	5.1 ± 0.4
4	0.9 ± 0.07	0.6 ± 0.1,

i.e. after a period length of one year and with fluctuations of monthly means exceeding 50% of the maxima, the rate of oxygen consumption of the fish returned to almost exactly its starting level. The standard deviations (SD) shown in this example are typical of all the 3-day experiments.

Temperature induces phase shifts of the circannual period. At 20° C the shape of the curve of monthly means of R_{rout} reflects the reproductive cycle of this population of roach as determined by Koch & Wieser (1983). The decreasing phase between April and July corresponds to the pre- and post-spawning period and the breakdown of the old gonads, whereas the increasing phase between September and December corresponds to the build-up of the new gonads. At the two lower temperatures the peaks and valleys of the circannual period shifted to later periods of the year, probably reflecting the general slowing-down of metabolic processes.

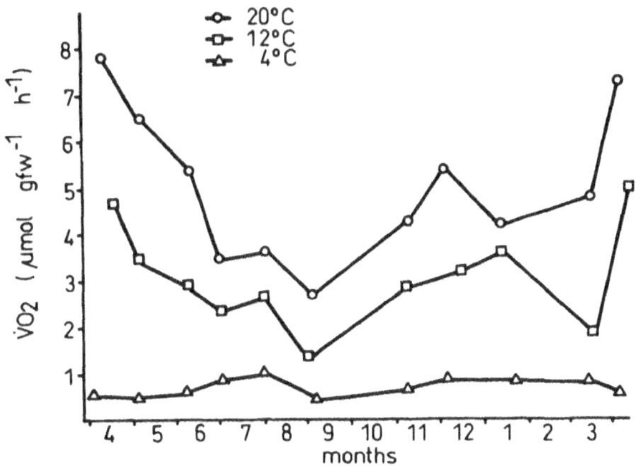

Fig. 4. Average routine rate of oxygen consumption of groups of *Rutilus rutilus* acclimated to 4, 12, or 20° C for a whole year, and measured at the same temperature every month for three days. For typical ranges of SD see text.

$\dot{V}O_2$ at 12° C of non-acclimated fish

Fish were caught at irregular intervals over the course of a year, transferred to the laboratory and brought to 12°C at a rate of 2°C per 3h. On the following day, oxygen consumption was measured at 12°C for 14 days. The fish used in this experiment weighed 20 g ± 10%. The average daily rate of metabolism declined slowly during the measuring period. This probably reflected both the adjustment of the fish to the experimental conditions and the effect of starvation. If the average rate (i.e. R_{rout}) over the whole measuring period is plotted against month of year a slightly U-shaped curve results (Fig. 5a). However, if the values of the first measuring day (Fig. 5b, circles) are removed, significant differences between monthly periods can no longer be detected. This again proves that no temperature compensation of metabolism had taken place in the field. On the other hand, the amplitude of the temperature change was directly proportional to locomotor activity, and thus to $\dot{V}O_2$, during the first day of the experiment. This resulted in the U-shaped curve which may be taken as a sign of temperature compensation of seasonal metabolism in poikilothermic animals (Wieser 1973). In this case it was clearly a sign of the relationship between locomotor activity and the magnitude of the temperature gradient between field and experiment to which the animals had been subjected.

Tissue respiration

Of several buffers tested the one used in the pre-

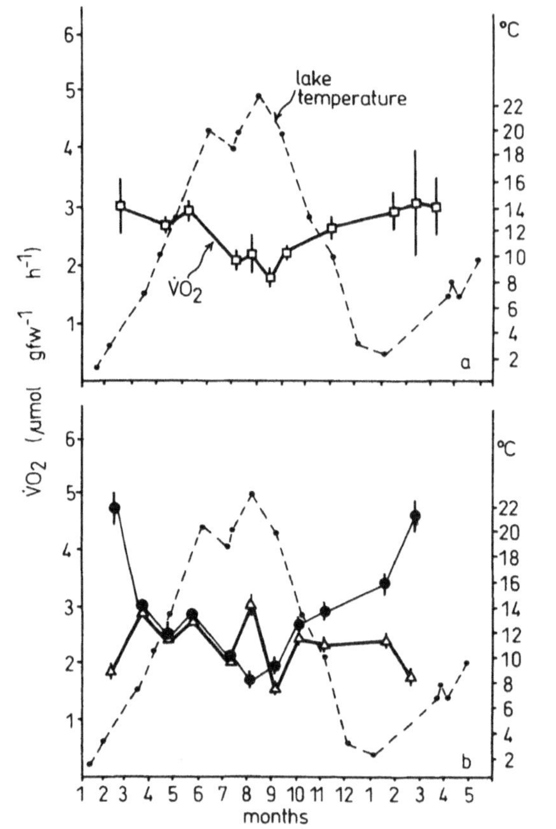

Fig. 5. Rate of oxygen consumption of groups of *Rutilus rutilus* caught at different times of year, transferred to (at a rate of ca 2° C $3\,h^{-1}$) and measured at 12° C for 14 days. Upper panel: average routine rate (means ± SD) of whole measuring period; lower panel: average rates of first (full circles) and last (open triangles) day of measurement, shown separately. Monthly lake temperatures at depth where fish had been caught are also shown.

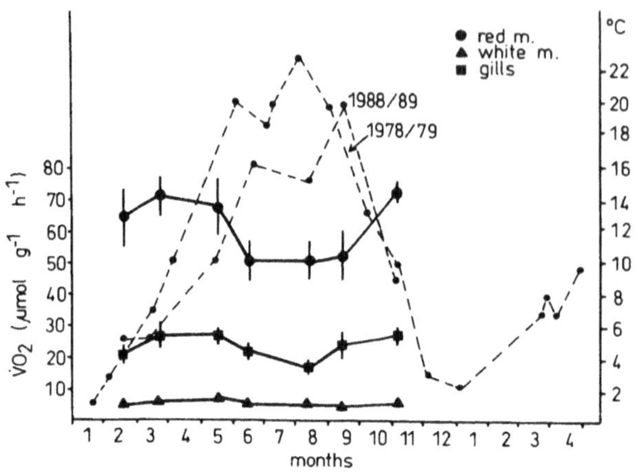

Fig. 6. Rate of oxygen consumption of red muscle, white muscle, and gill tissue 'brei' of *Rutilus rutilus* caught at different months of year and measured, within a few days, at 15° C. Lake temperatures for 1988/1989, when this series of measurements was carried out, and for 1978/1979 are shown.

sent investigation leads to the highest and most stable rates of oxygen consumption of the minced tissues. The addition of 100 mM sucrose is important in this connection since it stabilizes mitochondrial function. Our values at 15° C are about four times higher than those reported for carp tissues at 20° C by Itazawa & Oikawa (1983), indicating that the comparison of whole animal respiration with summated tissue respiration attempted by those authors is a questionable exercise. The time sequence of the rate of oxygen consumption of the three tissues as summarized in Figure 6 and Table 3 is very similar to that of whole animal respiration after long-term acclimation to constant temperatures (Fig. 4). The decrease of tissue respiration between May and June correlates with the breakdown of the old gonads, the increase between September and November with the synthesis of the new gonads (Koch & Wieser 1983), but, again, there is little indication of a general temperature compensation of metabolism. It is of great interest that all tissues measured displayed a seasonal cycle of energy turnover, even those, like the gills, which are not directly involved in metabolic rearrangements in connection with the reproductive cycle of the fish.

Enzyme activities

The seasonal activities of three enzymes characterizing three metabolic pathways in two tissues are listed in Table 4: PFK (glycolysis), Thiolase (β-oxidation) and COX (electron transport chain). In red muscle all three enzymes follow the same seasonal trend shown by the rate of oxygen consumption of this tissue (Table 3), i.e. high activities in spring, low activities in summer, and a sharp increase in autumn. In the gills this trend is less clearly expressed. A summer minimum (restricted to August) and a November maximum can be dis-

Table 3. Rate of oxygen consumption (μmol g^{-1} h^{-1}) at 15° C of tissue brei of *R. rutilus* collected at different months of the year. Means, SD, and number of fish measured are indicated. Significantly different from preceding value at 0.05 (*), 0.01 (**), and 0.001 (***) level.

Month	Red muscle	Gills	White muscle
Feb	63.9 ± 8.5 (13)	20.7 ± 2.4 (12)	5.1 ± 0.7 (11)
March	70.4 ± 5.9 (15)*	26.7 ± 3.5 (13)***	5.3 ± 1.2 (15)
May	67.7 ± 9.2 (10)	26.6 ± 2.6 (10)	6.7 ± 0.8 (10)**
Jun	50.3 ± 6.4 (10)***	21.6 ± 2.3 (10)***	4.6 ± 0.6 (9)***
Aug	50.0 ± 6.5 (7)	16.2 ± 2.3 (6)***	4.6 ± 1.4 (5)
Sep	52.6 ± 7.5 (10)	24.3 ± 3.9 (10)***	5.1 ± 0.8 (10)
Nov	72.3 ± 2.6 (5)***	27.0 ± 2.4 (5)	6.5 ± 1.7 (5)*

cerned for COX and Thiolase but not for PFK which displayed no seasonal trend at all.

These new findings can be compared with those from two previous investigations on the Na^+, K^+-ATPase of the kidney (Schwarzbaum et al. 1990) and the digestive enzymes in the gut (Hofer 1979a, b) carried out with the same population of roach (see below and Fig. 7).

Discussion

Whole animal respiration: temperature relationships

Our experimental design included long-term (up to one year) acclimation to constant and fluctuating temperatures as well as long-term (3 or 14 days) measurements of oxygen consumption. Thus we observed fish fully acclimated to environmental conditions in the holding tanks as well as in the respirometer and we assume that these data represent the metabolic responses of unstressed fish swimming in the routine mode. In this way we avoided the problems of short-term stress responses to temperature change which have hampered many interpretations of whole animal : temperature relationships (for discussions see Fry 1947, Wieser 1973, Cossins & Bowler 1987, Forstner & Wieser 1990).

None of the experimental designs employed supports the hypothesis that the energy metabolism of

Table 4. Activities (μmol min^{-1} g tissue $weight^{-1}$) of three enzymes in two tissues of *Rutilus rutilus* collected at different months of the year and measured at 20° C. Means, SD, and number of fish measured (n) are indicated. Asterisks indicate mean activities which were significantly different from preceding value at 0.05 (*) and 0.01 (**) level. n.d. = not determined.

Red muscle Month	COX	(n)	Thiolase	(n)	PFK	(n)
Feb	46.3 (8.7)	3	8.2 (0.8)	6	4.6 (1.6)	6
March	59.6 (14.2)	4	8.3 (0.9)	7	4.4 (0.9)	8
May	57.8 (9.7)	5	7.0 (0.8)	5*	2.9 (1.3)	5*
Jun	38.2 (4.7)	7**	5.2 (0.7)	5**	3.2 (1.9)	8
Aug	39.0 (10.1)	5	5.2 (2.3)	4	4.9 (1.3)	6*
Nov	64.9 (11.7)	4**	7.0 (1.1)	5*	7.0 (2.2)	5
Gills						
Feb	6.2 (1.1)	5	n.d.	n.d.		
March	6.8 (1.1)	7	0.9 (0.3)	3	1.2 (0.2)	4
May	9.8 (1.0)	4**	0.9 (0.2)	3	1.1 (0.1)	3
Jun	9.3 (1.5)	4	1.1 (0.2)	3	1.3 (0.2)	6
Aug	6.7 (1.4)	5*	0.8 (0.1)	3*	1.3 (0.2)	6
Nov	11.0 (3.1)	4*	1.3 (0.4)	3	1.3 (0.1)	3

roach during routine swimming activity is temperature compensated. This is made particularly clear by the fact that in groups of fish acclimated to and measured at 4, 12 or 20° C (corresponding to lake temperatures in winter, spring or autumn, and summer, respectively) the rate of oxygen consumption faithfully traced the curve of environmental temperature, with Q_{10} values ranging from ca. 5.6 (4 to 12° C) to ca. 1.9 (12 to 20° C). The same conclusion is suggested by experiments in which fish caught in the lake every month of the year were measured at 12° C without acclimation. With the exception of the first day, when the fish were still excited, the average rate of oxygen consumption at 12° C was independent of lake temperature from approximately 4 to 20° C. Absence of temperature compensation of oxygen consumption in the juveniles of this species has also been reported by Forstner & Wieser (1990).

On the other hand, our data provide evidence that the fish reacted sensitively to seasonal factors. Animals acclimated to an annual temperature sinus or to annual constant temperatures displayed characteristic cycles of oxygen consumption which correlated with their reproductive cycle. For example, in roach maintained for one year at 20° C and a constant photoperiod, $\dot{V}O_2$ decreased from April to July and increased again from September to December (Fig. 4), which corresponds to the annual cycle of gonad breakdown and resynthesis (Koch & Wieser 1983). In roach subjected to an annual temperature sinus a temperature of approximately 14° C constitutes a threshold above which the rate of oxygen consumption increases far above the constraints set by the temperature sensitivity in biochemical reactions per se, the Q_{10}-value across this range being as high as 8.2 (Table 2).

Thus temperature appears to set the average rate of energy turnover of spontaneously swimming roach according to a relationship with high Q_{10}-values in the lower temperature range and low Q_{10}-values in the upper range. This relationship may be altered by endogenous or seasonal factors. There is, however, no indication of a tendency towards a greater temperature *in*sensitivity of whole animal energy turnover.

Systemic versus cellular responses to climatic change

The absence of temperature compensation of whole animal respiration in the roach contrasts with the fact that cyprinids provide some of the most striking examples of temperature compensation of metabolism at the cellular and molecular level (Guderley & Blier 1988, Wieser 1990). Capacity adaptation has been reported for the respiratory rates of tissues as well as for the activities of oxidative and muscle enzymes in goldfish, crucian carp and common carp (Kanungo & Prosser 1959, Freed 1965, Caldwell 1969, Johnston 1979, Sidell 1980, Penney & Goldspink 1981). Common carp and goldfish have even been called 'the champion compensators' among freshwater teleosts (Guderley & Blier 1988). Much less information is available for other species of cyprinids, but thermal compensation has been reported for the myofibrillar ATPase of roach and tench (Heap et al. 1985) and for the Na^+, K^+-ATPase in the gills and kidney of roach (Schwarzbaum et al. 1990). On the other hand, a so-called 'inverse' acclimation effect (type 5 according to Precht et al. 1955) of striking magnitude has been documented for the activities of amylase and proteases in the gut of roach (Hofer 1979a, b).

Our data on whole animal respiration in roach intimate a discrepancy in temperature relationship between the systemic and the cellular level of metabolism in cyprinids. This may be interpreted in several ways. First of all, it should be pointed out that virtually no information exists on the temperature relationships of the more mesothermic and cold stenothermic species of European cyprinids (Wieser 1990). Thus it may be that the property of thermal compensation of metabolism is a characteristic of warm eurythermic species but is absent in species with lower temperature optima (Guderley & Blier 1988). Secondly, it might be argued that carefully conducted long-term studies with swimming cyprinids are rare, so that no definite conclusions can be reached as to the temperature relationships of their standard and active metabolism under equilibrium conditions. A third, and scientifically more attractive, hypothesis is that capacity adaptation on the cellular level may be compensated on the systemic level. This possibility was discussed by Roberts (1966) who found no, or an inverse, capacity adaptation in swimming carp, but a clear type 3 response (i.e. temperature compensation) in gill and muscle tissues. Roberts suggested that the whole animal response is under systemic control, which means that an increased energy turnover due to cold acclimation on the tissue level may be compensated on the systemic level. More circumstantial evidence for the existence of systemic control in this species has been presented by Koch & Wieser (1983) who showed that in summer and autumn the locomotor activity of roach decreases, thus partly compensating for the metabolic cost of gonad synthesis taking place at that time.

A final consideration is that the combined effects of season and temperature in nature might elicit responses in the cells and tissues of poikilotherms differing from those caused by acclimation to temperature alone in the laboratory. This possibility was also discussed by Roberts (1961, 1964). In our experiments the seasonal profiles of the metabolic activity of swimming fish (measured and acclimated to 20° C), red muscle tissue, and red muscle aerobic enzymes (from fish collected in Seefelder See at different times of the year) are very similar, following the seasonal profile of relative ovary weight in this population of roach (Koch & Wieser 1983). Some of the changes in metabolic activity observed can be interpreted as a result of temperature compensation (e.g. the increase of activity between September and November, which reflects the fall in temperature during this period), but other changes do not allow such an interpretation (e.g. the increase or constantly high level of activity between February and May). However, if viewed as a seasonal phenomenon all of these changes turn out to be correlated and linked to the reproductive cycle of the fish.

Thus acclimation to artificial temperature regimes in the laboratory may bring to light certain elements of the metabolic *capacity* of tissues and cells that may be suppressed, compensated or altered in fish living under a natural seasonal regime.

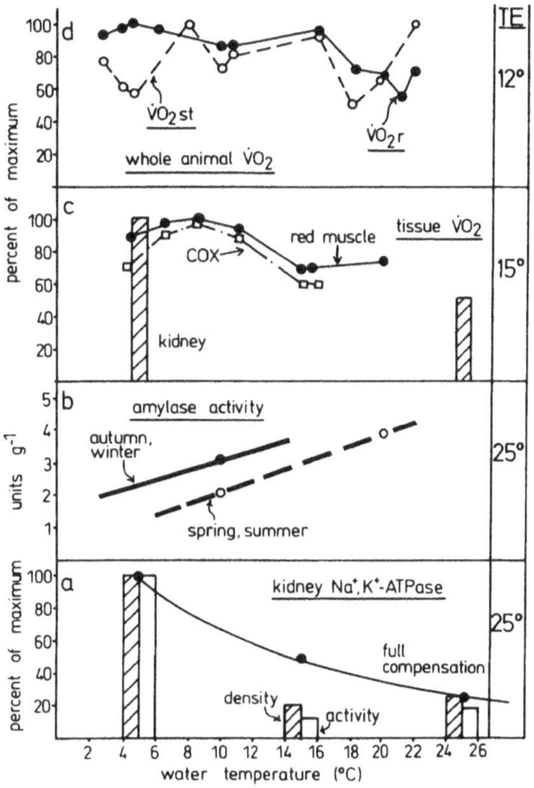

Fig. 7. Diagram summarizing temperature relationships of a series of physiological functions of *Rutilus rutilus* from Seefelder See. Relative or absolute rates are plotted against acclimation or lake temperature: *a* – relative activity, based on enzyme assays, and density, based on ouabain binding studies, of Na^+, K^+-ATPase in kidney of fish acclimated to 5, 15 or 25° C, measured at 25° C (after Schwarzbaum et al. 1990). The curve shown simulates a metabolic reaction with a Q_{10}-value of 2.0 which after acclimation to different temperatures shows full rate compensation when measured at an intermediate temperature. *b* – activity (in μmol min^{-1} g^{-1}) of amylase in the gut fluid of fish caught at different times of year, and measured at 25° C. Elevation of regression line for fish caught in autumn and winter significantly different ($p \leq 0.01$) from that for fish caught in spring and summer (after Hofer 1979a). *c* – relative activity of tissue respiration and COX activity in fish acclimated to 5 and 25° C (kidney) or caught at different times of year (red muscle tissue and COX), both measured at 15° C. *d* – routine ($\dot{V}O_2r$) and standard ($\dot{V}O_2st$) rates of oxygen consumption of fish caught at different times of year and measured at 12° C. Based on data shown in Figure 5.

A composite picture of the temperature relationships studied in one population of roach from a subalpine lake is presented in Figure 7. Here the water temperature of the lake at the time of collection, or to which the fish had been acclimated in the laboratory, is given on the abscissa. The ordinate presents relative (a, c, d) or absolute (b) activities of biological processes measured at a constant temperature (TE), on three levels of organization: enzymatic (a, b, and – partly – c), tissue (c), and whole organism (d). Clearly, very diverse responses to environmental temperature are exhibited by the metabolic systems of the roach, ranging from full temperature compensation (Na^+, K^+-ATPase of the kidney; compare theoretical curve of full compensation shown in a) to absence of compensatory reactions (whole animal respiration), or to inverse acclimatization (amylase activity in the gut). Tissue respiration and the activities of COX (Fig. 7c), Thiolase and PFK (Table 4) in red muscle convey the impression of combining inverse (4–8 and $\geq 15°$ C) and compensatory (8–15° C) strategies. However, this seemingly complex temperature relationship resolves into a simpler pattern if considered to express the seasonal aspect of the reproductive activity of this population (Koch & Wieser 1983).

Acknowledgements

The investigations reported in this review have been supported by the 'Fonds zur Förderung der wissenschaftlichen Forschung in Österreich', projects no. 3307 and S-35/04.

References cited

Brett, J. 1964. The respiratory metabolism and swimming performance of young sockeye salmon. J. Fish. Res. Board Can. 21: 1183–1226.

Brett, J.R. & T.D.D. Groves. 1979. Physiological energetics. pp. 280–352. *In:* W.S. Hoar, D.J. Randall & J.R. Brett (ed.) Fish Physiology, vol. 8, Academic Press, New York.

Bullock, T.H. 1955. Compensation for temperature in the metabolism and activity of poikilotherms. Biol. Rev. 30: 311–342.

Caldwell, R.S. 1969. Thermal compensation of respiratory enzymes in tissues of the goldfish (*Carassius auratus* L.) Comp. Biochem. Physiol. 31: 79–91.

Cossins, A.R. & K. Bowler. 1987. Temperature biology of animals. Chapman & Hall, London. 339 pp.

DuPreez, H.H., A. McLachlan & J.F.K. Marais. 1986a. Oxygen consumption of a shallow water teleost, the spotted grunter, *Pomadasys commersonni* (Lacepede, 1802). Comp. Biochem. Physiol. 84A: 61–70.

DuPreez, H.H., W. Strydom & P.E.D. Winter. 1986b. Oxygen consumption of two marine teleosts. *Lithognathus mormyrus* (Linnaeus, 1758) and *Lithognathus lithognathus* (Cuvier, 1830) (Teleostei: Sparidae). Comp. Biochem. Physiol. 85A: 313–331.

Forstner, H. 1983. An automated multiple-chamber intermittent-flow respirometer. pp. 111–126. *In:* E. Gnaiger & H. Forstner (ed.) Polarographic Oxygen Sensors, Springer Verlag, Heidelberg.

Forstner, H. & W. Wieser. 1990. Patterns of routine swimming and metabolic rate in juvenile cyprinids at three temperatures: analysis with a respirometer-activity-monitoring system. J. Comp. Physiol. 160B: 71–76.

Freed, J. 1965. Changes in activity of cytochrome oxidase during adaptation of goldfish to different temperatures. Comp. Biochem. Physiol. 14: 651–659.

Fry, F.E.J. 1947. Effects of the environment on animal activity. Univ. Toronto Stud. Biol. Ser. 55: 1–62.

Guderley, H. & P. Blier. 1988. Thermal acclimation in fish: conservative and labile properties of swimming muscle. Can. J. Zool. 66: 1105–1115.

Hazel, J. & C.L. Prosser. 1970. Interpretation of inverse acclimation to temperature. Z. vergl. Physiol. 67: 217–228.

Heap, S.P., P.W. Watt & G. Goldspink. 1985. Consequences of thermal change on the myofibrillar ATPase of five freshwater teleosts. J. Fish Biol. 26: 733–738.

Hinterleitner, S., J. Thurner-Flür, W. Wieser & N. El-Fiky. 1989. Profiles of enzyme activity in larvae of two cyprinid species with constrasting life styles (Cyprinidae: Teleostei). J. Fish Biol. 35: 709–718.

Hofer, R. 1979a. The adaptation of digestive enzymes to temperature, season and diet in roach, *Rutilus rutilus* L. and rudd *Scardinius eryhthrophthalmus* L. 1. Amylase. J. Fish Biol. 14: 565–572.

Hofer, R. 1979b. The adaptation of digestive enzymes to temperature, season and diet in roach, *Rutilus rutilus* and rudd, *Scardinius eryhthrophthalmus;* Proteases. J. Fish Biol. 15: 373–379.

Itazawa, Y. & S. Oikawa. 1983. Metabolic rates in excised tissues of carp. Experientia 39: 160–161.

Johnston, I. 1979. Calcium regulatory proteins and temperature acclimation of actomyosin ATPase from a eurythermal teleost (*Carassius auratus* L.). J. Comp. Physiol. 129: 163–167.

Kanungo, M.S. & C.L. Prosser. 1959. Physiological and biochemical adaptation of goldfish to cold and warm temperatures. I. Standard and active oxygen consumption of cold- and warm-acclimated goldfish at various temperatures. J. Cell. Comp. Physiol. 54: 259–263.

Koch, F. & W. Wieser. 1983. Partitioning of energy in fish: can

reduction of swimming activity compensate for the cost of reproduction? J. Exp. Biol. 107: 141–146.
Krogh, A. 1914. The quantitative relation between temperature and standard metabolism in animals. Intern. Z. physik. – Chem. Biol. 1: 491–508.
Mark, W., W. Wieser & C. Hohenauer. 1989. Interactions between developmental processes, growth, and food selection in the larvae and juveniles of *Rutilus rutilus* (L.) (Cyprinidae). Oecologia 78: 330–337.
Oikawa, S. & Y. Itazawa. 1983. Examination of techniques for manometric determinations of tissue respiration. Bull. Jap. Soc. Sci. Fish. 49: 23–26.
Penney, R.K. & G. Goldspink. 1981. Short term temperature acclimation in myofibrillar ATPase of a stenotherm *Salmo gairdneri* Richardson and an eurytherm *Carassius auratus*. J. Fish Biol. 18: 715–721.
Precht, H., J. Christophersen & H. Hensel. 1955. Temperatur und Leben. Springer Verlag, Berlin. 514 pp.
Prosser, C.L. 1958. General summary: the nature of physiological adaptation. pp. 167–180. *In:* C.L. Prosser (ed.) Physiological Adaptation, Amer. Physiol. Soc., Washington, D.C.
Roberts, J.L. 1961. The influence of photoperiod upon thermal acclimation by the crucian carp, *Carassius carassius* (L.). Zool. Anz. (Suppl.) 24: 73–78.
Roberts, J.L. 1964. Metabolic responses of fresh-water sunfish to seasonal photoperiods and temperatures. Helgoländer wiss. Meeresunters. 9: 459–473.
Roberts, J.L. 1966. Systemic versus cellular acclimation to temperature by poikilotherms. Helgoländer wiss. Meeresunters. 14: 451–465.
Schlieper, C. 1950. Temperaturbezogene Regulationen des Grundumsatzes bei wechselwarmen Tieren. Biol. Zentrbl. 69: 216–226.
Schwarzbaum, P., W. Wieser & H. Niederstätter. 1991. Strong effects of temperature on properties of Na^+, K^+-ATPase and on tissue respiration of kidney and gill in an eurythermic fish (*Rutilus rutilus* (L.); Cyprinidae). Comp. Biochem. Physiol. 98A: 483–489.
Sidell, B. 1980. Responses of goldfish *Carassius auratus* muscle to acclimation temperature alterations in biochemistry and proportions of different fiber types. Physiol. Zool. 53: 98–107.
Spärck, R. 1936. On the relation between metabolism and temperature in some marine lamellibranchs and its ecological and zoogeographical importance. Kgl. Dansk. Vidensk. Selsk. Biol. Medd. 13: 1–27.
Wieser, W. 1973. Temperature relations of ectotherms: a speculative review. pp. 1–23. *In:* W. Wieser (ed.) Effects of Temperature on Ectothermic Organisms, Springer Verlag, Heidelberg.
Wieser, W. 1991. Physiological energetics and ecophysiology. pp. 426–455. *In:* I.J. Winfield & J.S. Nelson (ed.) Cyprinid Fishes, Systematics, Biology and Exploitation, Chapman & Hall, London.
Wieser, W., H. Forstner, N. Medgyesy & S. Hinterleitner. 1988. To switch or not to switch: partitioning of energy between growth and activity in larval cyprinids (Cyprinidae: Teleostei). Functional Ecology 2: 499–507.
Winberg, G.G. 1961. New information on metabolic rate in fishes. Fish. Res. Board Can. Transl. Ser. 362: 1–38.

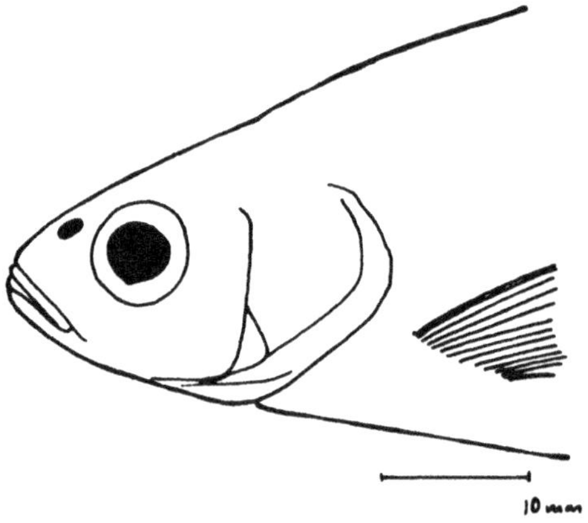
10 mm

Environmental Biology of Fishes **33**: 87–95, 1992.

Influence of temperature and ambient oxygen on the swimming energetics of cyprinid larvae and juveniles

Ruediger Kaufmann & Wolfgang Wieser
University of Innsbruck, Department of Zoology, Technikerstrasse 25, A-6020 Innsbruck, Austria

Received 16.3.1990 Accepted 12.10.1990

Key words: Chalcalburnus chalcoides, Fish larvae, Respirometry, Swimming efficiency, Scope for activity, Critical speed, Energy partitioning

Synopsis

The relationship between respiration and swimming speed of larvae and juveniles (2–100 mg fresh mass) of Danube bleak, *Chalcalburnus chalcoides* (Cyprinidae), was measured at 15° and 20° C under hypoxic (50% air saturation), normoxic, and hyperoxic (140% air saturation) conditions. In a flow-tunnel equipped with a flow-through respirometer the animals swam at speeds of up to 8 lengths · s^{-1}; speeds were sustained for at least two minutes. The mass specific standard, routine, and active respiration rates declined with increasing body mass at both temperatures. Metabolic intensity increased with temperature, but also the critical swimming speed (at which oxygen uptake reached its maximum) was higher at 20° than at 15° C by about 30%. Nevertheless, the oxygen debt incurred by the fish at the highest speeds was about 40%, and the net cost of swimming about 32%, lower at 20° than at 15° C. The standard metabolic rate was more strongly dependent on temperature (Q_{10} around 2.5) than the maximum active rate (Q_{10} below 2). Whereas standard and routine respiration rates were well regulated over the pO_2-range investigated (8.5–25.8 kPa), the active rates showed a conformer-like pattern, resulting in factorial scopes for activity between 2 and 4. Under hypoxia, the critical swimming speed was lower than under normoxia by about 1.5 l · s^{-1}, but the net cost of swimming was also lower by about 30%. On the other hand, hyperoxia neither increased the swimming performance nor did it lead to a further increase of the metabolic cost of swimming. The hypoxia experiments suggest that in response to lowered tensions of ambient oxygen maintenance functions of metabolism not directly related to swimming may be temporarily reduced, leading to increased apparent swimming efficiency under these conditions. The responses of the larvae of Danube bleak to low temperature and low ambient oxygen are discussed in terms of the metabolic strategies by which energy-limited animals meet the challenge of environmental deterioration.

Introduction

In a previous investigation (Kaufmann 1990) it was shown that in fish larvae both the standard metabolic rate (defined as the rate of oxygen consumption at zero swimming speed) and the cost of swimming are very high. The net cost of transport of the larvae of two species of cyprinids swimming at maximum speed proved to be about twice as high as that expected from extrapolating the cost of transport of adult fish (as established by Beamish 1978) to the larval body mass. These data have been obtained on fish swimming at different speeds under optimum conditions, i.e. air saturated water at 20° C, in a sensitive tunnel respirometer. How would these small fish with their low stamina respond to an additional environmental stress, like low temperature or lowered pO_2? Would they maintain their mechanic and metabolic performance levels by, for example, increasing their swimming efficiency? If not, which aspect of swimming and swimming metabolism would be affected most by a change of environmental parameters? Investigations of this kind are likely to provide new insights into the factors limiting maximum performance in small aquatic animals, and they may elucidate the mechanisms underlying the coupling of metabolic and mechanic functions.

In fish larvae there is the additional problem that changes in respiratory capacity can be expected, in view of the dramatic morphological and physiological changes accompanying the development of early life stages. For example, the swimming musculature differentiates only gradually after hatching (El-Fiky et al. 1987) and the oxygen supply initially relies on cutaneous diffusion since the gills do not become operative until later (De Silva & Tytler 1973, El-Fiky & Wieser 1988).

An extensive literature exists on the influence of environmental parameters such as temperature and oxygen tension on the routine respiration of larvae (reviewed by Rombough 1988), but few data are available concerning the responses of larvae at high levels of activity. The importance of assessing the cost of swimming under changing environmental conditions lies in the fact that activity is one of the dominant items in the energy budget: energy-limitation may be the decisive factor behind the high mortality of young fish in natural populations (Wieser 1991).

Material and methods

Animals

Eggs obtained from adult Danube bleak, *Chalcalburnus chalcoides mento* (Agassiz), from Mondsee, an oligotrophic, meromictic Austrian lake, were fertilized in the laboratory. The hatching fresh mass of the larvae was about 2 mg. At the rearing temperature of 20° C, swimming started around the sixth day. The hatchlings were kept at 20° C under normoxic conditions and fed *Artemia* nauplii. Rearing and maintenance followed the protocol given in Wieser et al. (1988).

In the first year of this investigation the influence of temperature was studied at 20° C and at 15° C with larvae of 3.3–27 mg fresh mass (10–26 days after hatching, total length 9–17 mm). The effect of ambient oxygen was investigated the following year on the same stock under hypoxic (50% air saturation), normoxic, and hyperoxic (140% air saturation) conditions at 20° C. Larvae and juveniles of 2–120 mg fresh mass were used (8–60 days of age, total length 8.5–25 mm).

The number of individuals used in a respirometric experiment varied according to body mass from 150 down to 5, so as to keep the total fresh mass in the range of 300–600 mg. Even with the large numbers of the smallest larvae the freedom of movement within the flow-tunnel was not impaired. The animals were transferred to the experimental temperature (for 15° C experiments) and deprived of food 8–12 h before introduction into the respirometer. When the oxygen tension was changed, this was done during the experiments after a normoxic adaptation phase. Thus, there was no long-term acclimation to either temperature or pO_2.

Respirometry

A small flow-tunnel (total volume 100 ml) closely resembling the construction of Brett (1964) was connected to a flow-through respirometer according to the design of Gnaiger (1983), which uses two polarographic sensors to measure oxygen tension at the inlet and the outlet of the flow-tunnel. This apparatus allowed continuous measurement of the oxygen uptake during swimming. To cope with the rapid respiration changes and the short swimming periods of the larvae, it was necessary to correct the slow response of the system by means of a differential equation containing two experimentally determined time constants. With this method an effective temporal resolution of about 30 s was achieved. Details of these respirometric techniques have been given elsewhere (Kaufmann et al. 1989, Kaufmann 1990).

For normoxic conditions the incoming water of the flow-tunnel was equilibrated with air; for hypoxic and hyperoxic conditions gas mixtures (prepared by a gas-mixing pump: type SA 27/4, Wösthoff OHG, Germany) of 10% O_2 in N_2 (approx. 50% air saturation) and 10% O_2 in air (approx. 140% air saturation) were used. The ambient oxygen tensions given (pO_2) refer to long-term averages of the respirometer's outlet and may be somewhat lower than the pO_2 of the inspired water (Hughes et al. 1983). Respiration rates (R) are given in units of μmol $O_2 \cdot h^{-1} \cdot g^{-1}$ fresh mass throughout this article.

Experimental protocol

In the flow-tunnel the larvae started swimming against the water current when the latter attained a speed of about $2\,cm \cdot s^{-1}$. This is similar to their routine speeds under natural conditions (Wanzenböck & Schiemer 1989) and was the lowest speed used in the experiments. Average duration of an experiment was three hours. After introduction into the respirometer the animals were exposed to a random variety of water velocities for one hour to allow handling effects to wear off. At this point, in all hypoxia and hyperoxia experiments, the new oxygen concentrations were initiated. The larvae were exposed to exercise speeds for periods of usually five minutes. At the highest speeds, when stamina declined rapidly, the periods were shortened and speed was further increased as long as it could be maintained for at least two minutes. Swimming trials alternated with resting periods of five minutes at routine speed; these were extended if, after strenuous swimming, the respiration rate remained elevated due to repayment of what in the previous investigation (Kaufmann 1990) was shown to be a true oxygen debt. In all cases the normal routine respiration rate was reattained within 15 minutes and thus the magnitude of the oxygen debt could be estimated. Every experiment comprised swimming trials at 5–10 different speeds.

Calculations and terminology

Routine respiration rates (R_r) are the averages of all periods during which the fish swam at speeds of $2\,cm \cdot s^{-1}$; active rates (R_a) represent the maximum oxygen uptake inducible by swimming at the highest speeds. If an oxygen debt occurred after the swimming trials its repayment was included to give the total oxygen uptake evoked by exercise (see Kaufmann 1990 for details of calculation).

The relationship between respiration rate (R(u)) and swimming speed (u) is described by functions of the form $R(u) = R_s + a \cdot u^b$, which consist of the metabolic standard rate (R_s) plus an activity-induced respiration expressed as a power function of the swimming speed u (discussed by Kaufmann 1990). The parameters R_s, a and b of these power-performance relationships were estimated by non-linear regression with the Levenberg-Marquardt algorithm (Press et al. 1986). The energetic net cost of swimming (i.e. cost of transport), giving the amount of energy required for a unit mass of animal to swim a unit distance ($J \cdot g^{-1} \cdot km^{-1}$), was calculated from the respiration induced by swimming assuming an oxycaloric equivalent of $450\,kJ \cdot mol^{-1}\,O_2$ (Brett & Groves 1979). As a measure of performance, two speeds were used: maximum speed (u_{max}) which the fish were able to maintain for two minutes, and critical speed (u_{crit}) at which the respiration rate R(u) reaches the maximum oxygen uptake R_a, above which stamina decreases rapidly and an oxygen debt builds up (showing that anaerobic processes are involved).

Results

Influence of temperature

The average power-performance relations of three size classes of *C. chalcoides* at 15° C and 20° C are shown in Figure 1. At both temperatures the mass

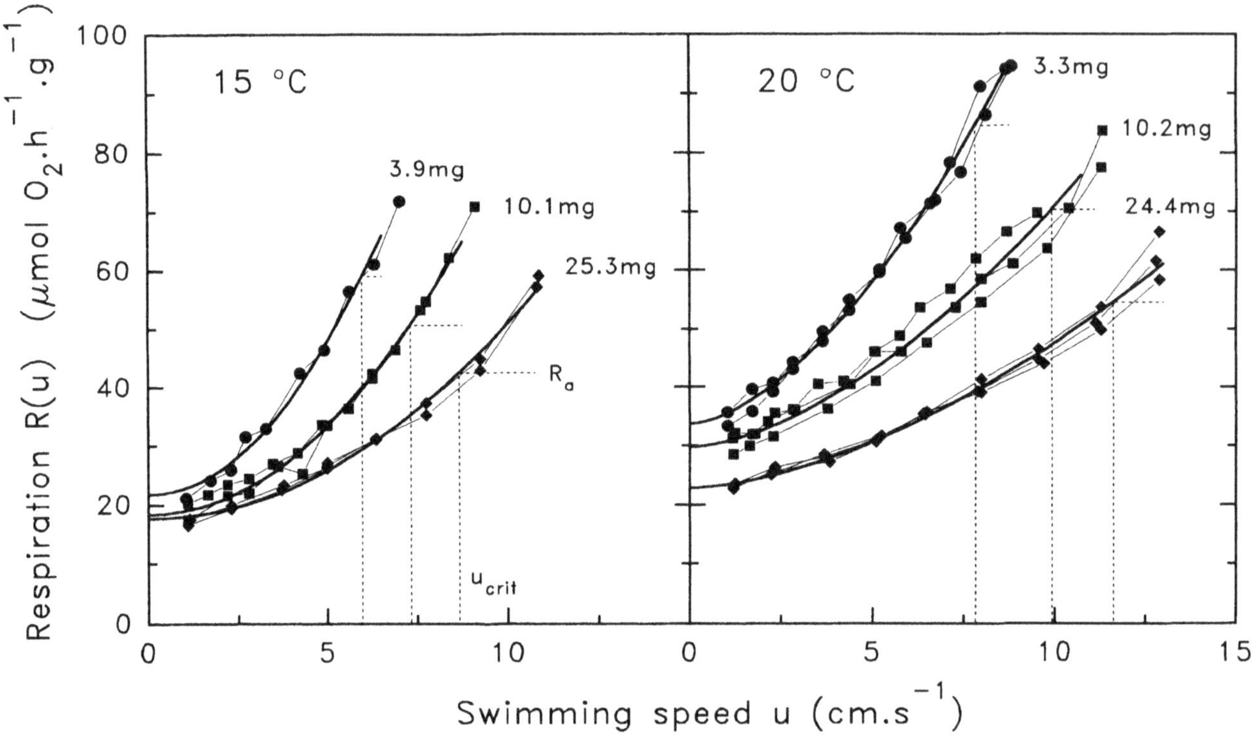

Fig. 1. Power-performance relationships between swimming speed and respiration of three size classes of *C. chalcoides* at 15° and 20° C. Given are individual values of the experiments listed in Table 1 and fitted curves of the form $R(u) = R_s + a \cdot u^b$. The thin dashed lines mark the maximum rate of respiration at the highest speeds (R_a) and the critical swimming speeds (u_{crit}). The segments of the curves beyond the maximum rate of oxygen consumption are measures of anaerobic energy production, calculated from the repayment of the oxygen debt at the end of each swimming trial (for method of calculation see Kaufmann 1990).

specific standard, routine and active respiration rates declined during growth in a similar manner. As expected, the metabolic intensity increased with temperature, but the swimming performance (i.e. maximum and critical speed) was also higher at 20° than at 15° C ($p < 0.001$, ANCOVA, with log M as covariate). Relative speed provides a mass-independent measure of this difference, u_{crit} being 5.6 ± 0.5 lengths · s^{-1} at 15° C and 7.6 ± 0.7 lengths · s^{-1} at 20° C (means ± s.d.).

The temperature dependence in terms of the Q_{10} was much less pronounced for the active rates R_a (Q_{10} = 1.6–2.1) than for the standard rates R_s (Q_{10} = 2.2–2.8), with the routine rates R_r somewhere in between. The essential metabolic and performance data at the two temperatures are summarized in Table 1.

A higher aerobic capacity and lower costs of swimming both contribute to the better performance of the fish at the higher temperature. From 15° to 20° C the aerobic scope for activity (R_a–R_s) increased by 27% ($p < 0.001$, ANCOVA on log-transformed data), while the oxygen debt incurred at the highest speeds decreased by 40% ($p = 0.015$, t-test). Swimming is most costly for the smallest larvae. In addition, the power-performance curves are steeper and more convex at 15° C (Fig. 1). This is mainly due to a higher exponent (*b*) ($p = 0.013$, t-test) in the relation $R(u) = R_s + a \cdot u^b$, which was 2.03 ± 0.25 and 1.66 ± 0.17 (means ± s.d.) at 15° and 20° C respectively. At low swimming speeds this does not seriously affect the net cost of swimming. With increasing speed, however, swimming costs become significantly higher at the lower temperature (Fig. 2). For larvae swimming at 6 lengths · s^{-1}, the cost of transport at 15° C was 32% above that at 20° C ($p < 0.001$, ANCOVA on log-transformed data).

Influence of ambient oxygen

The experiments dealing with the influence of pO_2, covered a wider weight range than the temperature experiments described above. The decrease of the metabolic rates and swimming costs during growth was similar in both the temperature and the pO_2 experiments.

Power-performance curves obtained at three levels of ambient oxygen tension: hypoxic 8.5 kPa, normoxic 18.5 kPa and hyperoxic 25.8 kPa, are shown in Figure 3. In all cases an oxygen debt occurred only above a well defined critical speed. This example clearly showed a performance reduction under hypoxia, with the critical speed (u_{crit}) being on an average 1.5 lengths · s^{-1} lower than under normoxic conditions ($p = 0.013$, ANCOVA/Scheffé-test). Increasing the pO_2 into the hyperoxic range, on the other hand, did not significantly improve swimming performance. In addition, respiratory rate and net cost of swimming increased with ambient oxygen. This change in metabolic costs is mainly due to the parameter (*a*) of the power-performance relationship $R(u) = R_s + a \cdot u^b$, the speed exponent (*b*) remaining constant at 1.78 ± 0.45 (mean ± s.d.), independent of pO_2. Thus the influence of oxygen tension on the net cost of swimming is the same at all speeds. Under hypoxia the net cost was 30% lower than under normoxia ($p = 0.014$, ANCOVA/Scheffé-test), whereas the difference between normoxia and hyperoxia was not statistically significant. There was no indication that under hypoxia the lack of oxygen was counteracted by an intensified employment of anaerobic pathways; in fact, quite to the contrary: the oxygen debt after strenuous exercise also decreased under hypoxia ($p = 0.029$, ANOVA/Scheffé-test). In summary, each of the three parameters: critical speed, swimming cost, and oxygen debt, increased with the availability of oxygen up to the normoxic level ($p = 0.001$–0.003, ANOVA/ANCOVA).

Whereas the standard (R_s) and routine (R_r) respiration rates were insensitive to pO_2, the availability of oxygen limited the maximum aerobic rate (R_a). As mentioned before, all metabolic levels decreased with the animals' fresh mass (Fig. 4). In addition, the active rates (R_a) were highly pO_2-dependent over the whole range of oxygen concentrations, from 8.5 to 25.8 kPa ($p < 0.001$, ANCOVA/Scheffé-tests on log-transformed data). Thus by normalizing the respiration rates with respect to the normoxic standard rate, the pO_2-characteristics shown in Figure 5 were obtained. The oxyregulatory behaviour at low energy demand, converted to a

Table 1. Summary of all temperature experiments, indicating average individual fresh body mass of fish (*M*), parameters of the functional relationship between the rate of oxygen consumption and swimming speed $R(u) = R_s + a \cdot u^b$, and the following levels of metabolic rate (*R*) and swimming speed (*u*): R_s = standard, R_r = routine, R_a = maximum; u_{crit} = critical, u_{max} = maximum.

		(mg)			(μmol $O_2 \cdot g^{-1} \cdot h^{-1}$)			(cm · s^{-1})	
		M	a	b	R_s	R_r	R_a	u_{crit}	u_{max}
	1	3.9	2.53	1.55	20.5	22.0	59.0	6.0	6.3
	2	10.0	0.33	2.26	19.5	21.5	51.0	7.4	8.4
15° C	3	10.2	0.48	2.13	17.2	18.5	50.4	7.2	9.1
	4	24.7	0.21	2.17	17.6	17.0	44.1	9.1	10.8
	5	25.9	0.31	2.04	16.9	16.5	40.9	8.3	10.9
	1	3.3	2.52	1.47	31.9	34.1	84.5	7.9	8.8
	2	3.4	1.70	1.66	33.5	37.9	84.7	7.7	8.7
	3	8.6	1.92	1.37	29.9	32.9	69.7	9.2	9.5
20° C	4	11.0	0.80	1.69	27.5	32.0	69.7	10.4	11.3
	5	11.2	0.59	1.82	30.4	34.3	71.8	10.2	11.4
	6	22.6	0.34	1.88	22.2	24.5	54.5	11.0	12.9
	7	23.3	0.35	1.82	23.0	25.2	53.6	11.4	12.8
	8	27.3	0.63	1.57	22.4	24.9	55.6	12.4	12.9

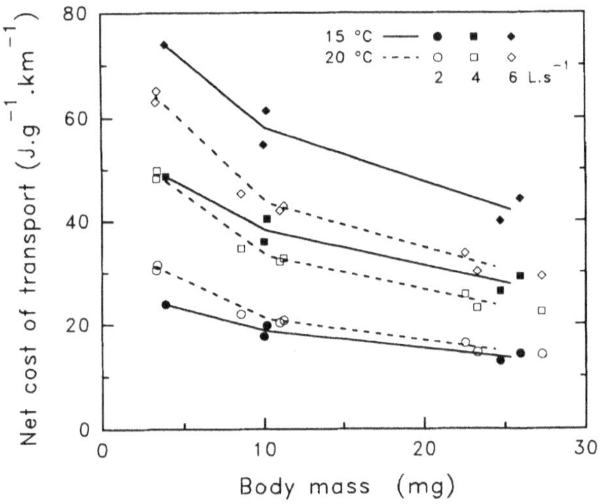

Fig. 2. Mass-dependence of net costs of transport at 15° C (full symbols, solid lines) and 20° C (open symbols, broken lines). Values estimated from the relationships shown in Figure 1 and fitted curves are given for swimming speeds of 2, 4 and 6 lengths $\cdot$ s^{-1}.

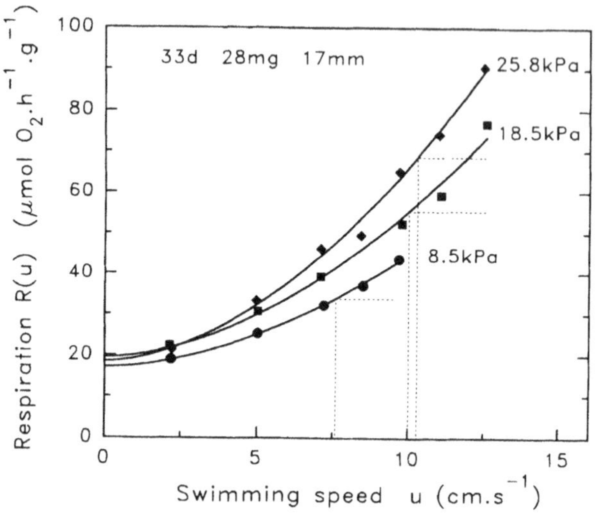

Fig. 3. Power-performance relationships at hypoxic (8.5 kPa), normoxic (18.5 kPa) and hyperoxic (25.8 kPa) pO_2 (20° C) for larvae of 28 mg fresh body mass. As in Figure 1 the horizontal dashed lines mark the maximum rate of oxygen consumption at the highest speeds, their intersection with the fitted regression curves characterizing the critical speed (u_{crit}). The segments of the curves beyond u_{crit} are measures of anaerobic energy production calculated from the repayment of the oxygen debt at the end of each swimming trial (see Kaufmann 1990).

conformer-like pattern as oxygen demand increased with activity.

Discussion

This study uncovered some rather subtle responses of swimming performance and swimming metabolism of young Danube bleak to changes in water temperature and ambient oxygen. As already stressed by Febry & Lutz (1987) in a comparable study, relations of this kind can be analyzed correctly only if the overriding influence of locomotor activity on metabolic rate is accounted for. This makes essential the use of tunnel respirometers, in which the oxygen consumption of fish forced to swim at known speeds can be measured.

Effects of temperature on swimming performance and metabolic rate

The reduction of water temperature from 20° to 15° C led to impairment of the swimming performance of the larvae of Danube bleak. Both maximum and critical speed decreased by about 30%, and at these speeds the metabolic cost of transport was about 32% higher at 15° than at 20° C (Fig. 2). This is a new finding. In adult fish the cost of transport has been reported to increase with increasing temperature or to remain constant (Brett 1964, Beamish 1970, 1981, Smit et al. 1971, Priede & Holliday 1980, Duthie 1982). The effect of water temperature on the metabolic rate of the larvae of whitefish and trout swimming at relatively low speed has been studied by Dabrowski (1986), but the methods employed did not permit the construction of a full power-performance profile.

The increase in metabolic cost of transport induced by low temperature in the larvae of *C. chalcoides* swimming at high speed was much larger than the expected influence of viscosity on hydrodynamic drag power, which according to formulae given by Smit et al. (1971) would be of the order of only 5%. Alternative explanations for the unexpectedly large temperature effect on active metabolic rate and swimming performance include changes of metabolic efficiency, or higher frictional energy losses in the muscles at low temperature.

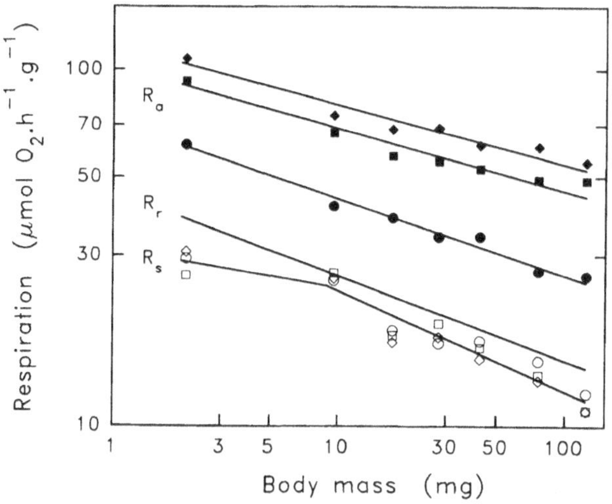

Fig. 4. Development during growth of standard (R_s), routine (R_r) and active (R_a) respiration rate. For R_a the data are given for experiments under hypoxic (8.5 kPa, full circles), normoxic (18.5 kPa, full squares) and hyperoxic (25.8 kPa, full diamonds) conditions. The allometric exponent is -0.18 at all three pO_2-levels. For pO_2-independent R_s and R_r individual data (open symbols) and fitted regressions are shown.

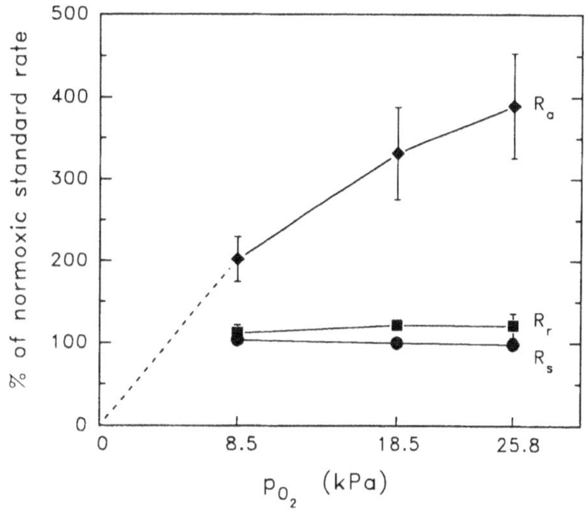

Fig. 5. pO_2-characteristic of standard (R_s), routine (R_r) and active (R_a) respiration rates obtained by referring the data to the normoxic (18.5 kPa) standard metabolism (100%). All size classes were pooled. Error bars indicate standard deviations (N = 7).

Moreover, the swimming style and thus the propulsive efficiency may depend on temperature (Rulifson 1977). However, visual observations of the larvae gave no clue to such changes, but kinematic data are lacking. At any rate, it appears that in fish weighing less than about 25 mg (Fig. 1), a drop in water temperature may seriously impair locomotor performance and constitute a severe drain on energy reserves if the fish are required to cover a given distance as fast as possible.

Cost of transport and energy partitioning at low oxygen partial pressure

Calculating the energetic cost of transport from respiration rates involves certain problems. Even if an oxygen debt is incurred and included in the calculations, it remains uncertain whether this gives an accurate estimate of the anaerobic energy component. The larvae seem to swim acrobically over a wide speed range since an oxygen debt builds up only after very strenuous swimming (Fig. 1, 3), whereas in adult cyprinids the anaerobic white muscles are active even at moderate speeds (Johnston & Goldspink 1973, Wokoma & Johnston 1983). Anaerobic processes have also been assumed to be involved in cases where no oxygen debt could be detected (Smit et al. 1971). The second problem is the common assumption that the energy demand for swimming can simply be added to a constant maintenance metabolism, an assumption which is not supported by experimental evidence and is rendered more questionable by the present results (see Wieser 1989 for a discussion of this problem).

The decrease in swimming costs at lower ambient oxygen content (Fig. 3) is unexpected if it is assumed that there are no unrecognized anaerobic contributions to the energy supply. Even after strenuous swimming under hypoxia the oxygen debt gave no indication of a higher anaerobic load than under normoxia. During subsequent swimming trials an impairment of performance, as would be expected if the aerobic status had not been fully restored, was not found. Moreover, there was no indication of a more efficient swimming pattern under hypoxia, which could have provided an explanation for the lower energy demand.

The interpretation remaining is that during swimming some of the normal maintenance functions are reduced to allow additional power for locomotion. The idea of such a shift in the partitioning of energy (Wieser 1989) is supported by the observation that in stressed adult fish, blood flow can be redirected from internal organs to the muscles (Randall & Daxboeck 1982). A lower cost of swimming under hypoxia has also been reported in goldfish (Smit et al. 1971), whereas it seems to remain constant in trout (Bushnell et al. 1984). A similar effect attributed to energy partitioning was observed in sablefish, in which the apparent cost of transport decreased when the standard rate of metabolism was elevated in the postfeeding state (Furnell 1987).

Strategies for dealing with a deteriorating environment

The responses of fast-swimming larvae to ecological change reflect two basic strategies of animals facing deterioration of their environment. In one case the fish reacted by mobilizing additional energy sources in order to meet the high cost of transport at the low temperature. However, since the increase in energy turnover was not matched by an increase in swimming performance, the energy status of the fish must have deteriorated. In the second case, i.e. low ambient oxygen, the fish appeared to react by slowing down non-essential physiological functions so that more energy could be channeled to muscle power. This caused an apparent increase of the efficiency of swimming which would be beneficial if, by fast swimming, the fish could escape from the source of stress. However, in the long run this strategy, too, should turn out to be detrimental, since redirecting the flow of energy from one physiological function to another provides only a temporary solution.

Continuity in pO_2-relations during development

It was expected that the dramatic developmental changes after hatching, particularly the replacement of the skin by the gills as the major pathway of oxygen to the tissues (Holeton 1971, De Silva & Tytler 1973, El-Fiky & Wieser 1988) and the delayed development of anaerobic muscle power (El-Fiky et al. 1987) would be reflected by a break in the pO_2-relations of the developing fish. However, no evidence for such a break was detected. As in juvenile and adult fish the ability to regulate oxygen consumption over a range of pO_2-values decreased with increasing energy demand (Ott et al. 1980, Hughes et al. 1983, Rombough 1988). In all fish tested so far active respiration rates (Fry 1971, Duthie & Hughes 1987) and swimming performance (Davis et al. 1963, Dahlberg et al. 1986, Beamish 1978, Bushnell et al. 1984) are sensitive to hypoxia and tend towards a maximum as the partial pressure of ambient oxygen increases.

Acknowledgements

This work was supported by the 'Fonds zur Förderung der wissenschaftlichen Forschung' of Austria, project no. S-35/04. Niki Medgyesy and Wolfgang Mark kept the fish larvae in good shape. We thank Joy Wieser for correcting the manuscript.

References cited

Beamish, F.W.H. 1970. Oxygen consumption of largemouth bass, *Micropterus salmoides*, in relation to swimming speed and temperature. Can. J. Zool. 48: 1221–1228.

Beamish, F.W.H. 1978. Swimming capacity. pp. 101–187. *In:* W.S. Hoar & D.J. Randall (ed.) Fish Physiology, Volume 7, Academic Press, New York.

Beamish, F.W.H. 1981. Swimming performance and metabolic rate of three tropical fishes in relation to temperature. Hydrobiologia 83: 245–254.

Brett, J.R. 1964. The respiratory metabolism and swimming performance of young sockeye salmon. J. Fish. Res. Board Can. 21: 1183–1226.

Brett, J.R. & T.D.D. Groves. 1979. Physiological energetics. pp. 279–352. *In:* W.S. Hoar, D.J. Randall & J.R. Brett (ed.) Fish Physiology, Volume 8, Academic Press, New York.

Bushnell, P.G., J.F. Steffensen & K. Johansen. 1984. Oxygen consumption and swimming performance in hypoxia-acclimated rainbow trout *Salmo gairdneri*. J. exp. Biol. 113: 225–235.

Dabrowski, K.R. 1986. Active metabolism in larval and juve-

nile fish: ontogenetic changes, effect of water temperature and fasting. Fish Physiology and Biochemistry 1: 125–144.
Dahlberg, M.L., D.L. Shumway & P. Doudoroff. 1986. Influence of dissolved oxygen and carbon dioxide on swimming performance of largemouth bass and coho salmon. J. Fish. Res. Board Can. 25: 49–70.
Davis, G.E., J. Foster, C.E. Warren & P. Doudoroff. 1963. The influence of oxygen concentration on the swimming performance of juvenile pacific salmon at various temperatures. Trans. Amer. Fish. Soc. 92: 111–124.
De Silva, C.D. & P. Tytler. 1973. The influence of reduced environmental oxygen on the metabolism and survival of herring and plaice larvae. Neth. J. Sea Res. 7: 345–362.
Duthie, G.G. 1982. The respiratory metabolism of temperature-adapted flatfish at rest and during swimming activity and the use of anaerobic metabolism at moderate swimming speeds. J. exp. Biol. 97: 359–373.
Duthie, G.G. & G.M. Hughes. 1987. The effect of reduced gill area on the oxygen consumption and swimming speed of rainbow trout. J. exp. Biol. 127: 349–354.
El-Fiky, N., S. Hinterleitner & W. Wieser. 1987. Differentiation of swimming muscles and gills, and development of anaerobic power in the larvae of cyprinid fish (Pisces, Teleostei). Zoomorphology 107: 126–132.
El-Fiky, N. & W. Wieser, 1988. Life styles and patterns of development of gills and muscles in larval cyprinids (Cyprinidae; Teleostei). J. Fish Biol. 33: 135–145.
Febry, R. & P. Lutz. 1987. Energy partitioning in fish: the activity-related cost of osmoregulation in a euryhaline cichlid. J. exp. Biol. 128: 63–85.
Fry, F.E.J. 1971. The effect of environmental factors on the physiology of fish. pp. 1–98. *In:* W.S. Hoar & D.J. Randall (ed.) Fish Physiology, Volume 6, Academic Press, New York.
Furnell, D.J. 1987. Partitioning of locomotor and feeding metabolism in sablefish (*Anoplopoma fimbria*). Can. J. Zool. 65: 486–489.
Gnaiger, E. 1983. The twin-flow microrespirometer and simultaneous calorimetry. pp. 134–166. *In:* E. Gnaiger & H. Forstner (ed.) Polarographic Oxygen Sensors, Springer-Verlag, Berlin.
Holeton, G.F. 1971. Respiratory and circulatory responses of rainbow trout larvae to carbon monoxide and to hypoxia. J. exp. Biol. 55: 683–694.
Hughes, G.M., C. Albers, D. Muster & K.H. Götz. 1983. Respiration of the carp, *Cyprinus carpio* L., at 10 and 20° C and the effects of hypoxia. J. Fish Biol. 22: 613–628.
Johnston, I.A. & G. Goldspink. 1973. A study of the swimming performance of the crucian carp *Carassius carassius* (L.) in relation to the effects of exercise and recovery on biochemical changes in the myotomal muscles and liver. J. Fish Biol. 5: 249–260.
Kaufmann, R. 1990. Respiratory cost of swimming in larval and juvenile cyprinids. J. exp. Biol. 150: 343–366.
Kaufmann, R., H. Forstner & W. Wieser. 1989. Respirometry – methods and approaches. pp. 51–76. *In:* C.R. Bridges & P.J. Butler (ed.) Techniques in Comparative Respiratory Physiology, Society for Experimental Biology Seminar Series, Cambridge University Press, Cambridge.
Ott, M.E., N. Heisler & G.R. Ultsch. 1980. A re-evaluation of the relationship between temperature and the critical oxygen tension in freshwater fishes. Comp. Biochem. Physiol. 67A: 337–340.
Press, W.H., B.P. Flannery, S.A. Teukolsky & W.T. Vetterling. 1986. Numerical recipes – the art of scientific computing. Cambridge University Press, Cambridge. 817 pp.
Priede, I.G. & F.G.T. Holliday. 1980. The use of a new tilting tunnel respirometer to investigate some aspects of metabolism and swimming activity of the plaice (*Pleuronectes platessa* L.). J. exp. Biol. 85: 295–309.
Randall, D.J. & C. Daxboeck. 1982. Cardiovascular changes in the rainbow trout (*Salmo gairdneri*) during exercise. Can. J. Zool. 60: 1135–1142.
Rombough, P.J. 1988. Respiratory gas exchange, aerobic metabolism, and effects of hypoxia during early life. pp. 59–161. *In:* W.S. Hoar & D.J. Randall (ed.) Fish Physiology, Volume 11A, Academic Press, San Diego.
Rulifson, R.A. 1977. Temperature and water velocity effects on the swimming performances of young-of-the-year striped mullet (*Mugil cephalus*), spot (*Leiostomus xanthurus*), and pinfish (*Lagodon rhomboides*). J. Fish. Res. Board Can. 34: 2316–2322.
Smit, H., J.M. Amelink-Koutstaal, J. Vijverberg & J.C. Von Vaupel-Klein. 1971. Oxygen consumption and efficiency of swimming goldfish. Comp. Biochem. Physiol. 39A: 1–28.
Wanzenböck, J. & F. Schiemer. 1989. Prey detection in cyprinids during early development. Can. J. Fish. aquat. Sci. 46: 995–1001.
Wieser, W. 1989. Energy allocation by addition and by compensation: an old principle revisited. pp. 98–105. *In:* W. Wieser & E. Gnaiger (ed.) Energy Transformations in Cells and Organisms, Georg Thieme Verlag, Stuttgart.
Wieser, W. 1991. Limitations of energy acquisition and energy use in small poikilotherms: evolutionary implications. Funct. Ecol. (in press).
Wieser, W., H. Forstner, N. Medgyesy & S. Hinterleitner. 1988. To switch or not to switch: partitioning of energy between growth and activity in larval cyprinids (Cyprinidae: Teleostei). Funct. Ecol. 2: 499–507.
Wokoma, A. & I.A. Johnston. 1983. Anaerobic metabolism during activity in the rainbow trout (*Salmo gairdneri*). Experientia 39: 1366–1367.

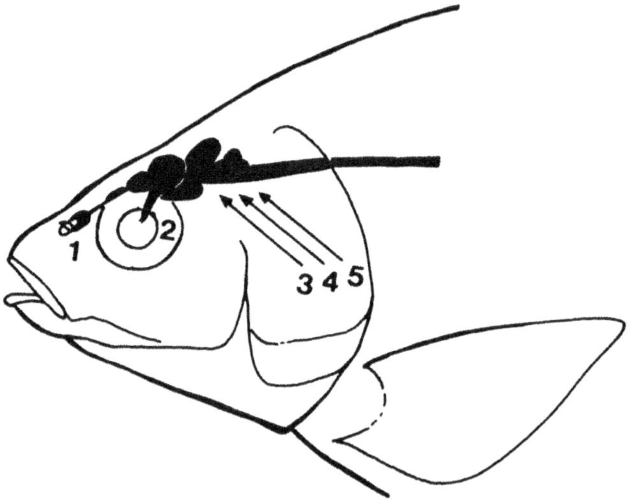

1
2
3
4
5

Environmental Biology of Fishes **33**: 97–104, 1992.

Quantitative fine structural diversification of red and white muscle fibres in cyprinids

Alexandra M. Sänger
Institute of Zoology, University of Salzburg, Hellbrunnerstrasse 34, 5020 Salzburg, Austria

Received 15.1.1990 Accepted 14.8.1990

Key words: Morphometry, Intermediate muscle fibre type, Mitochondrial and lipid content, Species grouping, Ostariophysan fishes

Synopsis

Based on the results of investigations into the fine structure of red and white muscle of roach, *Rutilus rutilus,* and chub, *Leuciscus cephalus* (Sänger et al. 1990), analysis of axial muscle has been extended to include nine other cyprinid species. Quantitative comparisons were made of myofibrils, mitochondria, lipid and subsarcolemmal cytoplasm in relation to muscle fibre type and species. A comparison of these variables between the species shows that for red fibres there are significant differences in all measured variables in at least some cases. The most striking difference is in lipid content, with Danube bleak, *Chalcalburnus chalcoides mento,* having the highest amount. For white fibres, there were significant differences only in intermyofibrillar mitochondria and myofibrils, which were most significant in Danube bleak. As a preliminary study, intermediate fibres have been examined qualitatively and compared with the other two fibre types using the above mentioned variables. The results allow cyprinids to be arranged according to their scores from the different measurements. These groupings are discussed in relation to the different life-styles of the fish.

Introduction

Fish axial muscle shows anatomically distinct zonation of fibre types. Aside from other fish, in cyprinids two main types of myotomal muscle, red and white muscle, and subtypes can be distinguished. They have different biochemical, histochemical, physiological and morphological properties (Akster 1983, 1985, Johnston et al. 1977, Sänger et al. 1988, 1989).

In an earlier investigation intraspecific comparisons were made of the fine structure of red and white muscle fibres and the effect of habitat and season (Sänger et al. 1990). The aim of the present study is to make quantitative interspecific comparisons of the fine structure of muscle fibre types of a range of cyprinids and to relate the findings to the life-styles and swimming behaviour of the species.

Materials and methods

Cyprinid species were collected by gill-netting from three localities, the backwaters of the river Danube near Vienna (*Alburnus alburnus* (Aa), the bleak; *Abramis ballerus* (Aba), the blue bream; *Abramis brama* (Abr), the common bream; *Aspius aspius* (Asp), the asp; *Blicca bjoerkna* (Bb), the white bream; *Chondrostoma nasus* (Cn), the nase; *Leuciscus cephalus* (Lc), the chub; *Rutilus rutilus* (Rr), the roach), the Seefelder See near Innsbruck (*Rutilus rutilus*) and the Mondsee near Salzburg (*Chal-*

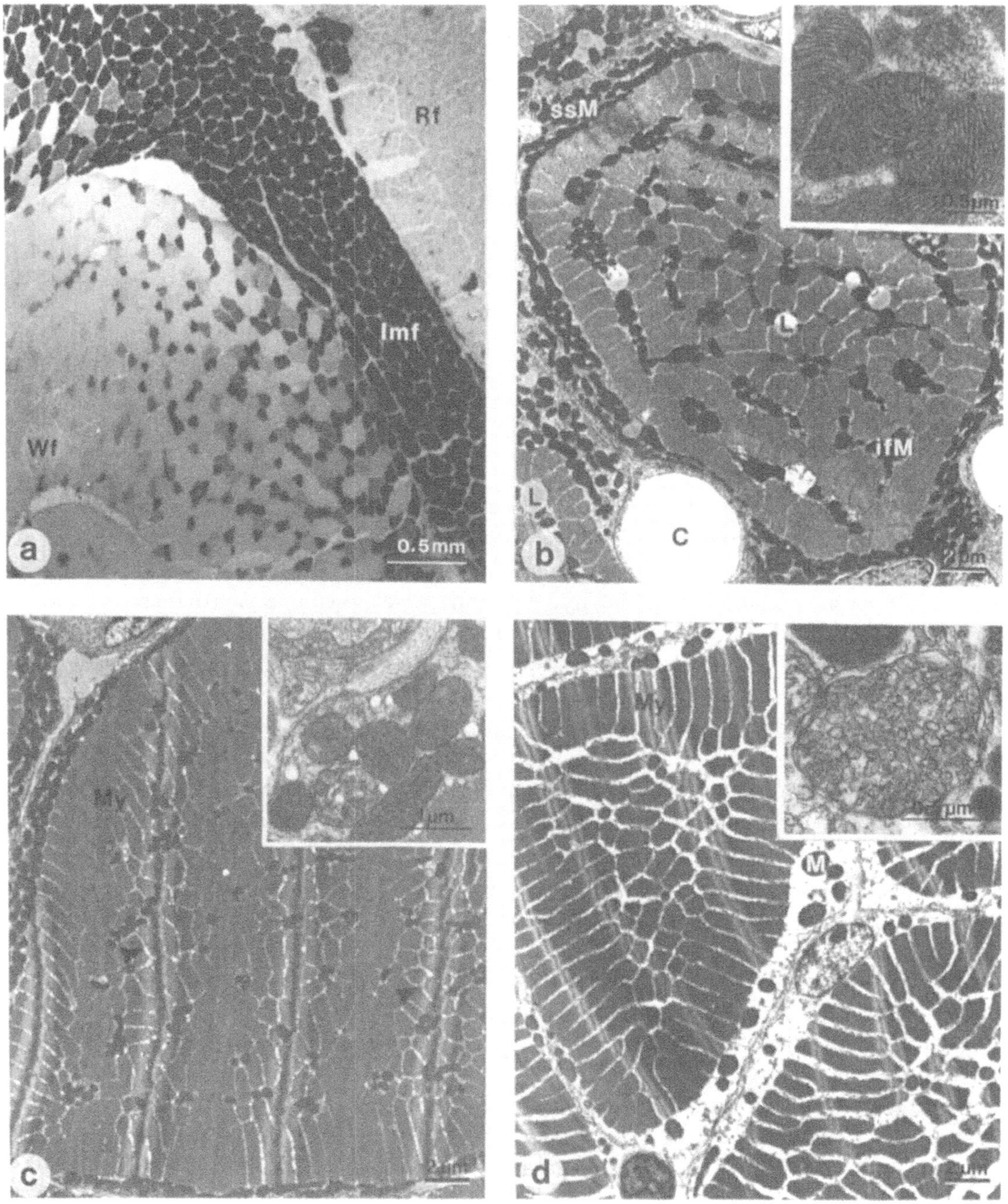

Fig. 1. a – Histochemically treated muscle sample (mATPase reaction, preincubated at pH 10.35) in *Chalcalburnus chalcoides mento* with red (Rf), intermediate (Imf) and white muscle fibres (Wf). b – Transverse section of a red muscle fibre in *Chondrostoma nasus*. Note well-developed capillarisation (C) and the large number of intermyofibrillar (ifM), subsarcolemmal (ssM) mitochondria and lipid (L). The insert shows the typical lamellar shape of red fibre mitochondria. c – Transverse section of an intermediate muscle fibre in *Chondrostoma nasus* with fewer mitochondria than red muscle fibres, almost no lipid but a well developed sarcoplasmic reticulum (arrowheads) and myofibrils (My). The insert shows both lamellar and tubular types of mitochondria. d – Transverse section of white muscle fibres in *Aspius aspius* with a large myofibrillar component (My), few mitochondria (M) and no lipid. The insert shows the typical tubular shaped mitochondria of white muscle fibres.

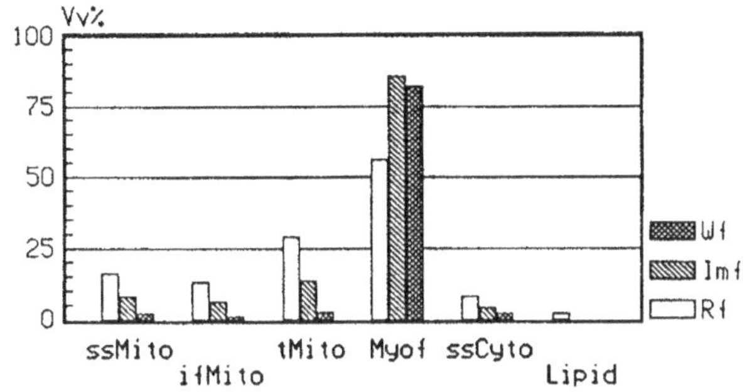

Fig. 2. Comparison of the three fibre types (red = Rf, intermediate = Imf, white = Wf) by volume densities (V_V%) of subsarcolemmal (ssMito), intermyofibrillar (ifMito) and total amount (tMito) of mitochondria, myofibrils (Myof), subsarcolemmal cytoplasm (ssCyto) and lipid.

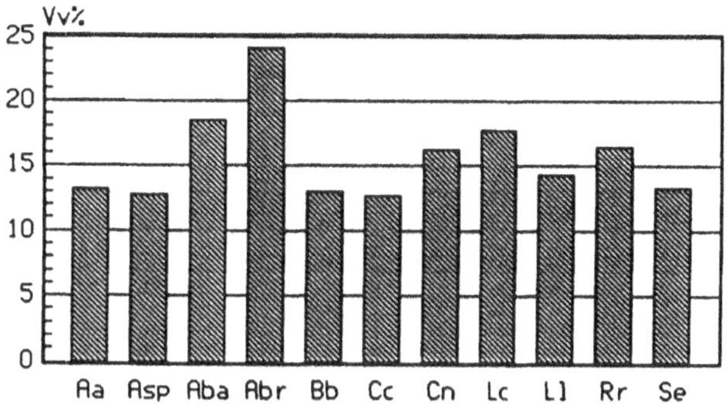

Fig. 3. Mean values of the volume densities (V_V%) of subsarcolemmal mitochondria of red muscle fibres for the various cyprinid species (Aa = *Alburnus alburnus;* Asp = *Aspius aspius;* Aba = *Abramis ballerus;* Abr = *Abramis brama;* Bb = *Blicca bjoerkna;* Cc = *Chalcalburnus chalcoides mento;* Cn = *Chondrostoma nasus;* Lc = *Leuciscus cephalus;* Ll = *Leuciscus leuciscus;* Rr = *Rutilus rutilus;* Se = *Scardinius erythrophthalmus*).

calburnus chalcoides mento (Cc), the Danube bleak; *Leuciscus cephalus, Leuciscus leuciscus* (Ll), the dace; *Scardinius erythrophthalmus* (Se), the rudd). The number (n) of individuals of the different species, respectively, are: n(Aa) = 5, n(Asp) = 9, n(Aba) = 2, n(Abr) = 2, n(Bb) = 2, n(Cc) = 9, n(Cn) = 3, n(Lc) = 4, n(Ll) = 2, n(Rr) = 16, n(Se) = 2. For the analysis of red muscle fibres all 11 cyprinid species were used, for analysis of white fibres only 5 species provided suitable material. Preparation of muscle samples for electron microscopy and morphometrical analyses were described by Sänger et al. (1990).

The variables measured were subsarcolemmal mitochondria (ssM), intermyofibrillar mitochondria (ifM), total amount of mitochondria (tM), myofibrils (Myof), subsarcolemmal cytoplasm (ssCyto) and lipids (Lip). They are expressed in percentages as volume densities per muscle fibre (V_V%). Significance of group differences (e.g. the differences between species) was assessed with one-way analysis of variance (ANOVA). In addition two pairwise multiple comparison tests (Tukey and Bonferroni method, respectively) were used to identify which group differences account for a significant overall F-value. Each species was tested against all other species. In the case of unequal group variances (tested with Levene's test), Welch and Brown-Forsythe procedures were used instead of standard ANOVA. Normal distribution of data was tested after Lilliefors (1967).

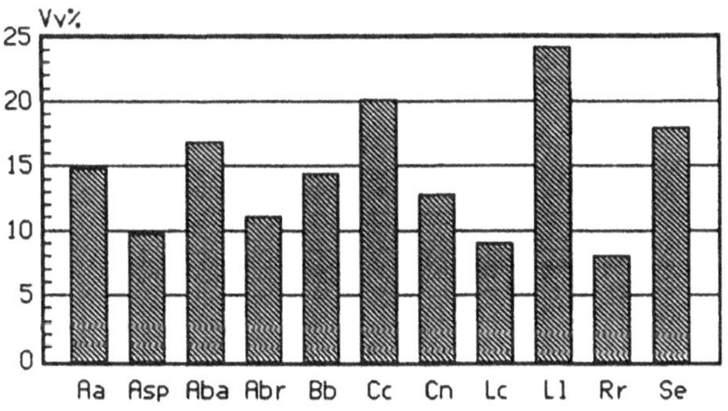

Fig. 4. Mean volume densities of intermyofibrillar mitochondria of red muscle fibres (for abbreviations see Fig. 3).

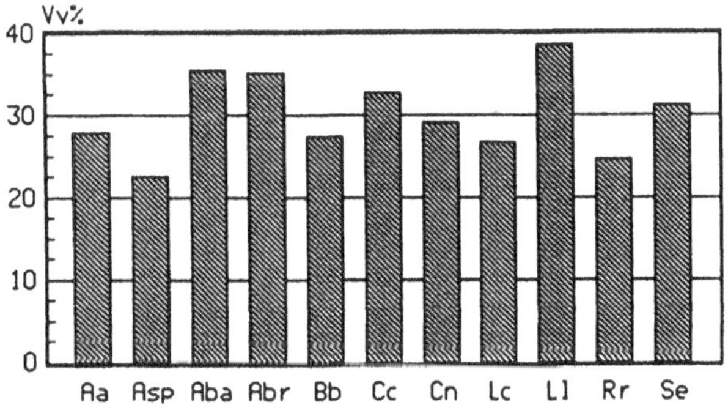

Fig. 5. Mean volume densities of the total amount of mitochondria of red muscle fibres (for abbreviations see Fig. 3).

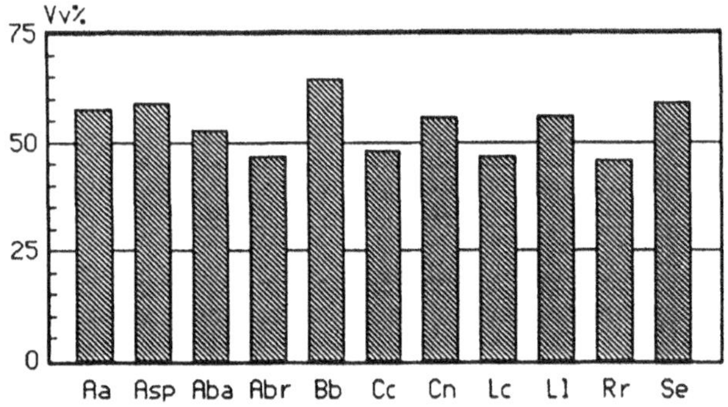

Fig. 6. Mean volume densities of myofibrils of red muscle fibres (for abbreviations see Fig. 6).

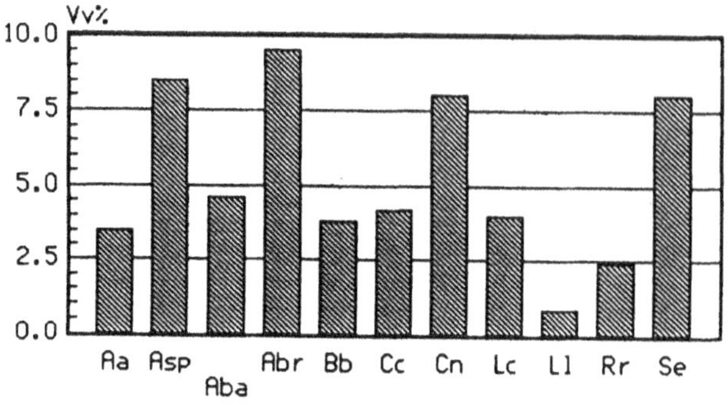

Fig. 7. Mean volume densities of subsarcolemmal cytoplasm of red muscle fibres (for abbreviations see Fig. 3).

Results

The morphological characteristics of the intermediate muscle fibres (Fig. 1a) are between those of red and white muscle fibres. For example, there is a well developed sarcoplasmic reticulum and the numbers of subsarcolemmal (ssM) and intermyofibrillar mitochondria (ifM) are between those values of the other two fibre types (Fig. 1b, c, d). In contrast to red muscle fibres, where we can observe only lamellar mitochondria (Fig. 1b) and white muscle fibres with tubular mitochondria (Fig. 1d), intermediate fibres show tubular and lamellar mitochondria (Fig. 1c). Figure 2 shows that also the quantitative ultrastructure of intermediate fibres is between that of red and white fibres with the exception of myofibrils, for which the intermediate fibres show the greatest numbers.

The overall ANOVA for the ssM of red fibres does not show a significant difference among spe-

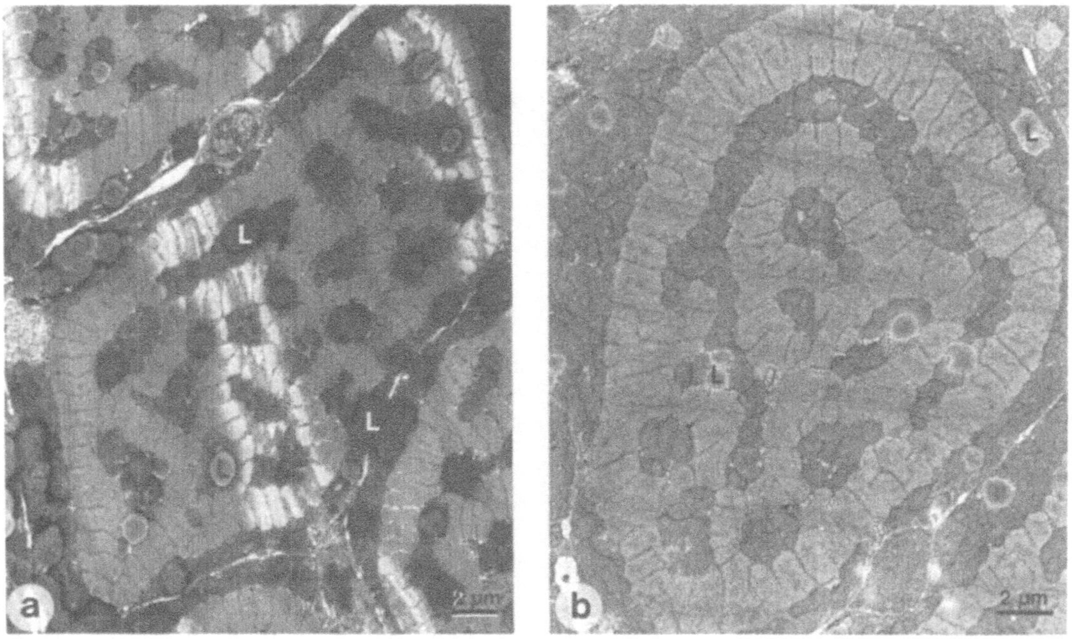

Fig. 8. a – Transverse section of a red muscle fibre in *Chalcalburnus chalcoides mento.* Note the large amount of lipid (L). b – Transverse section of a red muscle fibre in *Leuciscus leuciscus* with the second greatest lipid content (L) of the studied cyprinids.

cies. However, the Bonferroni method (which is more robust when group variances are unequal) indicates some significant pairwise differences: roach (Rr) and asp (Asp) at the 0.1% level and Danube bleak (Cc) at the 1% level when compared with the common bream (Abr) and asp (Asp) compared with the blue bream (Aba) at the 5% level (Fig. 3).

For the ifM there is a highly significant ($p \leq 0.01$) overall F-value, due mainly to the group differences of roach (Rr), chub (Lc) and asp (Asp) compared with Danube bleak (Cc) and dace (Ll) at the 1% level, and roach (Rr) compared with blue bream (Aba) at the 5% level (Fig. 4).

The group differences of Danube bleak (Cc) compared with asp (Asp) (1% level) and roach (Rr) (5% level) are accountable for the highly significant overall F-value for the volume density of the total amount of mitochondria (Fig. 5).

Again, the difference between Danube bleak (Cc) and asp (Asp) (5% level) is due to the highly significant overall F-value for myofibrils together with the difference between roach (Rr) and asp (Asp) at the 1% level (Fig. 6).

Regarding ssCyto (subsarcolemmal cytoplasm) we have several group differences for the highly significant overall F-value. These are asp (Asp) compared with Danube bleak (Cc), bleak (Aa) and roach (Rr) at the 1% level and compared with chub (Lc) at the 5% level and roach (Rr) compared with nase (Cn) at the 5% level (Fig. 7).

The ANOVA indicates highly significant variations in lipid content, due to Danube bleak (Cc) (Fig. 8a), which has the highest lipid content of all studied cyprinid species. With Tukey's multiple comparison test all group differences are significant at the 1% (Danube bleak (Cc) compared with roach (Rr), chub (Lc), white bream (Bb), blue bream (Aba), bleak (Aa), nase (Cn) and asp (Asp)) respectively 5% level (Danube bleak (Cc) compared with rudd (Se) and common bream (Abr)) with the exception of dace (Ll) (Fig. 8b) which has the second-largest lipid content (Fig. 9).

In white muscle fibres we could observe only significant overall F-values for ifM and myofibrils. Regarding ifM (Fig. 10), there are group differences between Danube bleak (Cc) compared with asp (Asp), roach (Rr) and chub (Lc) (at the 1% level) and with nase (Cn) (at the 5% level) together causing the highly significant overall ANOVA (Fig. 11). Regarding myofibrils, roach (Rr) differed significantly from asp (Asp) at the 5% level using Bonferroni method (Fig. 12).

The visual view of the mean values from the histograms probably suggests other group differences as significant than the test does. This is due to the number of individuals which is considered with the test but not with histograms.

From these analyses it is possible to order cyprinid species according to their high or low values of the measured variables. This arrangement of fish species is nearly the same for red and white muscle fibres:

Red muscle fibres:

	high values	low values
Total amount of mitochondria	Cc, Abr, Se, Cn, Ll, Aa, Aba	Asp, Bb, Lc, Rr
Lipid	Cc, Lc, Rr, Ll, Aa	Asp, Bb, Abr, Se, Cn, Aba
Myofibrils	Asp, Bb, Se, Cn, Ll, Aa, Aba	Cc, Lc, Abr, Rr

White muscle fibres:

	high values	low values
Total amount of mitochondria	Cc, Cn	Asp, Lc, Rr
Myofibrils	Cc, Asp	Cn, Lc, Rr

Discussion

The morphometric data observed for intermediate muscle fibres are similar to those collected for pink muscle fibres in the carp by Akster (1985). The comparison of the three fibre types shows the posi tion of the intermediate muscle fibres being between red and white fibres with the exception of myofibrillar content. This higher density of myo-

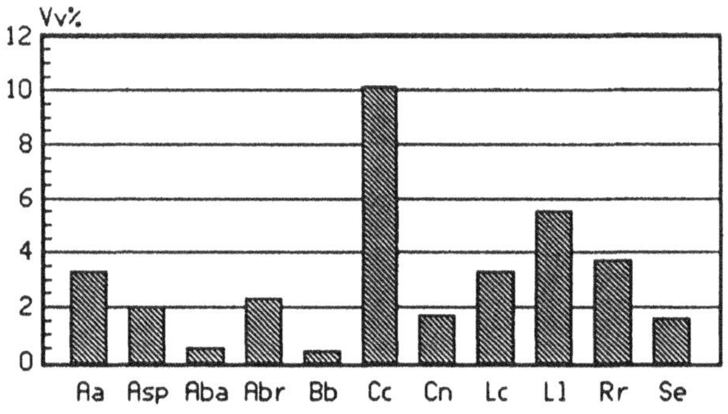

Fig. 9. Mean volume densities of lipid of red muscle fibres (for abbreviations see Fig. 3).

fibrils in the intermediate fibres is probably due to a difference in intermyofibrillar space, that might be related to a better developed sarcoplasmic reticulum in white muscle fibres and thus a better rate of tension development (Johnston 1982a).

Summarizing the findings of the quantitative ultrastructure of red muscle fibres, it is striking that for all measured variables except ssM (subsarcolemmal mitochondria) there is a significant group difference between Danube bleak (Cc) and asp (Asp) (Fig. 3, 4, 5, 6, 7, 9). This may be related to the different and contrary life-style and swimming behaviour of these two species: Danube bleak is considered to be a sustaining swimmer and plankton feeder in contrast to the piscivorous asp which, partly performs burst swimming activities. Starting from the high values of ifM (intermyofibrillar mitochondria) in red and also white muscle fibres and the very high red fibre lipid content for Danube bleak one could suppose a special life-style compared with the other cyprinids studied. According to Moyes et al. (1989) red and white muscle differ not only in mitochondrial abundance but also in mitochondrial properties. Whereas red muscle mitochondria appear to be able to utilize a wide range of metabolic fuels (fatty acids, pyruvate, amino acids but not ketone bodies), white muscle mitochondria appear to be specialized for oxidation of pyruvate and, to a lower extent, fatty acids. The resulting high values for lipid, thus may indicate,

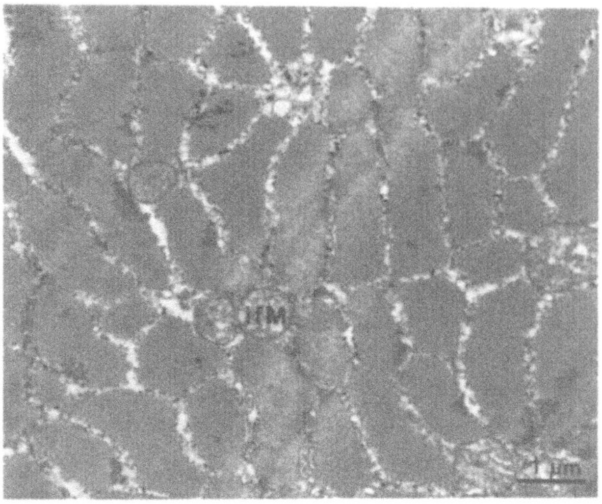

Fig. 10. Transverse section of a white muscle fibre in *Chalcalburnus chalcoides mento* with its high intermyofibrillar mitochondrial content (ifM).

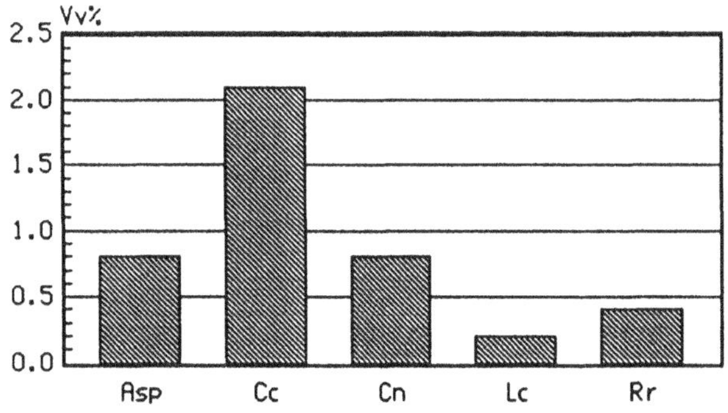

Fig. 11. Mean volume densities (V_V%) of intermyofibrillar mitochondria of white muscle fibres (Asp = *Aspius aspius;* Cc = *Chalcalburnus chalcoides mento;* Cn = *Chondrostoma nasus;* Lc = *Leuciscus cephalus;* Rr = *Rutilus rutilus).*

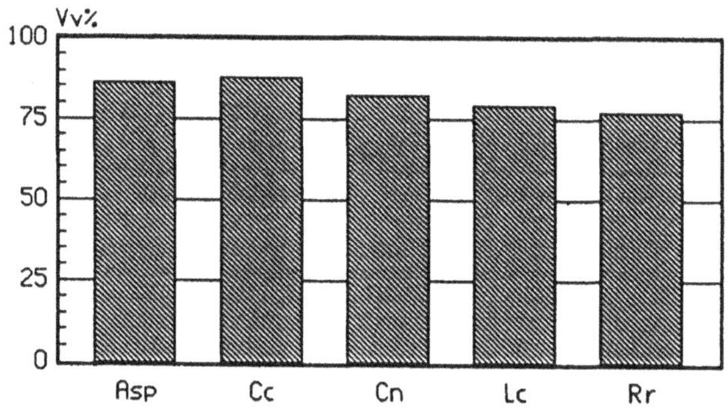

Fig. 12. Mean volume densities of myofibrils of white muscle fibres (for abbreviations see Fig. 11).

that the ifM of the red muscle fibres of Danube bleak use mainly lipid for energy supply or oxygen transport. Because of the better solubility of oxygen in lipid than in aqueous cytosol, Londraville (1988) suggests that lipid is facilitating oxygen transport from capillaries to mitochondria within the muscle fibre in those fish with a relatively low capillary density but high volume density of lipid. Morphometry on capillary density has not been done up to now but would be necessary for strengthening this hypothesis on the possible function of intracellular lipid in the cyprinids studied.

In summary, on the basis of muscular energy supply, Danube bleak represents metabolically the most aerobic cyprinid species. It is well known that environmental factors like temperature (Johnston & Maitland 1980, Johnston 1982b, Egginton & Sidell 1989), hypoxia (Johnston & Bernard 1982, Sänger et al. 1990) and food availability (Johnston 1981) influence ultrastructural features of muscle fibres. Acclimation from 25 to 5° C results in an increase in total mitochondrial volume of red and white muscle fibres as well as intracellular lipid droplets in red fibres (Egginton & Sidell 1989). This effect can be excluded in our material as water temperature was nearly the same for all original habitats. Nevertheless, to eliminate that the resulting differences in the measured parameters, particularly those involving Danube bleak, are due to habitat influences, further studies using laboratory-reared specimens are required.

In contrast to the single species, the arrangement of fish species according to their volume densities of mitochondria, lipid and myofibrils, is difficult to explain according to life-style and swimming behaviour. Mosse & Hudson (1977) showed the connection between proportion of muscle type and swimming behaviour. Red muscle is metabolically adapted for sustained swimming whereas white muscle is needed for more vigorous activities. If we stay within a muscle type it should be possible to connect ultrastructural features with swimming behaviour of a fish. Regarding red muscle fibres the resulting high values for mitochondria and lipid of bleak (Aa) and Danube bleak (Cc) may be connected with their continuous swimming. The adult asp (Asp) as a piscivore, shows lower mitochondrial content with higher amount of myofibrils which may be connected with a more vigorous swimming activity.

To get further information about this grouping, cluster analysis may help and, since this grouping is based on fine structure of muscle, the information obtained has to be integrated with enzymatic, respirometric and taxonomic relationships.

Acknowledgements

The author thanks Alfred Goldschmid, John Haslett, Hans Hoppeler, Franz Maier and Ferdinand Österreicher for valuable and critical comments on the manuscript and Adda Mänhardt, Rudolf Hametner and Christian Ortner for excellent technical work. The investigation was supported by the 'Fond zur Förderung der wissenschaftlichen Forschung in Österreich', project No. S-35 and by the Forschungsinstitut Gastein-Tauernregion, project No. K-18.

References cited

Akster, H.A. 1983. A comparative study of fibre type characteristics and terminal innervation in head and axial muscle of the carp (*Cyprinus carpio* L.): a histochemical and electronmicroscopical study. Neth. J. Zool. 33: 164–188.

Akster, H.A. 1985. Morphometry of muscle fibre types in the carp (*Cyprinus carpio* L.). Relationships between structural and contractile characteristics. Cell Tiss. Res. 241: 193–201.

Egginton, S. & B.D. Sidell. 1989. Thermal acclimation induces adaptive changes in subcellular structures of fish skeletal muscle. Amer. J. Physiol. 256: R1–R9.

Johnston, I.A. 1981. Quantitative analysis of muscle breakdown during starvation in the marine flatfish. *Pleuronectes platessa.* Cell Tissue Res. 214: 369–386.

Johnston, I.A. 1982a. Biochemistry of myosins and contractile properties of fish skeletal muscle. Mol. Physiol. 2: 15–29.

Johnston, I.A. 1982b. Capillarisation, oxygen diffusion distances and mitochondrial content of carp muscles following acclimation to summer and winter temperatures. Cell Tissue Res. 222: 325–337.

Johnston, I.A. & L.M. Bernard. 1982. Ultrastructure and metabolism of skeletal muscle fibres in the tench: effects of long term acclimation to hypoxia. Cell Tissue Res. 227: 179–199.

Johnston, I.A. & B. Maitland. 1980. Temperature acclimation in crucian carp, *Carassius carassius* L., morphometric analyses of muscle fibre ultrastructure. J. Fish Biol. 17: 113–125.

Johnston, I.A., W. Davison & G. Goldspink. 1977. Energy metabolism of carp swimming muscles. J. comp. Physiol. 114: 203–216.

Lilliefors, H.W. 1967. On the Kolmogorov-Smirnov test for normality with mean and variance unknown. J. Amer. Statist. Assoc. 62: 399–402.

Londraville, R.L. 1988. Quantitative ultrastructural differences in aerobic muscle from two antarctic fishes. *Trematomus newnesi* and *Notothenia gibberifrons*. Amer. Zool. 28: 60A.

Mosse, P.R.L. & R.C.L. Hudson. 1977. The functional roles of different muscle fibre types identified in the myotomes of marine teleosts: a behavioural, anatomical and histochemical study. J. Fish Biol. 11: 417–430.

Moyes, C.D., L.T. Buck, P.W. Hochachka & R.K. Suarez. 1989. Oxidative properties of carp red and white muscle. J. exp. Biol. 143: 321–331.

Sänger, A., H. Claassen & H. Adam. 1988. The arrangement of muscle fiber types in the axial muscle of various cyprinidae (Teleostei). Zool. Anz. 221: 44–49.

Sänger, A., A. Goldschmid & H. Adam. 1989. Muscle fiber type distribution of various cyprinids. Progr. Zool. 35: 561–563.

Sänger, A.M., Z.S. Kim & H. Adam. 1990. The fine structure of muscle fibres of roach *Rutilus rutilus* (L.) and chub *Leuciscus cephalus* (L.); Cyprinidae, Teleostei: intraspecific differences and effects of habitat and season. J. Fish Biol. 36: 205–213.

Environmental Biology of Fishes **33**: 105–124, 1992.

Structure, development and function of the branchial sieve of the common bream, *Abramis brama*, white bream, *Blicca bjoerkna* and roach, *Rutilus rutilus*

Coen van den Berg, Ferdinand A. Sibbing, Jan W.M. Osse & Wim Hoogenboezem
Department of Experimental Animal Morphology and Cell Biology, Agricultural University, Marijkeweg 40, 6709 PG Wageningen, The Netherlands

Received 5.10.1990 Accepted 18.1.1991

Key words: Cyprinids, Eutrophication, Zooplankton, Filter feeding, Gill rakers, Retention ability, Capacity, Comb model, Channel model, Energy ratio

Synopsis

The filter feeding organ of cyprinid fishes is the branchial sieve, which consists of a mesh formed by gill rakers and tiny channels on the gill arches. In order to establish its possible role during growth we measured the following morphological gill raker parameters over a range of sizes in three cyprinid fishes, bream, white bream and roach: inter raker distance, bony raker length, raker width, cushion length and channel width. At any given standard length common bream has the largest inter raker distance, roach the lowest and white bream is intermediate. In the 'comb model' of filter feeding the inter raker distance is considered to be a direct measure of the mesh size and retention ability (= minimal size of prey that can be retained) of a filter. For the three species under study there is a conflict between the comb model and experimental data on particle retention. Lammens et al. (1987) found that common bream has a large retention ability whereas roach and white bream have a much smaller one. A new model, the 'channel model' (Hoogenboezem et al. 1991) has been developed for common bream; in this model the lateral gill rakers can regulate the mesh size of the medial channels on the other side of the gill slit. The present data indicate that this model is not appropriate for white bream and roach. At any given standard length white bream and roach only reach 70% of the raker length of common bream, which means that in this model the gill slits should to be very narrow during filter feeding. The gill rakers consist of a bony raker and a fleshy cushion. The bony rakers have a rather long needle-like part outside the cushion in bream, but not in white bream and roach which have blunt gill rakers. Blunt gill rakers are not suited to reduce the diameter of the medial channels. The comb model seems more appropriate for white bream and roach, but doubts about the validity of this simple model remain. The sum of the areas of the medial channels is an approximation of the area through which water flows in the filter. This channel area therefore gives an impression of the capacity or flow rate of the filter. With this capacity estimation and an estimation of energy consumption we calculated an energy ratio of filter feeding. The energy ratio decreases with increasing standard length with an exponent close to the expected exponent of − 0.40. The energy ratio is highest in bream, intermediate in white bream and lowest in roach.

Introduction

In Tjeukemeer, a shallow, eutrophic fresh water lake in the Netherlands, the omnivorous common bream, *Abramis brama*, and the piscivorous pike perch, *Stizostedion lucioperca*, are strongly dominant (Lammens 1986, Lammens 1989). The population sizes of white bream, *Blicca bjoerkna*, and roach, *Rutilus rutilus*, are small (Lammens 1986, Lammens et al. 1987).

In this lake zooplankton and chironomid larvae are the major food source for fishes (de Nie et al. 1980, Vijverberg & Richter 1982a, 1982b, Lammens 1986, de Nie 1987). On an annual basis the diet of bream consists for one half of zooplankton and for one half of chironomid larvae (Vijverberg & Richter 1982a).

Small food, like zooplankton is retained by the branchial sieve, which is a filter formed by gill rakers on the branchial arches (Zander 1906, Hoogenboezem et al. 1991). The two main properties of a filter are its retention ability (the minimal size of particles that can be retained) and its capacity (the volume of water that can be filtered per unit time). There is evidence that bream has a better retention ability than white bream and roach; small food particles are still found in the intestines of freshly caught bream with a standard length at which these particles are absent in the other two species (Lammens et al. 1987; Fig. 11).

Since zooplankton is a major food source in Tjeukemeer it is hypothesized that success in eutrophic water is coupled to a good retention ability and a large capacity of the branchial sieve.

The first hypothesis to be tested is whether differences in filter functioning (retention ability and capacity) between the three species can be related to differences in structure and growth of the branchial sieve. A second hypothesis is whether the functioning of the filter is related to success in zooplankton-rich environments. These relations have been investigated for bream by Hoogenboezem et al. (1990, 1991).

In this paper the relevance for the studied species of two filter feeding models, the comb model and the channel model, is evaluated with the present data. In the comb model of filter feeding the inter raker distance is considered to be a direct measure of the mesh size and retention ability (= minimal size of prey that can be retained) of a filter. The channel model has been proposed for bream by Hoogenboezem et al. (1991; Fig. 12b) In this model water and prey flow parallel to the arches and turn 90° into the medial channels where the prey is retained; the water flows into the gill slit towards the gills. The curvature of the channels in bream (Fig. 3a) could possibly help to guide the water current into the channels. Apart from this a mechanism to reduce the mesh size is introduced (Hoogenboezem et al. 1991). The gill rakers on the lateral side of an arch can be depressed into the medial channels on the opposite side of the gill slit. Evidence for this is provided by X-ray films of filter feeding bream, which show that during filter feeding the maximal inter arch distance is small enough to allow the gill rakers to reach the other side of the gill slit (Hoogenboezem et al. 1991). In this way the mesh size of the medial channels is reduced by at least 50%. In other words, in this model the mesh size is adjustable using either the channel width or the reduced channel width.

The assumption is made that the branchial sieve acts as a mechanical sieve and does not have a sticky mucus layer so that the mechanism of prey retention is simple sieving (Rubinstein & Koehl 1977). This assumption is based on the observation that no small, free living algae are found in the mucus on the branchial sieve of bream in Tjeukemeer although such algae are the dominant particles in the lake (Hoogenboezem, personal communication). Simple sieving means that no particles smaller than the mesh size will be retained. This does not necessarily mean that retention curves must be perfect step functions because variations in mesh size within the filter may occur (see below, 'variations within the branchial sieve'). A second assumption is that the only selection phase occurs in the branchial sieve.

For this paper Hoogenboezem measured a number of branchial sieve parameters in an ontogenic series of bream; later van den Berg measured similar parameters for white bream and roach and some additional parameters for the three species and worked out the results.

Material and methods

The eleven white breams (SL 101–232 mm) and fifteen roaches (SL 48–281 mm) were collected a

b

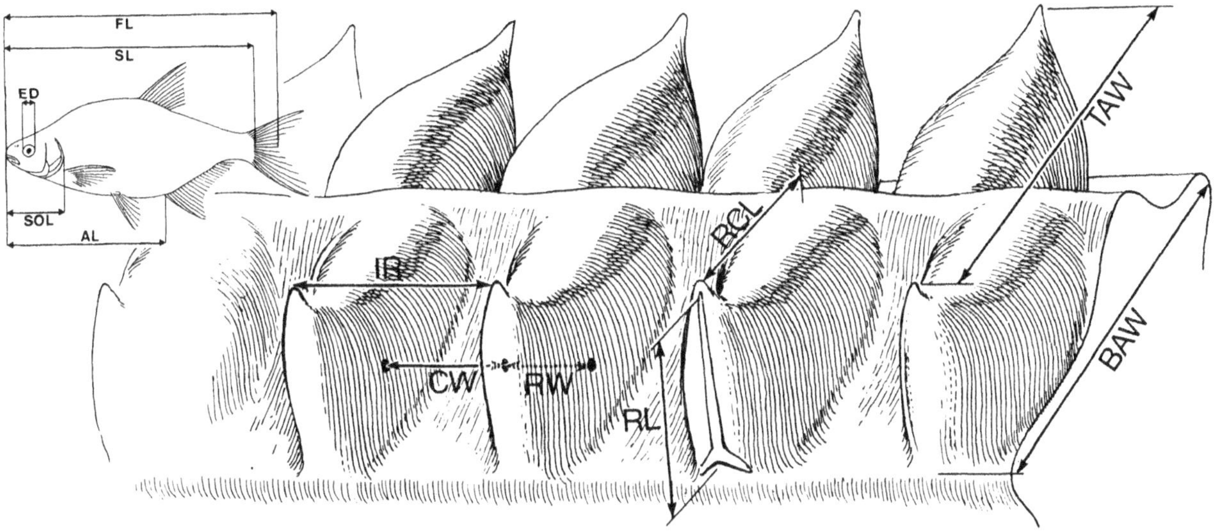

Fig. 1. a – the definition of the standard parameters: standard length (SL), fork length (FL), eye diameter (ED), head length (SOL) and anal length (AL). b – part of a gill arch of white bream; indicated are the definitions of the measurements of bony raker length (RL), inter raker distance (IR), channel width (CW), raker cushion width (RW), raker cushion length (RCL), tip arch width (TAW) and basal arch width (BAW). The position of measurement of CW and RW is indicated with large dots; this position is located at a height of 0.5 RL and about 0.25 RCL deep in the channel.

from a number of Frysian lakes, but not from Tjeukemeer. The thirty-one breams (SL 30–430 mm) were captured in Tjeukemeer. All fish were captured with trawlnets and immediately killed and stored in Bouin's fluid (Romeis 1968). In order to estimate the effect of tissue deformation due to fixation we measured one fresh specimen of each species, stored it in Bouin's fluid and measured it again two weeks later. The inter raker distance and the raker length are almost unaltered (about 4% of change) in the three species. The cushions become lower and broader and due to this effect the channels become narrower. The trends of this change, but not the magnitude (up to 40% of change), are similar in the three species. We did not correct the measured values for the effect of tissue fixation.

We measured three groups of parameters: standard parameters, gill arch parameters and gill raker parameters. The first group (Fig. 1a) consists of standard length (SL), fork length (FL), anal length (AL), head length (SOL, snout operculum length), eye diameter (ED) and body weight (W). We measured these parameters with Vernier calipers (to the next 0.05 mm) or a spring rule (to the next 0.5 mm) depending on fish size. The orientation terms used below are defined in the section 'symbols'.

The gill arch parameters are: the number of gill rakers on the medial side of the ceratobranchials (NR); the length of each ceratobranchial (LCB), defined as the distance between the first and the last raker on the medial side of the ceratobranchial; the arch width from raker tip to raker tip (TAW) and the basal arch width of each arch (BAW), without the gill rakers (Fig. 1b); both widths were measured at the middle of the ceratobranchials. We measured at 30 × magnification using an ocular micrometer (to the next 0.03 mm). The length of the ceratobranchials of the larger fish were measured with Vernier calipers (to the next 0.1 mm). We measured these parameters on the four gill arches of one side.

We measured the gill raker parameters (Fig. 1b) at 30 × magnification with an ocular micrometer (to the next 0.03 mm). In a lateral view of the gill arch, perpendicular to the long axis of the bony rakers, the following parameters were measured: bony raker length (RL), inter raker distance (IR)

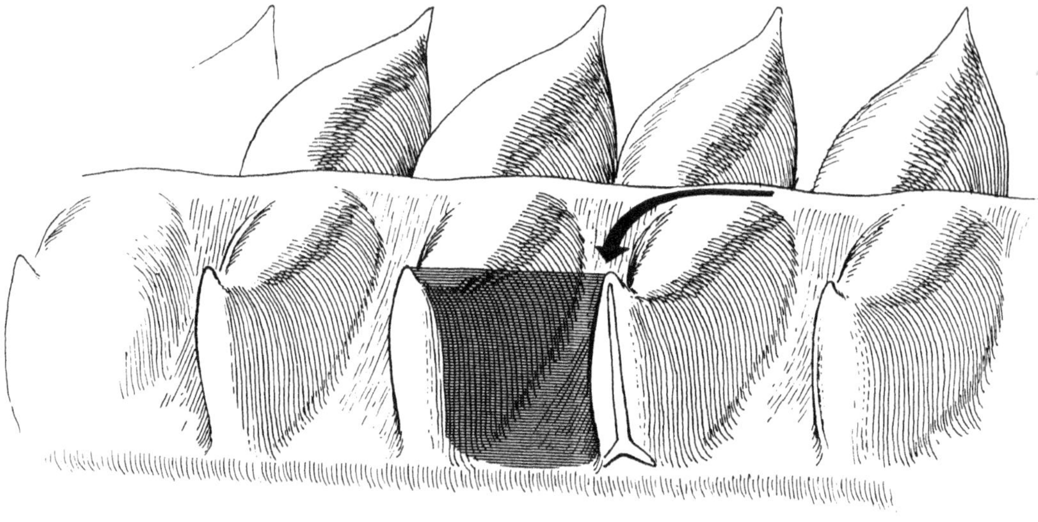

Fig. 2. Part of a gill arch; the channel area is the sum of the areas of the medial channels (one of which is indicated by horizontal hatching). These areas are calculated as the product of bony raker length and channel width; as indicated here they are situated at the end of the channel, in the plane of the gill rakers. The expected water flow is indicated with an arrow, the areas of the medial channels are at right angles with the expected flow.

and at half of the bony raker length the width of the raker cushion (RW) and the width of the channel (CW). We measured the same parameters in bream except for the inter raker distance, which we calculated however as the sum of cushion width and channel width. We measured the raker cushion length (RCL) in a top view of the gill arch. In bream we measured channel length instead of cushion length. We measured the gill rakers from the fifth to the ninth raker counting from anterior (which is approximately the middle of the ceratobranchial) and only on the medial side of the first arch, the lateral and medial side of the second arch and the lateral side of the third arch, in other words, the parameters of the second and third gill slit. In bream we measured the medial side of the first and the lateral side of the second arch, in other words the parameters of the second gill slit.

In order to quantify the variation of the gill raker parameters within a complete filter we measured all the rakers on all the ceratobranchials of one side of the branchial sieve for one specimen of each species. In this way we obtained information about variations within each arch and also about differences between the arches.

With the data of ceratobranchial lengths and widths we calculated the gill arch area (A_a), defined as the sum of the length width products of the first three arches multiplied by two to account for both sides:

$$A_a = 2 \sum_{i=1}^{3} LCB_i \times BAW_i \quad .$$

A more interesting area for capacity estimations is the filter area or the total cross sectional area where water passes the filter. Two different areas can be defined depending on the filter model that is used. The channel area (A_c, Fig. 2) is the sum of all the medial channel diameters of both sides, defined as raker length times channel width or, in other words, the diameter at the end of the channel; this area is based on the channel model (Hoogenboezem et al. 1991; Fig. 12b):

$$A_c = F \sum_{i=1}^{5} RL_i \times CW_i \quad ,$$

where F is a factor relating the sum of the five channel diameters of the standard measurements with the total medial channel area. We calculated this factor with the aid of the complete filter measurements (see above). The gill slit area, which depends on the distance between the arches, is based on the interdigitating model (Sibbing 1991; Fig. 12a). There are no data of inter-arch distances during filter feeding for white bream and roach so the gill slit area could not be calculated.

We analysed the data on a Macintosh II with a statistical program Statworks™. We calculated the relations between pairs of data sets with the model I type of linear regression rather than with model II assuming that the standard length was measured with a much higher degree of accuracy than the branchial sieve parameters (Sokal & Rohlf 1969). We tested the significance of differences between the species with the non-parametric Mann Whitney U-test, which assumes identical distributions of the three independent sets of data. The ratio of the y-value and the x-value (SL) is the test parameter (e.g. IR divided by SL); in the case of an area the test parameter is the ratio area per SL^2. The value of y/x has to be rather constant per species to be able to use the Mann Whitney U-test properly. If the relation between the parameters is linear and if the regression line tends to go through the origin, the value of y/x will be rather constant. In the present data the deviations from going through the origin are never very large and corrections were not necessary.

Anatomy of the branchial sieve

The filter feeding apparatus of cyprinids is the branchial sieve (Fig. 3). The cerato- and epibranchials bear gill rakers on the lateral and medial sides, pointing to the gill slits. The gill rakers filter the water flowing through the gill slits and retain particles larger than the local mesh size. Two possible functions of the branchial sieve are protection of the gills from damaging objects and retention of food particles (Zander 1906).

The fifth gill arches are the pharyngeal jaws (Sibbing 1982, 1988). They bear short rakers on the lateral side only. The ceratobranchials of the four gill bearing gill arches lie parallel in the floor of the pharynx and run latero caudally; they are slightly curved (Zander 1906). Rostrally they are connected in a complex of small hypobranchialia and basibranchialia, which is covered by the postlingual organ (Sibbing & Uribe 1985). Caudally they connect with the epibranchials which contribute to the roof of the pharyngeal cavity. The epibranchials are much shorter than the ceratobranchials and have smaller gill rakers. The palatal organ covers most of the roof of the pharyngeal cavity (Sibbing & Uribe 1985, Sibbing et al. 1986); it plays an important role in food processing. The pharyngeal cavity in cross section has the shape of a horizontal slit; the palatal organ is right above the branchial sieve and probably plays an important role in the functioning of the branchial sieve (Hoogenboezem et al. 1991).

The gill rakers consist of a bony raker (branchiospine) from which a fleshy cushion runs to the middle of the arch surface, which in bream is elevated to a distinct central ridge (Fig. 3a–c). In bream the rakers have a clear needle-like point, whereas in white bream and roach they are blunt (Zander 1906); in bream the top 1/3 of the bony raker is covered only by a thin layer of tissue, whereas in white bream and roach the bony rakers are almost completely embedded in thick cushion tissue. Between two adjacent gill rakers a channel is present. In bream the channels start at the central ridge, first run almost parallel to the arch axis and then turn 90° towards the rim of the arch (Fig. 3a); this curvature becomes less distinct from anterior to posterior on the arches. In white bream and roach all the channels are straight and run perpendicular to the arch axis (Fig. 3b, c).

Gill rakers on opposite sides of a gill slit interdigitate so that a slit might form a functional sieving unit; for bream evidence for this was provided by Hoogenboezem et al. (1991).

Results

Variations within the branchial sieve

We measured the gill raker parameters of all the rakers of one half of the branchial sieve of one bream (SL 310 mm), one white bream (SL 232 mm) and one roach (SL 141 mm). In Fig. 4 the hemi-

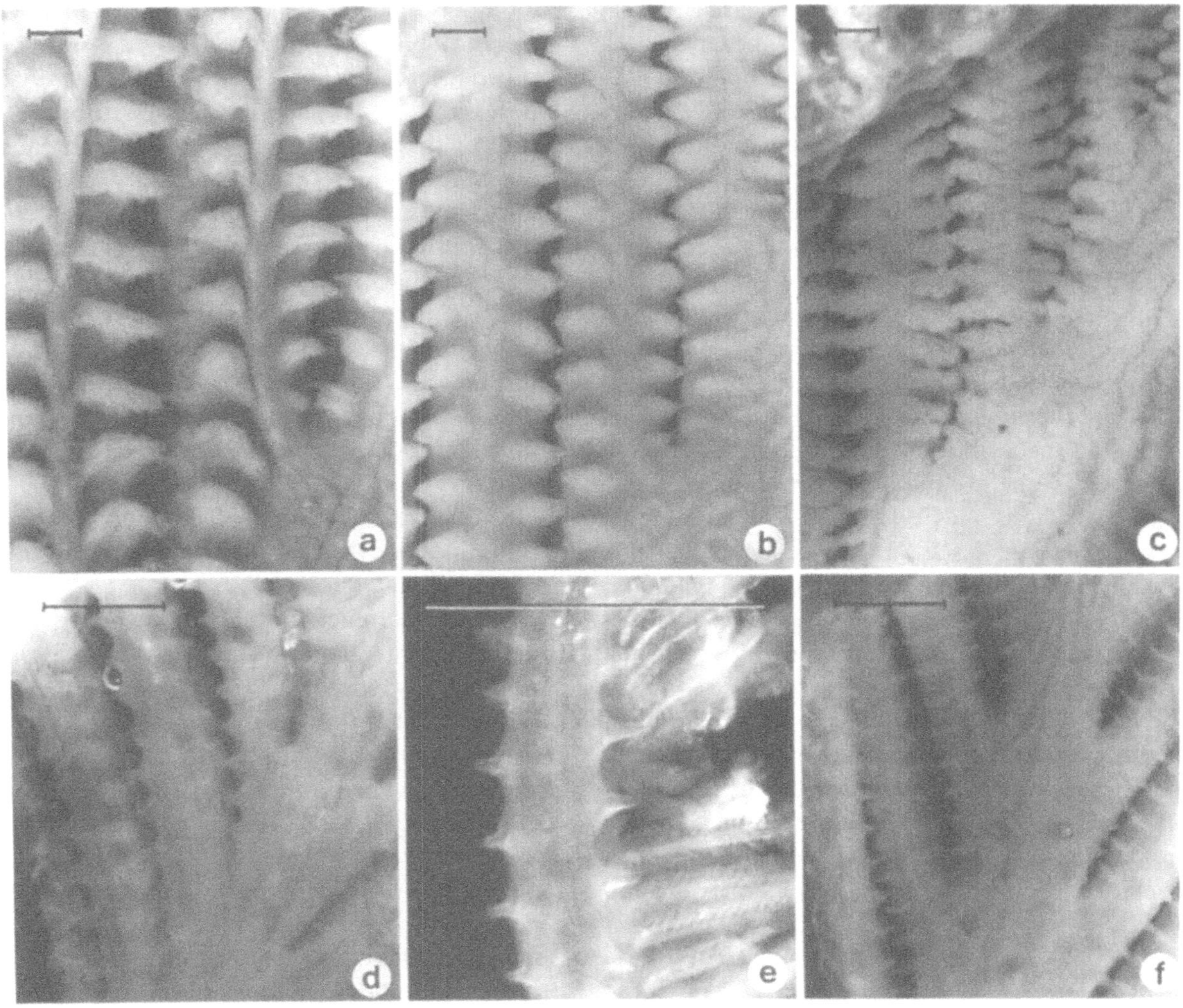

Fig. 3. Photos of branchial sieves of fresh fish; the bars represent 1 mm. a – bream with SL 230 mm; the second, third and part of the fourth gill arch are visible and in the right bottom corner a part of the postlingual organ is visible. The bony raker tips protrude quite far from under the raker cushion. Notice the curvature of the cushions and channels and the distinct medial ridge. b – white bream with SL 204 mm; same orientation as bream in (a). The bony raker tips hardly protrude from the fleshy cushion. The channels are straight and a clear medial ridge is absent. c – roach with SL 254 mm; same orientation as bream in (a). The bony raker tips do not protrude far from the fleshy cushion. The channels are straight and a clear medial ridge is absent. d – bream with SL 55 mm. The raker cushions are small and straight, the bony rakers are already developed. e – roach with SL 35 mm; the second gill arch in a top view. The bony rakers are present but the raker cushions have not yet developed. f – roach with SL 70 mm. The bony rakers and raker cushions are developed now and a medial ridge has appeared. Notice the resemblance with the adult form (c).

branchs are indicated on the x-axis from rostral (first arch, lateral side) to caudal (fifth arch, lateral side). The gill raker parameters are each indicated with a symbol. For each hemibranch the average parameter values of all the rakers on that hemibranch are indicated as well as the extreme parameter values (maximum and minimum). The heavy lines connect the average values of the hemibranchs and the bands indicate the range of values between the extremes.

The lateral side of the first arch has rather strongly deviating gill rakers. Especially in bream, where these rakers are about two times longer than the other rakers and have a flattened blade-like shape. Hoogenboezem et al. (1991) suggest that in bream these gill rakers are modified to be able to seal off the large first gill slit in cooperation with the palatal organ. This seems less likely for white bream and roach because the gill rakers on the lateral side of their first arch do not have such a strongly deviating length as those of bream.

Generally, the parameters tend to become smaller posteriorly. The raker cushion length of white bream and roach clearly shows an asymmetry of the gill arch surface; the medial cushions are longer than the lateral cushions (Fig. 4b, c).

A general conclusion is that these branchial sieves are not entirely homogeneous, both within the arches and between the arches there are differences in gill raker parameters. The standard measurements were done at the middle of the arches. The parameter values are about maximal there, in other words the measurement area represents the maximal branchial sieve values. It can be expected therefore that smaller particles than predicted from the standard measurements will be retained to a certain degree.

Another aspect of the complete filter measurements is the number of gill rakers of the species. We measured this parameter in five white breams and roaches and in seven breams. The number of gill rakers of bream is larger than that of white bream and roach (Table 1). The number of gill rakers is constant above a standard length of about 60 mm. There is some literature evidence of intraspecific variation in these species. Zander (1906) found slightly higher gill raker numbers for roach and much higher ones for bream. Goldschmid et al. (unpublished) found intraspecific differences in the gill raker number of bream and roach populations from different mountain lakes in Austria and relates high gill raker numbers to abundance of planktonic food in certain lakes. Lindsey (1981) found intraspecific variation in gill raker numbers in *Coregonus* sp. and gives a review of literature on the subject. He found indications that the availability of food types has an influence on the number of gill rakers of a population after several generations. The relatively low number of gill rakers in bream and roach from the Frysian lakes might be explained by the fact that the eutrophication of these lakes and the resulting dominance of plankton only started in the sixties (Lammens 1986, p.68)

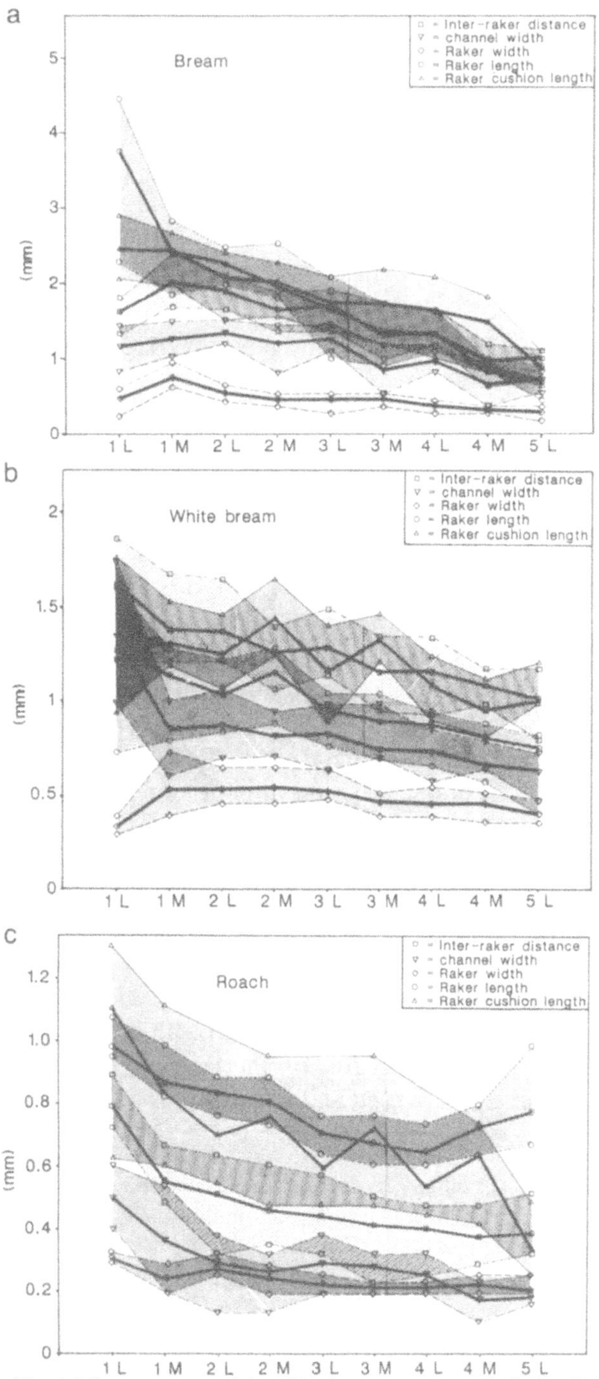

Fig. 4. Measurements of the gill raker parameters of all the gill rakers of one side of the branchial sieve of a 310 mm SL bream, a 232 mm SL white bream and a 141 mm SL roach. The hemibranchs are indicated on the x-axis, from rostral to caudal. The points connected by heavy lines indicate the average values of all the gill rakers on each hemibranch, the bands surrounding them indicate the minimum and maximum value on each hemibranch. Note the different scales on the y-axis. 1L = lateral side of the first arch, etc. Further explanations in the text.

Table 1. The number of rakers as measured by Zander (1906) (between brackets) and by us. The data by us are the averages of seven breams, five white breams and five roaches.

number of rakers	arch 1	arch 2	arch 3	arch 4
bream	16.1 (22)	15.5 (21)	14.2 (20)	13.3 (14)
white bream	12.8	12.8	12.0	9.0
roach	12.8 (13)	12.8 (14)	11.8 (14)	9.2 (8)

and that these populations have not adapted yet to the new food circumstances.

The differences between second and third gill slit can be treated in more detail for white bream and roach because the second and third slit of these species were measured in the developmental series. In Table 2 the ratio of parameters from the first and the second arch is given; it is evident that the parameters are some 10% smaller in the third than in the second gill slit in both species. This is in support of the complete filter measurements of one specimen (Fig. 4b, c) where we already noted the decrease of the parameters from the first to the fifth arch. The branchial sieves are not homogeneous.

Comparative structure and development of the branchial sieve

The parameters of the second slit are used throughout in order to make comparison with bream possible; the raker lengths on the lateral side of the second arch are used and for the other parameters the medial side of the first arch is used. This was done to be able to apply the channel model (Hoogenboezem et al. 1991). All results are plotted versus the standard length (SL) allowing comparison within and between the species.

Head length (SOL) and weight (W)

The differences between the species in head length are small though significant ($p = 0.05$, except white bream/bream).

Since weight and standard length are expected to have a cubic relation log W is plotted versus log SL. The exponent of the increase in weight of the three species is slightly higher than the isometric value 3; the exponent is 3.11, 3.03 and 3.13 for bream, white bream and roach, respectively; in bream and roach the difference from 3 is significant ($p = 0.01$, $p = 0.05$). The differences between the species are very small, despite the apparent clear difference in body shape between white bream/common bream and roach (in particular, the height of the back). Roach and bream are not significantly different ($p = 0.05$) but each is significantly different from white bream

Table 2. The difference of typical parameters between the third and the second gill slit is expressed as the ratio of the parameter in the second slit and that in the third slit. In this way the measurements of the complete filters can be verified partly with data of more specimens per species. 1 M = medial side of the first arch, etc.

		roach		white bream	
IR	1M/2M	1.11 ± 0.13[a]	; $p < 0.005$[b]	1.11 ± 0.17	; $p < 0.005$
IR	2L/3L	1.12 ± 0.12	; $p < 0.005$	1.08 ± 0.14	; $p < 0.005$
CW	1M/2M	1.15 ± 0.26	; $p < 0.005$	1.17 ± 0.25	; $p < 0.005$
RL	2L/3L	1.12 ± 0.11	; $p < 0.005$	1.13 ± 0.12	; $p < 0.005$
LCB	1/2	1.08 ± 0.08	; $p = 0.05$	1.05 ± 0.05	; $p = 0.025$
NR	1/2	0.98 ± 0.06	; n.s.	0.98 ± 0.06	; n.s.

[a] = mean $\pm$ standard deviation.
[b] = difference from 1 tested with sign-test.
; n.s. = not significant ($p > 0.05$).

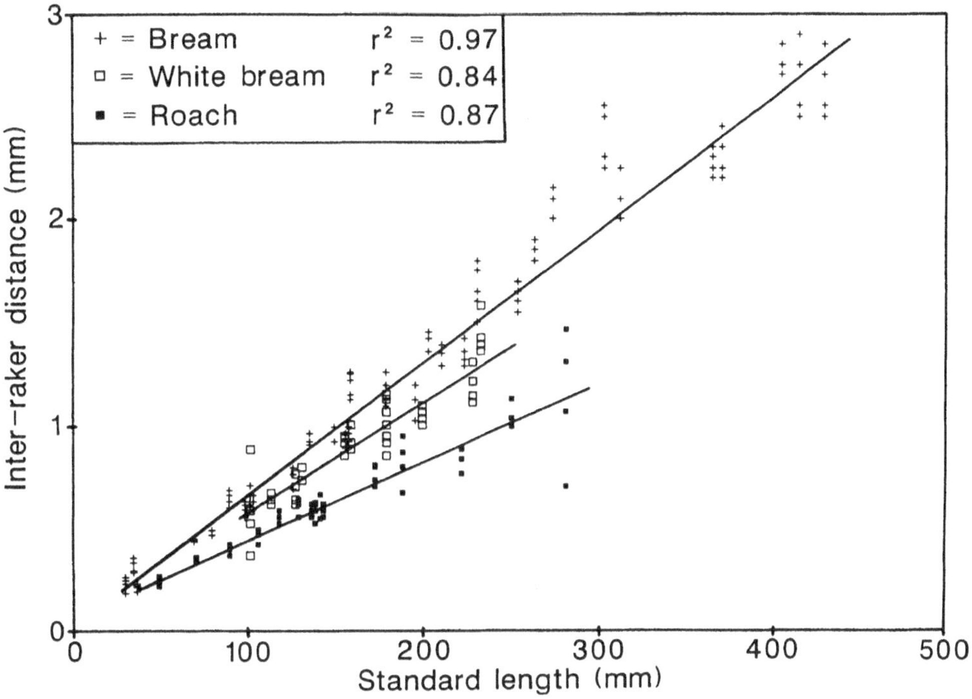

Fig. 5. Inter raker distance versus standard length; notice that bream reaches larger lengths than the other two species.

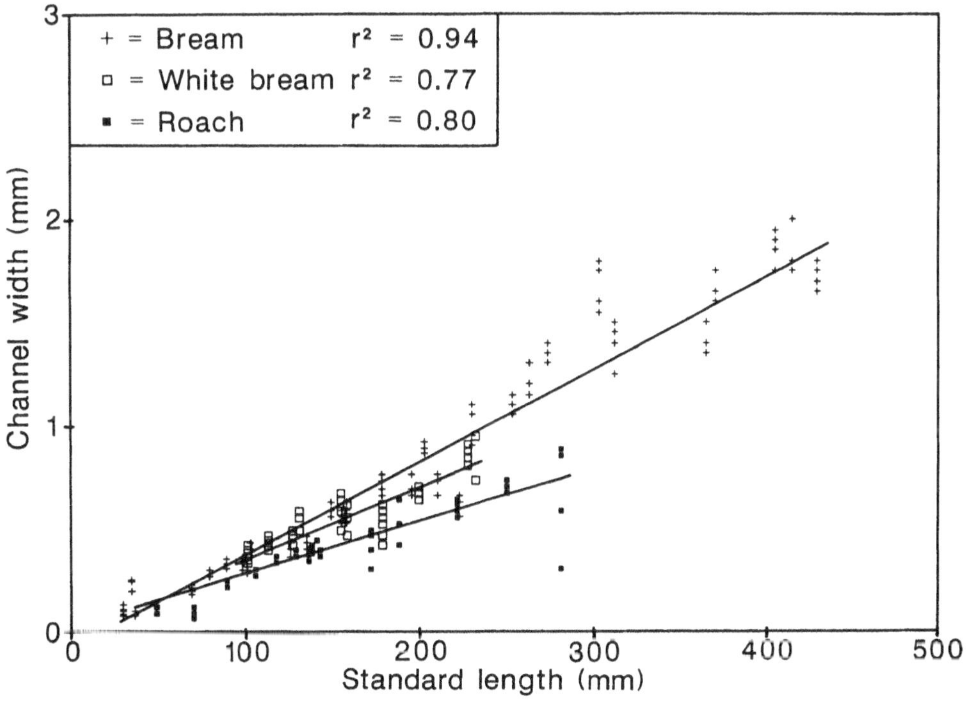

Fig. 6. Channel width versus standard length.

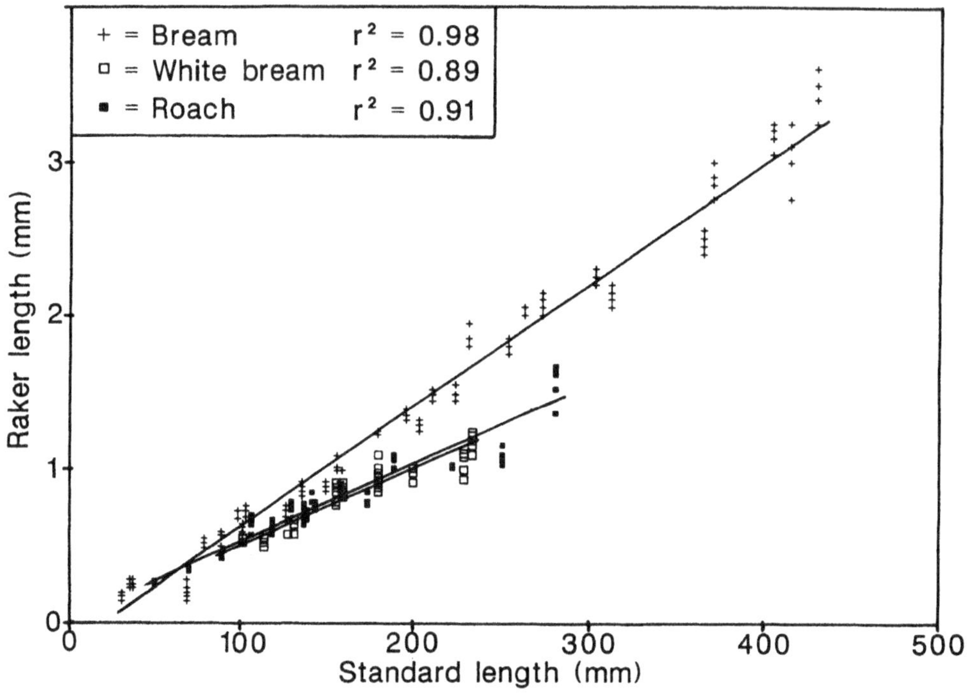

Fig. 7. Raker length versus standard length.

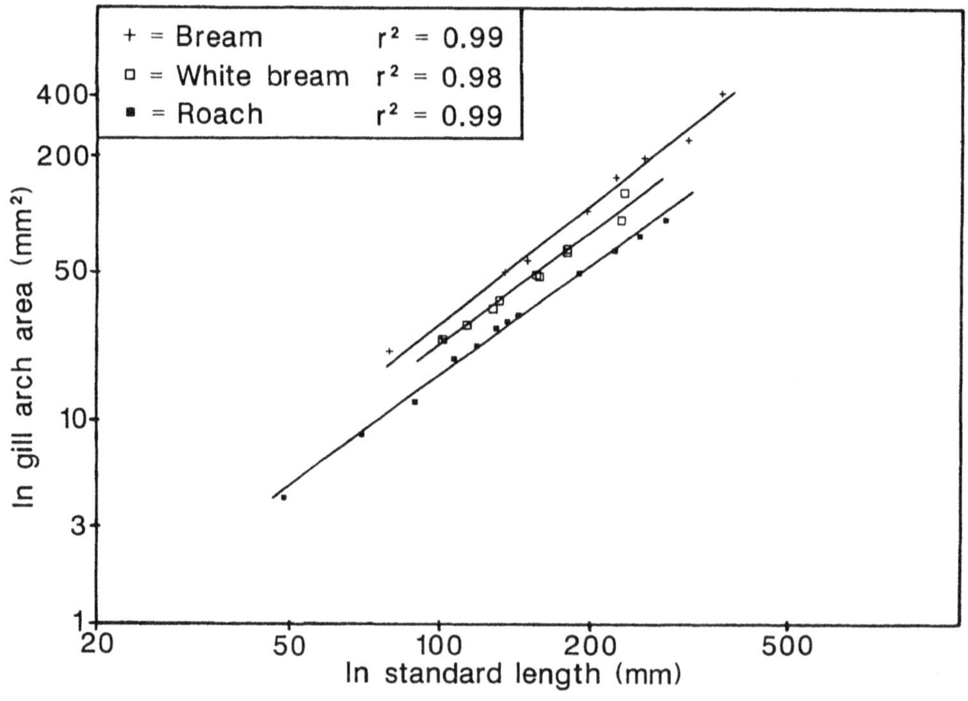

Fig. 8. Gill arch area versus standard length plotted on a double logarithmic scale.

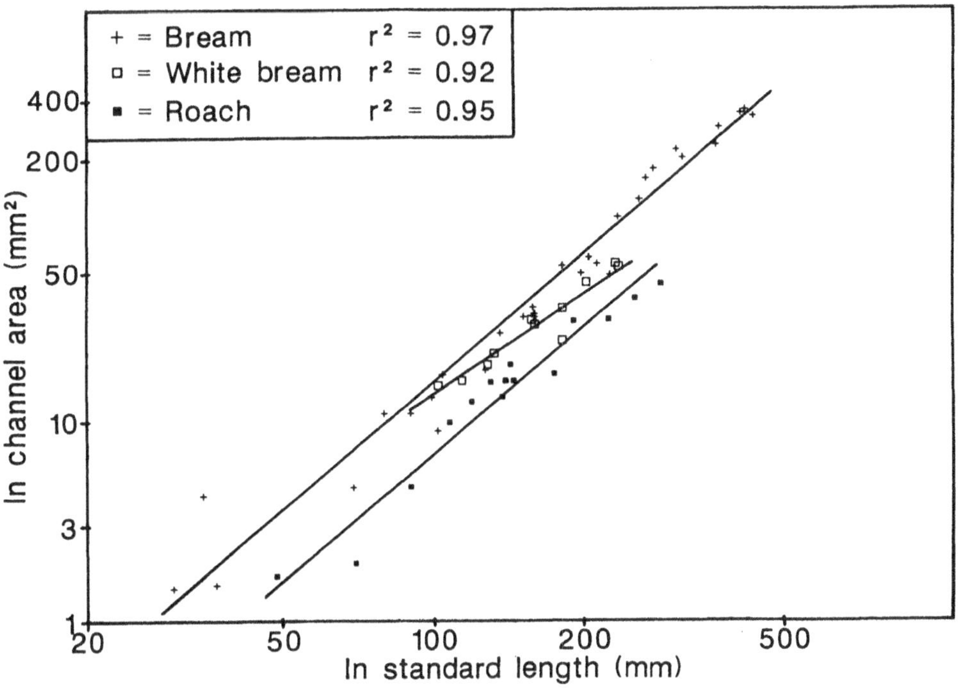

Fig. 9. Channel area versus standard length on a double logarithmic scale.

($p = 0.005$). The effect of fixation is probably large here, the fresh weight is higher than the presently found values.

The results below, which are always plotted versus standard length, would not become really different if they were plotted versus head length or weight.

Inter raker distance (IR, Fig. 5)
The inter raker distances increase isometrically in all the species and the regression lines almost go through the origin, which means that the isometry can possibly be extrapolated back to standard length zero. The value for bream at a given length is larger than that for white bream, which in turn is larger than the roach value. The ratio bream, white bream, roach is about 1 : 0.85 : 0.60. The differences between the species are significant ($p \ll 0.001$).

Channel width (CW, Fig. 6) and raker cushion width (RW)
The sum of CW and RW is the inter raker distance. These parameters show a different pattern of

Table 3. The exponents of standard length in the relation with two filter areas and the energy ratio (Figs. 8, 9, 13). The exponents are derived from the linear regression relations which have the shape: ln (area) = ln (constant) + exponent × ln (standard length).

	Exponent in gill arch area relation	Exponent in channel area relation	Exponent in energy ratio relation
bream	1.97 ± 0.08[a]	2.11 ± 0.07	−0.40 ± 0.07
white bream	1.86 ± 0.09	1.72 ± 0.16	−0.71 ± 0.19
roach	1.79 ± 0.04	2.05 ± 0.14	−0.44 ± 0.12

[a] = estimation ± standard error.

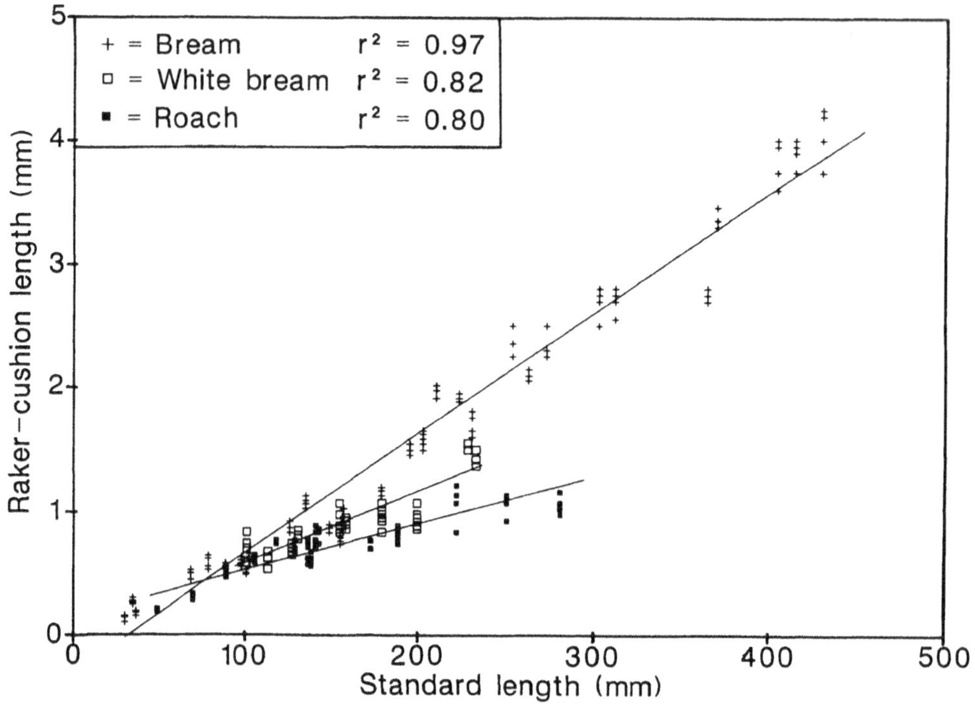

Fig. 10. Raker cushion length versus standard length.

growth than the inter raker distance. The extrapolated regression lines of white bream and roach pass the origin closely, but the one of bream not at all. The explanation for this may be an allometric growth in the early stages of development. Extrapolating a line from the smallest measured SL to the origin gives an idea of early growth. Judging from this approximation in bream the growth of CW accelerates after the early stages whereas the growth of RW decelerates. A consequence of this bend in the curve is that at SL 120 mm the channel widths of bream and white bream are about the same whereas at SL 250 mm the ratio bream, white bream, roach is about 1 : 0.85 : 0.60, which is the same ratio as for the inter raker distance. Appar-

Table 4. The approximate ratios of gill raker parameters of common bream, white bream and roach; A – the ratio of the parameter values at SL 150 mm, B – at SL 250 mm. The values were determined with the regression data.

	A	Common bream	White bream	Roach	B	Common bream	White bream	Roach
inter raker distance (IR)		1	0.85	0.65		1	0.85	0.60
channel width (CW)		1	0.90	0.65		1	0.80	0.65
raker width (RW)		1	0.80	0.65		1	0.95	0.65
raker length (RL)		1	0.75	0.75		1	0.70	0.70
gill arch area (A_a)		1	0.75	0.50		1	0.70	0.50
channel area (A_c)		1	0.75	0.45		1	0.60	0.45
raker cushion length (RCL)		1	0.80	0.65		1	0.70	0.50
energy ratio		1	0.65	0.45		1	0.60	0.45

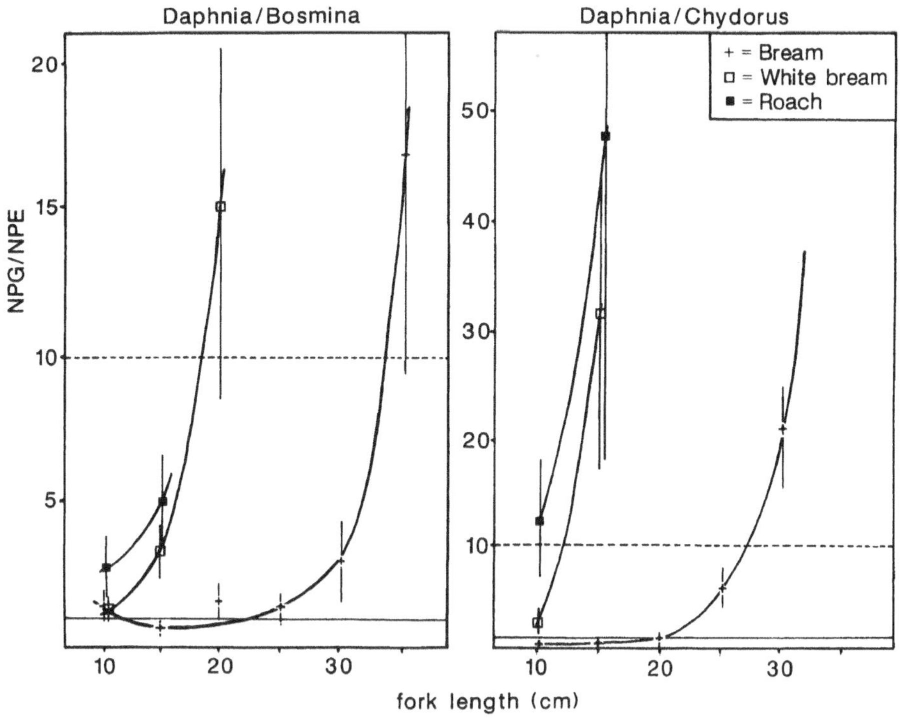

Fig. 11. Retention curve based on gut contents of freshly caught fish; the average lengths in Tjeukemeer of the three mentioned cladocerans are: *Daphnia hyalina* 1.11 mm, *Bosmina coregoni* 0.43 mm and *Chydorus sphaericus* 0.28 mm (Lammens et al. 1987). Vertically plotted is the ratio of the numbers of two plankton species in the gut (NPG) divided by the same ratio in the environment (NPE); a value of one means that both species are caught to the same extent, a higher or lower value means that one of the two prey species is preferred or selected; horizontally plotted is the fork length; the horizontal line at 10 is the (arbitrary) limit above which we defined the retention of the smaller species to have stopped being significant. The underlying assumption is that the larger species (*Daphnia hyalina*) is still retained completely (100%) at that point whereas merely 10% of the smaller species is retained. After Lammens et al. (1987).

ently the early development differs in the three species under study.

Allometric growth in the early stages of fish development is very common and is well described in literature, e.g. for carp *Cyprinus carpio* (Hoda & Tsukahara 1971, Osse et al. 1986). Roughly,

Table 5. The critical standard lengths in this table are based on the data of Lammens et al. (1987) in Fig. 11 and the average prey lengths on data from the same paper; the average prey length was determined with plankton samples from Tjeukemeer. The critical channel width was calculated from the critical standard length using the regression parameters.

	Zooplankton species	Average prey length (mm)	Critical channel width (mm)	Critical standard length (mm)
bream	*Chydorus sphaericus*	0.28 ± 0.028[a]	1.05	250
	Bosmina coregoni	0.43 ± 0.07	1.31	306
white bream	*Chydorus sphaericus*	0.28 ± 0.028	0.40	109
	Bosmina coregoni	0.43 ± 0.07	0.57	164
roach	*Chydorus sphaericus*	0.28 ± 0.028	0.23	85
	Bosmina coregoni	0.43 ± 0.07	0.43	160

[a] = mean ± standard deviation.

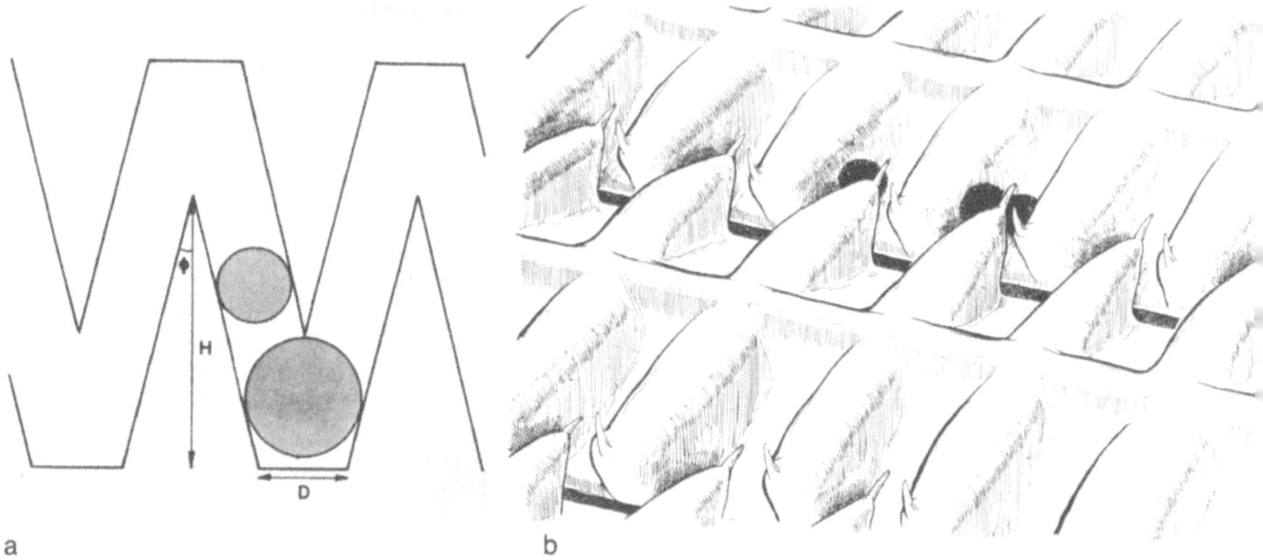

Fig. 12. Two filter feeding models predicting the mesh width of the branchial sieve. a– The saw tooth model or interdigitating model. Notice the importance of gill raker shape and spacing (top angle (Φ) and length (H) of the triangular gill raker, distance (D) between the gill rakers) and of the distance between the arches. b– Schematic drawing to illustrate the channel model. The lateral gill rakers are depressed here, in this position they reduce the mesh size of the medial gill raker channels by at least 50%.

growth of the carp can be described in two quite discreet phases, an allometric one up to about 20 mm SL and an isometric one above 20 mm. We only have indirect information about the first phase since all measurements were done with fishes above about 50 mm SL.

The differences in channel width between the species are significant ($p = 0.001$ and $p = 0.02$ for white bream/bream).

Raker length (RL, Fig. 7)

The extrapolated linear regression lines do not go through the origin, for any of the three species. The same approximation as described above is used here. Bream shows an acceleration of growth after the earliest stages. The other two species show a slight deceleration of growth. At 70–80 mm SL the three species have approximately the same raker length, but at 250 mm SL bream differs clearly from the other two species; the ratio bream to white bream and roach is about 1:0.70. Roach and white bream do not differ significantly ($p = 0.05$) but both species differ significantly from bream ($p \ll 0.001$).

Gill arch area (A_a, Fig. 8)

Zander (1906) states that roach has a simple branchial sieve with a relatively small area and that bream on the other hand has a relatively large area because of curvature and elongation of the ceratobranchials. Our data confirm this. Apart from elongation of the ceratobranchials we found wider arches in bream. In Fig. 8 the gill arch areas are plotted on a logarithmic scale. The slope of the curves represents the exponent in the growth curve (Table 3). The exponent of roach differs significantly from the other two (isometry; $p = 0.001$). The gill arch area values of roach are lower than those of white bream and bream. Bream and white bream start to diverge at a SL of 100–120 mm. At 150 mm SL the ratio bream, white bream, roach is about 1 : 0.75 : 0.50 and at 250 mm SL 1 : 0.70 : 0.50. The differences between the three sets of values of area divided by SL^2 are all significant ($p = 0.001$).

Channel area (A_c, Fig. 9)

The factor F relating standard measurements of five channel diameters to the total medial channel area is 14.2 for bream, 16.5 for white bream and 12.6 for roach. The differences are a consequence

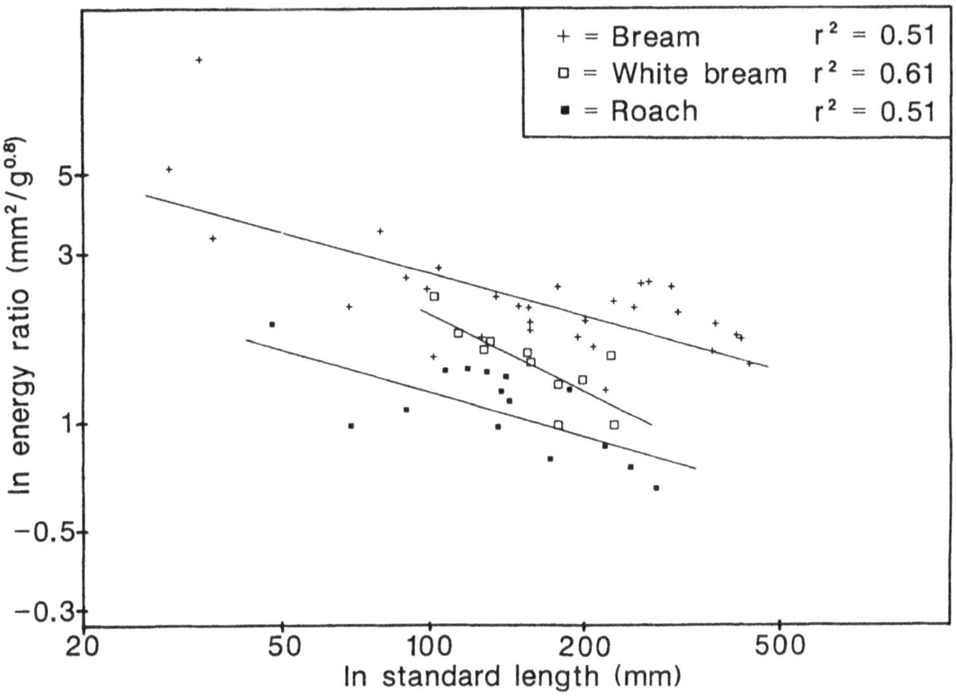

Fig. 13. Energy ratio versus standard length on a double logarithmic scale. The energy ratio equals the channel area divided by body weight to the power 0.80; for bream the body weights were not measured but calculated with a separately determined metric relation.

of differences in branchial sieve construction (Fig. 14). The exponent of the growth in white bream is low, 1.72 (Table 3); the difference from two is not significant. At 150 mm SL the ratio bream, white bream and roach is about 1: 0.75: 0.45 and at 250 mm SL the ratio is 1: 0.60: 0.45. The differences between the species are comparable to those of the arch area, but the channel area is smaller than the arch area. The differences between the species are significant ($p = 0.001$).

Raker cushion length (RCL, Fig. 10)

With the same reasoning as above we find an indication of accelerating growth in bream. At 80–90 mm SL the values are equal for the three fishes and at 250 mm SL the ratio is 1 : 0.70 : 0.50 (bream, white bream, roach). The accelerating growth rate of bream is in agreement with the observation that the channels are straight at a SL of 55 mm (Fig. 3d); the accelerating growth could be caused by increasing curvature of the channels. The differences between the species are significant ($p = 0.001$).

Summary of the results

The pattern of the species differences in gill raker parameter values is the same for all measurements (Table 4). In bream the values are about 20% higher than in white bream and about 40% higher than in roach. In the area values the differences are even more pronounced. Furthermore, the branchial arches of bream have more gill rakers than those of white bream and roach. Evidently the branchial sieve of bream is more strongly developed than that of white bream and roach; in other words, bream allocates more room inside its head for the branchial sieve. This is an indication that bream is more adapted to filter feeding than white bream and roach.

We have seen that quite a number of regression lines do not pass the origin closely and we have explained this with differences in early growth of the three species. In fact we do not know a lot about the development of the branchial sieve between SL 10 and 50 mm. In Fig. 3e the second gill arch of a roach with SL 35 mm is shown. The bony rakers are

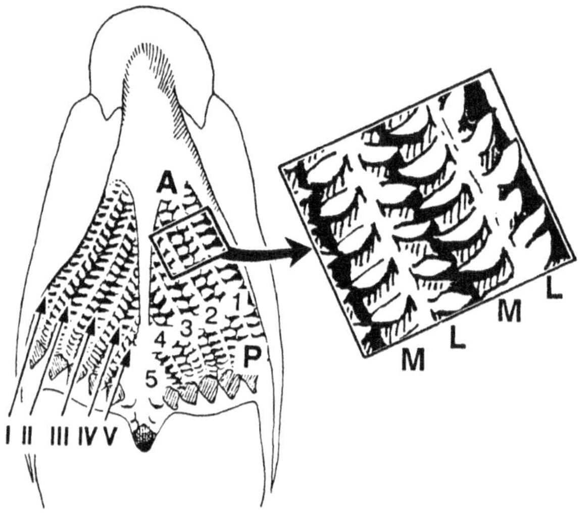

I–V	gill slit 1 to 5	1–5	gill arch 1 to 5
A	anterior side of gill arch	P	posterior side of gill arch
inset:			
L	lateral side of gill arch	M	medial side of gill arch

Fig. 14. Orientation terms; horizontal section of bream head showing the floor of the mouth cavity and the branchial sieve.

fully developed but a fleshy cushion is absent. At a SL of 70 mm the raker cushions have emerged and the branchial sieve already looks a lot like the adult form (Fig. 3f, c).

Discussion

Filter feeding models: comb model, channel model

The functional anatomy of the branchial sieve has not often been investigated; often only the retention (Durbin & Durbin 1975, Janssen 1976) or the anatomy (Beveridge et al. 1988) is investigated. MacNeill & Brandt (1990) however used a similar approach as in this paper, combining morphology, ontogeny and prey capture efficiency. A model that is often used (sometimes implicitly) as a functional description of mechanical sieving is the comb model or inter raker distance model. In this simple view of filter feeding the inter raker distance or the channel width is a direct measure for the mesh size and retention ability of a filter.

The comb model seems appropriate for clupeid and coregonid filter feeders (e.g. Drenner et al. 1984). In clupeids and coregonids the branchial sieves consist of very long and thin gill rakers mainly on the first arch, very much like a comb (Gibson 1988). However, experimental data do not always fit in with the comb model. Wright et al. (1983) and Gibson (1988) do not find an agreement between the comb model and retention experiments. Drenner et al. (1987) finds that filter feeding of the cichlid *Sarotherodon galilaeus* is not hampered by removal of gill rakers and microbranchiospines. The branchial sieves of the cyprinids studied in this paper lack the long specialized branchiospines typical of clupeid and coregonid filter feeders and their systems of branchiospines, cushions and channels are complex; the influence of inter arch distance on the mesh size could be large.

The ratio of inter raker distances in bream, white bream and roach is 1 : 0.85 : 0.60. This suggests that, applying the comb model, bream has the largest mesh size (lowest retention ability), roach the smallest and white bream is intermediate. This is clearly in conflict with ecological data (Fig. 11). Lammens et al. (1987) counted zooplankton in gut contents of bream, white bream and roach, which were all caught in lake Tjeukemeer. With these data he calculated the numerical proportion (NPG) of pairs of food species in the gut. These proportions he compared to the numerical proportions of the same pairs of food species in the environment (NPE). The ratio of NPG and NPE is a measure for selection of either of the two food species; a value of 1 means that there is no preference for either of the two food species. The food species pairs in Fig. 11 consist of a large cladoceran (*Daphnia hyalina*) and a small cladoceran (*Chydorus sphaericus* and *Bosmina coregoni*). An increase of the ratio in Fig. 11 means that the smaller species starts to slip through the branchial sieve. In Fig. 11 an arbitrary limit of 10 is drawn; at this line, where we assume that 10% of the smaller species is retained and 100% of the larger species, we call the retention of the smaller species critical; the standard length at which this occurs we call the critical standard length.

The guts of bream contained a critical proportion of *Bosmina* at a fork length of 330 mm (SL 306), whereas for white bream the critical FL is 180 mm (SL 164) and for roach 170 mm (SL 160) (Fig. 11); so white bream and roach have a comparable retention ability but bream has a much higher one. The order of retention ability is the reverse of the order mentioned above in connection with the comb model. The differences in abundance between these species in eutrophic water (Lammens 1986) are in agreement with the retention data of Lammens; bream is very successful in eutrophic lakes, roach and white bream usually are not. However, Lammens data should be used with care because it is not known whether the ingested cladocerans were taken in with gulping or with particulate feeding, the latter involving visual selection (Lammens et al. 1987). Therefore the proportions in the gut do not necessarily reflect the environmental proportions precisely and the retention abilities of the branchial sieves could be higher.

If one compares the critical channel width (defined as the channel width at the critical SL) with the actual prey length (Table 5) we find that bream retains prey about three times smaller than its critical channel width, which clearly is not compatible with the comb model. Roach has a critical channel width in the order of magnitude of the prey size, data compatible with the comb model. White bream however has a critical channel width 30–40% wider than the prey size. Because of the above we have doubts about the validity of the comb model in its simplest form for white bream and roach as much as for bream.

The fact mentioned above that the prey size is in the order of magnitude of the mesh size (as predicted by either of the two present models) is a further indication that the branchial sieves of these species are simple sieves (not sticky) as defined by Rubinstein & Koehl (1977), since the critical prey size of sticky sieves will usually be one order of magnitude lower than the physical mesh size (here: channel width or reduced channel width). A second indication for simple sieving is provided by the shape of Fig. 11. All the curves become very steep beyond the critical point; in other words, small food particles disappear completely from the diet soon after the critical standard length has been reached, which is typical for a simple sieve action, but not for sticky filters.

The channel model (Hoogenboezem et al. 1991) includes a mechanism to reduce the channel width by means of the raker from the opposite side of the gill slit (Fig. 12b). This mechanism is in good agreement with the finding in bream that the critical channel width is about 3 × the prey length (Table 5); the factor 3 instead of 2 could account for the thickness of the lateral raker. The results of white bream and roach are not in agreement with the channel model (with neither channel width nor reduced channel width). The present data provide some evidence that this way of filter feeding is unlikely for white bream and roach. Firstly, the raker length of these species is smaller than that of bream (ratio 1: 0.70, Fig. 7). This means that during filter feeding the gill slits of these species should be 30% narrower than those of bream in order to use the mechanism described above, which would cause a large decrease of filter capacity. Secondly there are differences in the shape of the rakers. In bream the rakers have a clear needle-like point, whereas in white bream and roach the rakers are blunt; in bream the top 1/3 of the bony raker is only covered by a thin layer of tissue, in white bream and roach this is not the case. A needle-like point is a good adaptation to reduce the mesh size of channels without plugging them off completely.

The proper functioning of the channel model depends on the relative position of two arches; the lateral rakers on one side of the slit must reach exactly in the channels on the medial side. This interdigitation can be disturbed by (small) movements of the arches. This factor provides a possible explanation for the relatively large inter raker distance in bream.

A model that might be useful for white bream and roach is mentioned by Sibbing in Nelson & Winfield (1991); the interdigitating or sawtooth model (Fig. 12a). In this planar model the gill rakers are represented as triangles; several parameters describing the shape of the gill slit determine the prey sizes that can be retained. Data about variations in inter-arch distance during filter feeding are needed to define the type of filter feeding and the mesh size of white bream and roach. X-ray films are being made in our laboratory to measure these parameters.

Capacity, filtering area and energy ratio

The capacity of a filter (volume per unit time) is an important property determining (together with the retention ability) the amount of prey that can be retained per unit time. The capacity of a filter is determined by the velocity of the water in the filter and the area through which water flows in the filter (the filtering area).

We want to estimate the ratio of energy intake per unit time through filter feeding and total energy consumption per unit time, in other words, we want to express the filter feeding energy intake as a fraction of total energy consumption. With energy is meant energy per fixed unit of time. When the ratio is high enough the fish can survive on a diet of zooplankton (as far as energy is concerned).

If the concentration and composition of zooplankton in the environment and the retention ability of the filter are constant, then the capacity of the filter is a direct measure of zooplankton and energy intake. If the water velocity in the filter is constant then the capacity is proportional to the filter area. There is some indirect evidence for this. Drenner et al. (1982) found that the filtering rate (= capacity) in gizzard shad, *Dorosoma cepedianum* is proportional to $SL^{2.08}$ and Smith (1989) found that the filtering rate in silver carp, *Hypophthalmichthys molitrix*, is proportional to $W^{0.713}$. Both values suggest a scaling of capacity with an area. All these points lead to the assumption that energy intake is proportional to the filtering area.

The energy consumption is proportional to the metabolic rate, which is very difficult to measure in fishes, the value of the exponent varies between 0.75 and 0.90 depending on author and taxonomic group (Schmidt-Nielsen 1984). Winberg (1960) found the following relation for cyprinids at 20° C:

$$\text{metabolic rate} = 0.47\ W^{0.80},$$

so in cyprinids the energy consumption is proportional to $W^{0.80}$.

Combining the above findings we find that the energy ratio is proportional to the filtering area divided by $W^{0.80}$. If the filtering area and the weight increase isometrically then the energy ratio is proportional to $SL^{-0.40}$:

$$\frac{A}{W^{0.80}} = \frac{L^2}{(L^3)^{0.80}} = L^{-0.40}.$$

The ratio decreases with increasing length, so filter feeding will become increasingly less suited as the sole feeding mode. It should be stressed here that the calculated values of the energy ratio only have a comparative meaning.

In Fig. 13 the energy ratio (calculated with the channel area) can indeed be seen to decrease with increasing SL. The exponent of this decrease (Table 3) is -0.40 for bream and -0.44 for roach, close to the expected value -0.40. For white bream however the exponent is -0.71; this deviation (non-significant, notice the low r^2 values) is caused by the non isometric growth of the channel area in white bream (Fig. 9). At 200 mm SL the ratio bream, white bream, roach is 1 : 0.60 : 0.45. This means that white bream can take in 40% less energy per unit time than bream and roach 55% less. The differences between the species are significant ($p = 0.001$).

One of the future research subjects is to measure the capacity directly in filter feeding fish in order to test the above ideas.

It seems that the success of bream in eutrophic environments can be explained by a superior ability to exploit zooplankton, i.e. by the efficiency of its branchial sieve. First, there is a strong indication that its capacity is larger than that of white bream and roach and second according to Lammens et al. (1987) the retention ability of its branchial sieve is much higher than that of white bream and roach.

List of symbols

A_a	= gill arch area
A_c	= sum of the channel diameters on the medial sides of the four arches
AL	= anal length
BAW	= basal arch width
CW	= channel width
ED	= eye diameter
F	= factor relating standard channel diameter measurements to total medial channel area
FL	= fork length
IR	= inter raker distance
LCB	= length of ceratobranchial
NPE	= numerical proportion of a pair of food species in the environment
NPG	= numerical proportion of a pair of food species in the gut
NR	= number of gill rakers
RCL	= raker cushion length
RL	= bony raker length
RW	= raker width
SL	= standard length
SOL	= snout operculum length, head length
TAW	= arch width from raker tip to raker tip
W	= body weight

Acknowledgements

We want to thank Yvette van Vugt and Peter Klinkhamer for doing most of the measurements. Most figures were made by Jos van den Boogaart and Wim Valen. This research is funded by the Dutch organization for Scientific Research (NWO); Department BION, project no. 811-428-265.

References cited

Beveridge, M.C.M., M.R.P. Briggs, M.E. Northcott & L.G. Ross. 1988. The occurrence, structure, and development of microbranchiospines among the tilapias (Cichlidae: Tilapiini). Can. J. Zool. 66: 2564–2572.

Drenner, R.W., W.J. O'Brien & J.R. Mummert. 1982. Filter feeding rates of gizzard shad. Trans. Amer. Fish. Soc. 111: 210–215.

Drenner, R.W., J.R. Mummert, F.Jr. de Noyelles & D. Kettles. 1984. Selective particle ingestion by a filter-feeding fish and its impact on phytoplankton community structure. Limnol. Oceanogr. 29: 941–948.

Drenner, R.W., G.L. Vinyard, K.D. Hambright & M. Gophen. 1987. Particle ingestion by *Tilapia galilaea* is not affected by removal of gill rakers and microbranchiospines. Trans. Amer. Fish. Soc. 116: 272–276.

Durbin, A.G. & E.G. Durbin. 1975. Grazing rates of Atlantic menhaden. Mar. Biol. 33: 265–277.

Gibson, R.N. 1988. Development, morphometry and particle retention capability of the gill rakers in the herring, *Clupea harengus* L.. J. Fish Biol. 32: 949–962.

Hoda, S.M.S. & H. Tsukahara. 1971. Studies on the development and relative growth in the carp (*Cyprinus carpio* (Linne)). J. Fac. Agriculture, Kyushu University 16: 387–509.

Hoogenboezem, W., F.A. Sibbing, J.W.M. Osse, J.G.M. van den Boogaart, E.H.R.R. Lammens & A. Terlouw. 1990. X-ray measurements of gill-arch movements in filterfeeding bream, *Abramis brama* (Cyprinidae). J. Fish Biol. 36: 47–58.

Hoogenboezem, W., J.G.M. van den Boogaart, F.A. Sibbing, E.H.R.R. Lammens, A. Terlouw & J.W.M. Osse. 1991. A new model of particle retention and branchial sieve adjustment in filter-feeding bream (*Abramis brama* (L.), Cyprinidae). Can. J. Fish. Aquat. Sci. 48: 7–18.

Janssen, J. 1976. Selectivity of an artificial filter feeder and suction feeders on calanoid copepods. Amer. Midland Natur. 95: 491–493.

Lammens, E.H.R.R. 1986. Interactions between fishes and the structure of fish communities in Dutch shallow, eutrophic lakes. Ph.D. Thesis, Agricultural University, Wageningen. 100 pp.

Lammens, E.H.R.R. 1989. Causes and consequences of the success of bream in Dutch eutrophic lakes. Hydrobiol. Bull. 23: 11–18.

Lammens, E.H.R.R., J. Geursen & P.J. McGillavry. 1987. Diet shifts, feeding efficiency and coexistence of bream (*Abramis brama*), roach (*Rutilus rutilus*) and white bream (*Blicca bjoerkna*) in eutrophicated lakes. Proc. V Congr. Europ. Ichtyol., Stockholm: 153–162.

Lindsey, C.C. 1981. Stocks are chameleons: plasticity in gill rakers of coregonid fishes. Can. J. Fish. Aquat. Sci. 38: 1497–1506.

MacNeill, D.B. & S.B. Brandt. 1990. Ontogenetic shifts in gill-raker morphology and predicted prey capture efficiency of the alewife, *Alosa pseudoharengus*. Copeia 1990: 164–171.

Nie, H.W. de. 1987. The decrease in aquatic vegetation in Europe and its consequences for fish populations. EIFAC/CEC-PI Occasional paper 19. 52 pp.

Nie, H.W. de, H.J. Bromley & J. Vijverberg. 1980. Distribution patterns of zooplankton in Tjeukemeer, the Netherlands. J. Plankton Res. 2: 317–334.

Osse, J.W.M., M.R. Drost, J. Vial & R. Aldunate. 1986.

Growth and feeding in larval carp, *Cyprinus carpio* (L). Ann. Mus. Roy. Afr. Centr., Sc. Zool. 251: 67–70.

Romeis, B. 1968. Mikroskopische Technik. Oldenbourg, München. 695 pp.

Rubinstein, D.I. & M.A.R. Koehl. 1977. The mechanisms of filter feeding: some theoretical considerations. Amer. Natur. 111: 981–994.

Schmidt-Nielsen, K. 1984. Scaling; why is animal size so important? Cambridge University Press, Cambridge. 241 pp.

Sibbing, F.A. 1982. Pharyngeal mastication and food transport in the carp (*Cyprinus carpio* L.): a cineradiographic and electromyographic study. J. Morph. 172: 223–258.

Sibbing, F.A. 1988. Specializations and limitations in the utilization of food resources by the carp, *Cyprinus carpio*: a study of oral food processing. Env. Biol. Fish. 22: 161–178.

Sibbing, F.A. 1991. Food capture and oral processing. pp. 377–412. *In:* I.J. Winfield & J.S. Nelson (ed.) Cyprinid Fishes. Systematics, Biology and Exploitation, Chapman & Hall, London.

Sibbing, F.A. & R. Uribe. 1985. Regional specializations in the oropharyngeal wall and food processing in the carp (*Cyprinus carpio* L.). Neth. J. Zool. 35: 377–422.

Sibbing, F.A., J.W.M. Osse & A. Terlouw. 1986. Food handling in the carp (*Cyprinus carpio*): its movement patterns, mechanisms and limitations. J. Zool. Lond. (A)210: 161–203.

Smith, D.W. 1989. The feeding selectivity of silver carp, *Hypophthalmichthys molitrix* Val.. J. Fish Biol. 34: 819–828.

Sokal, R.R. & F.J. Rohlf. 1969. Biometry: the principles and practice of statistics in biological research. W.H. Freeman, San Francisco. 776 pp.

Vijverberg, J. & A.F. Richter. 1982a. Population dynamics and production of *Daphnia hyalina* (Leydig) and *Daphnia cucullata* (Sars) in Tjeukemeer. Hydrobiologia 95: 235–259.

Vijverberg, J. & A.F. Richter. 1982b. Population dynamics and production of *Acanthocyclops robustus* (Sars) and *Mesocyclops leuckarti* (Claus) in Tjeukemeer. Hydrobiologia 95: 261–274.

Winberg, G.G. 1960. Rate of metabolism and food requirements of fishes. Fisheries Research Board Translation Series 194: 1–202.

Wright, D.I., W.J. O'Brien & C. Luecke. 1983. A new estimate of zooplankton retention by gill rakers and its ecological significance. Trans. Amer. Fish. Soc. 112: 638–648.

Zander, E. 1906. Das Kiemenfilter der Teleoster. Zeitschrift für Wissenschaftliche Zoologie 84: 619–713.

Environmental Biology of Fishes **33**: 125–134, 1992.

Density and distribution of external taste buds in cyprinids

Andreas Gomahr, Margit Palzenberger & Kurt Kotrschal
Zoologisches Institut der Universität Salzburg, Hellbrunnerstraße 34, A-5020 Salzburg, Austria

Received 2.11.1989 Accepted 24.10.1990

Key words: Chemosense, Comparative morphology, Neuroecology, Fish, Sensory compensation, Silver nitrate, Skin, Cyprinidae, Teleostei

Synopsis

The aim of the present qualitative and quantitative study was to survey the density and distribution of external taste buds (TB) in 10 common European cyprinid species. TB pores were stained with silver nitrate and counted in 7 sample areas on the body and 3 on the fins. TB densities decrease from rostral to caudal and from ventral to dorsal. Highest densities were encountered in the gular region of the minnow (297 per mm^2), white bream (291) and vimba (285), or at the forehead, in white bream (147). Of the fins, the pectoral showed highest densities (150 per mm^2 in vimba, 138 in the minnow). The overall TB density rank from high to low is: minnow *Phoxinus phoxinus*, white bream *Blicca bjoerkna*, crucian carp *Carassius carassius*, vimba *Vimba vimba*, bream *Abramis brama*, roach *Rutilus rutilus*, bleak *Alburnus alburnus*, sun bleak *Leucaspius delineatus*, sabre carp *Pelecus cultratus* and bitterling *Rhodeus sericeus*. This is also a fair rank of life styles from benthivory to open water planktivory and surface feeding. A significant positive correlation exists between densities of external taste buds and the relative sizes of the brain stem facial lobes. An apparent compensatory relationship between vision and taste, but not lateral line, is discussed.

Introduction

Taste buds (TB) are compound epidermal chemosensory and tactile organs (Davenport & Caprio 1982, Jakubowski & Whitear 1990, Reutter 1986, Weber 1827). In most fishes, TBs are not confined to the oropharyngeal cavity, but also occur at the body surface (Adam 1964, Atema 1971, Bardach & Atema 1971, Herrick 1901, 1904, Kapoor et al. 1975, Leydig 1851, Moore 1950, Reutter 1986, Schemmel 1967). These external TBs are particularly abundant in the ostariophysan fishes, like siluroids and cyprinids. External TBs are innervated from the facial nerve (Finger 1983, 1988, Freihofer 1963, Herrick 1901, Kiyohara et al. 1980) and show a polar arrangement at the body surface, with highest densities at the lips, the gular region, barbels (if present) and along other suitable appendages, like pectoral and pelvic fins (Atema 1971, Bardach & Atema 1971, Caprio 1988, Gomahr et al. 1988, Jakubowski & Whitear 1990, Kiyohara et al. 1980, Moore 1950).

Quantitative studies on densities and distribution of external TBs are rare (Atema 1971, Kiyohara et al. 1980, Schemmel 1967). Even less abundant are comparative investigations (Campos 1969, Schemmel 1967, Sinha & Moitra 1978). We therefore presently compare distribution patterns and densities of external TBs within a group of common European cyprinids and relate these data with aspects of sensory biology and life style.

Brain patterns allow prediction of sensory types in fishes (Davis & Miller 1967, Evans 1931, 1932, 1935, Kotrschal & Junger 1988, Kotrschal et al.

1991). We test whether external TB densities are correlated with relative sizes of the brain stem facial lobes, to elucidate if TB densities are predictive for brain morphology (Kotrschal & Palzenberger 1992) and vice versa.

Materials and methods

Most fish used for quantitative work were gill-netted in the Stopfenreuther Au, a Danube backwater area in the east of Austria. Additional species (*Carassius carassius, Leucaspius delineatus, Rhodeus sericeus*) were purchased at a local pet shop. For a list of species investigated see Table 1. The species considered are indifferent to currents and represent different feeding types as well as forming a gradient with respect to the preferred habitats, from the bottom to the water surface. For ecological information see Schiemer (1985, 1988).

For scanning electron microscopy (SEM), mainly laboratory-reared specimens were used (the same specimens used for quantification of solitary chemosensory cells, see Kotrschal et al. 1991). Fish were fixed at least for 2 days in a mixture of buffered aldehydes after deep anaesthesia with MS 222. After washing in buffer, whole fish or skin parts were digested for 45 min at 37° C in a buffered (phosphate, pH 7.4) 0.1% hyalonuridase solution. After 15 sec of ultrasonic treatment, specimens were washed in distilled water and dehydrated in a graded series of ethanol. After CO_2 critical point drying, specimens were sputtered with gold and observed in a Cambridge 250 Stereoscan at 10–20 kV.

For taste bud (TB) counts, only fish with relatively intact epidermis were used. Fish were kept in tanks for at least 24 hours prior to staining to avoid TB pores being covered by excess mucus, as may be the case in stressed fish. After prolonged narcosis in MS 222 (1 : 10000), fish were rinsed in MS-free water and body silhouettes were drawn on mm^2

Table 1. List of species investigated for taste bud (TB) densities, ranked according to the overall body mean TB densities (see Fig. 4 for mean values), with common names of species, numbers of specimens investigated (n), ranges of standard lengths (SL) in mm and TB numbers per mm^2 at selected sampling areas on the body and fins. Single values show no sd; - = missing values; 0 = no TBs present. Abbreviations used: caud = caudal, foreh = forehead, operc = operculum, pect = pectoral, pelv = pelvic, vent = ventral.

Species common names	n	SL mm	body areas means ± sd					fin areas means ± sd		
			gular	foreh	operc	vent	lateral	pect	pelv	caud
1 *Phoxinus phoxinus*	4	45– 55	297	95	106	150	63	138	83	–
minnow			± 75	± 77	± 70	± 86	± 24		± 64	
2 *Blicca bjoerkna*	17	85–160	291	147	64	36	7	120	91	28
white bream			± 74	± 59	± 31	± 25	± 5	± 38	± 43	
3 *Carassius carassius*	4	85– 95	162	82	95	–	61	90	92	–
crucian carp			± 24	± 14	± 55		± 37		± 21	
4 *Vimba vimba*	4	112–145	285	93	19	–	–	150	–	–
vimba			± 71	± 35				± 13		
5 *Abramis brama*	9	45–175	190	82	43	10	14	125	140	33
bream			± 91	± 28	± 27			± 81	± 132	
6 *Rutilus rutilus*	16	75–235	178	36	41	21	8	89	65	–
roach			± 57	± 19	± 10	± 8	± 3	± 42	± 26	
7 *Alburnus alburnus*	17	120–145	145	44	42	18	8	64	45	–
bleak			± 57	± 21	± 24	± 8	± 1	± 20		
8 *Leucaspius delineatus*	4	43– 58	18	15	0	0	0	0	0	0
sun bleak			± 3	± 1						
9 *Pelecus cultratus*	4	165–190	11	17	0	0	0	0	0	0
sabre carp			± 5	± 4						
10 *Rhodeus sericeus*	4	49– 58	11	10	0	0	0	0	0	0
bitterling			± 4	± 3						

grid paper for estimating body area on a digitizer. Total and standard lengths as well as weight were recorded. Fish were sprayed with 2% $AgNO_3$ solution (Whitear personal communication), followed by a reduction of the silver nitrate under sunlight or by aid of a halogen-film lamp. When the skin surface turned light tobacco-brown (after 1–2 minutes of light exposure) the fish was rinsed with distilled water. Whole fish or parts were fixed for 15 minutes in a 10% buffered formaldehyde solution, wrapped in bandage tissue, soaked with the fixative and stored deep-frozen. Thawed specimens were investigated and documented by drawing under a stereomicroscope, equipped with a camera-lucida. TBs appear as dark-brown dots on the light brown epidermis and can easily be distinguished from superficial neuromasts by shape, size and position. This quick and reliable method was superior to the use of other stains for labelling the TB pores.

Epidermal surfaces were projected to a digitizing tablet and TB counts were directly fed into an Apple IIe-computer. A standard 3 × 3 mm (9 mm^2) sampling area was searched for TBs at the following sites; gular, forehead, operculum (left/right), belly, lateral trunk and palatal organ (not considered within the present contribution), pectoral and pelvic fins (both left and right), and one side only of dorsal, caudal and anal fins (Fig. 1). Other areas like the lips and the circumorbital region were observed for qualitative comparison.

All TB densities throughout this paper are given per 1 mm^2. For all species comparisons, means of the different sample areas were used. This was possible as there appears to be no intraspecific difference in TB density in fish of different body sizes (see Discussion). One way ANOVA and Scheffe's procedure served to detect significant differences between species. Least square regression was used to test for possible growth-related changes in TB densities. For testing the significance of the correlation between TB densities and relative sizes of brain lobes the Spearman's rank test was employed.

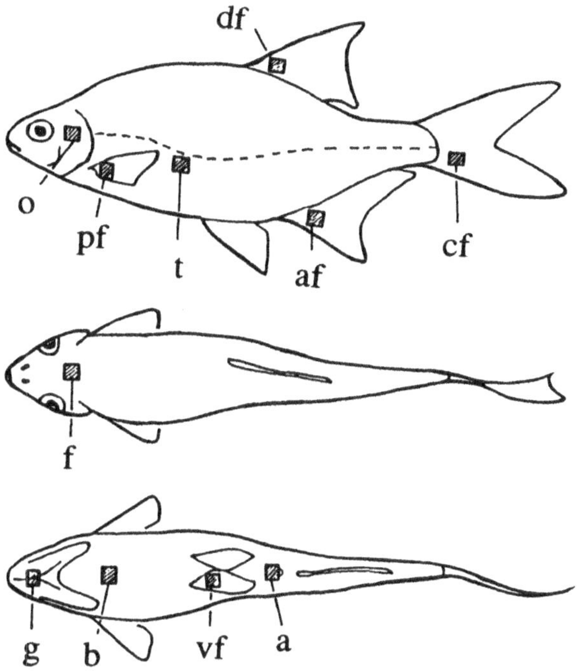

Fig. 1. Position of 9 mm^2 sampling areas on body and fins: anal fin (af), anal papilla (a), belly (b), caudal fin (cf), dorsal fin (df), forebrain (f), gular (g), lateral trunk (t), operculum (o), pectoral fin (pf), pelvic fin (vf).

Results

Scanning electron microscopy

Within and between the investigated cyprinid species, taste buds (TB) differ in size of the apical pore and in position relative to the surrounding skin.

Relatively small TBs were found in roach (Fig. 2a), with an apical pore diameter of 4–6 μm and containing 10–15 gustatory cells, which are easily recognized by their relatively large, single apical microvillus as compared with the multiple, small villi of the supporting cells. TBs in nase (Fig. 2b) also showed small pores, approximately 5 μm in diameter, with 10–15 gustatory apices. In other cyprinids, like bleak or chub (Fig. 2c, d) the TB pore may be 6–8 μm in diameter and contain up to 30 gustatory cell apices. Cyprinid TB pores approach 10 μm in diameter only occasionally.

In most cases, TB pores at the body surface were on a level with the surrounding squamous epider-

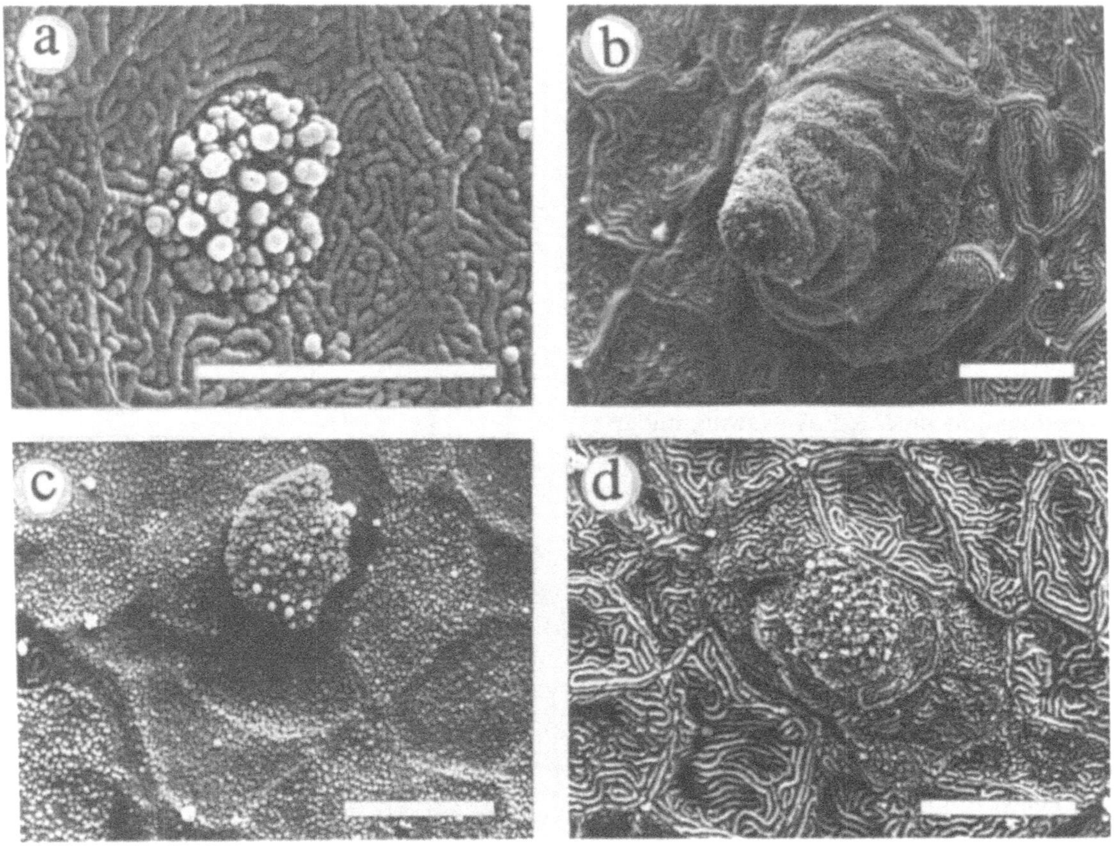

Fig. 2. Scanning electron micrographs of taste bud pores (all scale bars 10 μm): a – taste bud from the forehead of roach, b – hillocked taste bud from close to the anterior nostril in nase, c – taste bud from the nasal area in bleak, d – gular taste bud in chub.

mal cells (Fig. 2a, c, d). In the nase, *Chondrostoma nasus*, the TBs may be positioned on distinct hillocks (Fig. 2b), which surround the mouth, the nostrils and can also be found within the frontal oral cavity at the oral valve.

Taste bud distribution

TBs decrease in density from rostral to caudal and from ventral to dorsal (Table 1). Only in the open water or surface dwelling sun bleak and sabre carp and in the bitterling, were there approximately as many taste buds on the dorsal as on the ventral side of the head.

In areas lacking scales, TBs were scattered irregularly (Fig. 3a). In other areas, TBs were arranged according to underlying dermal structures. In unscaled areas with regular skin folds, like the gular, TBs stood in rows, following the crests of the folds (Fig. 3b). In epidermal areas overlying scales, TBs avoided the margins of the scales, and were restricted to the concerns formed by adjacent scales, where the epidermis is relatively deep (Fig. 3c). There, TBs may occur in rows or irregular groups, oriented perpendicularly to the body axis, often interspersed between neuromasts.

At the base of the pectoral fin, the scattered arrangement of TBs merged into rows on the fin. Such rows may either follow the fin rays, as in the pectoral and anal, or be situated on the fin web, as in the caudal fin, where TBs may alternate with neuromasts (Fig. 3d).

Taste bud density

A rank of species according to decreasing TB den-

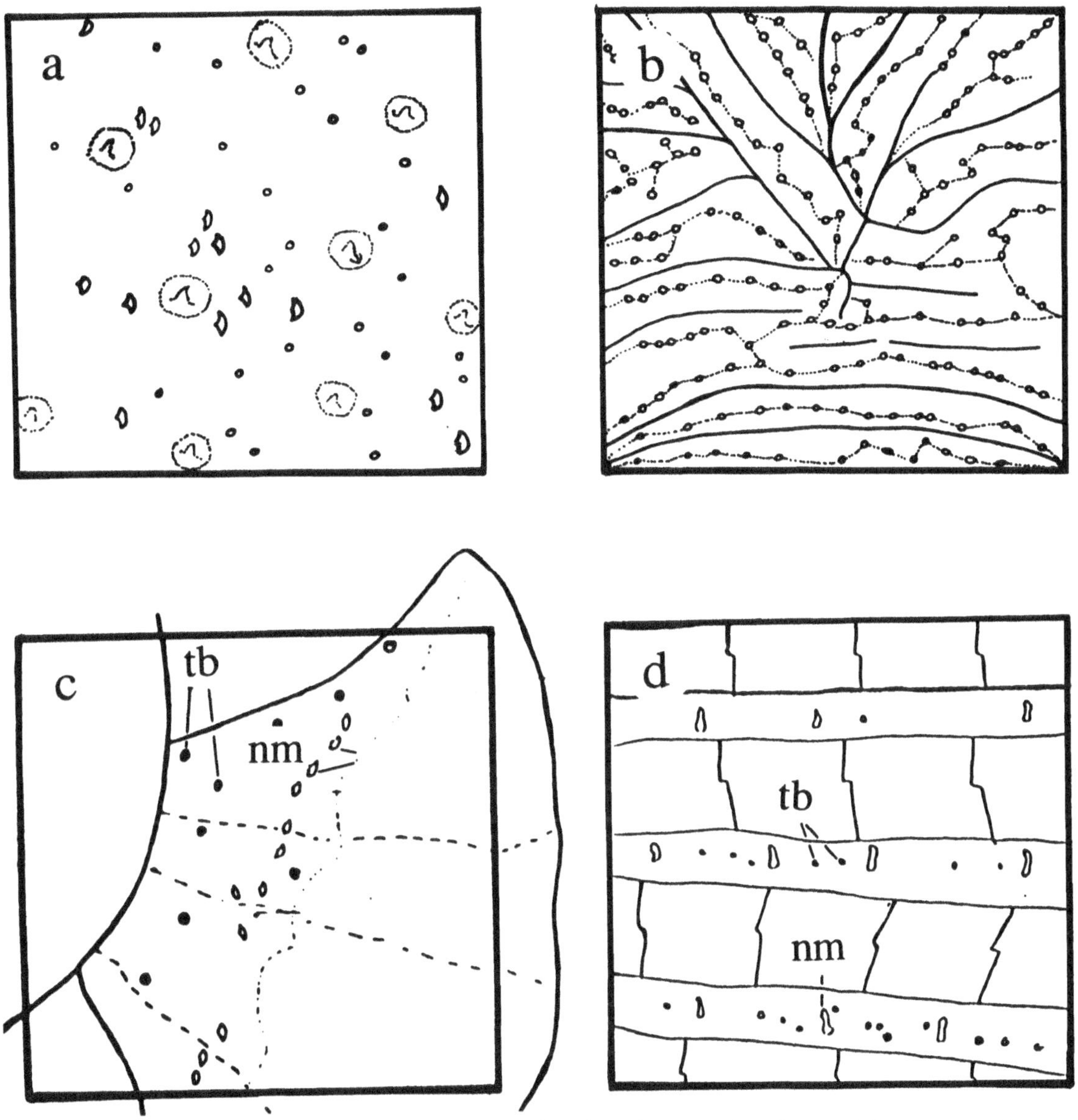

Fig. 3. Camera lucida drawings of representative sampling areas after silver nitrate staining, showing different taste bud arrangements (all quadrats represent an area of 9 mm^2): a – Opercular surface of a bleak. Thorns are the horny unculi, the nuptial dress in male cyprinids. Taste buds (circles) and neuromasts (elongated structures) shown. b – Taste buds (circles) on gular folds in white bream. The fold crests are indicated by stippled lines. c – Taste buds (tb), neuromasts (nm) and scales on the trunk of roach. Most of the taste buds and free neuromasts located between adjacent posterior scale margins. d – Alternating taste buds (tb) and neuromasts (nm) in rows on the caudal fin web near the caudal peduncle in bream.

sity (Table 1) can be loosely associated with the degree of benthic habitat and benthic feeding (Schiemer 1985) (Fig. 4). The minnow may feed from the bottom and also shelters there when threatened. White bream, crucian carp and vimba are predominantly benthic feeders. Bream, like the crucian carp, is a benthic oropharyngeal sorter, but may switch to feed on plankton. Roach is a generalized feeder, sun bleak and sabre-carp are mainly surface feeders. The bitterling revealed the lowest TB counts despite its loose association with the substrate. The relatively high TB density in bleak may correspond to its ability to switch to benthivory (Schiemer 1985) despite its predominant surface

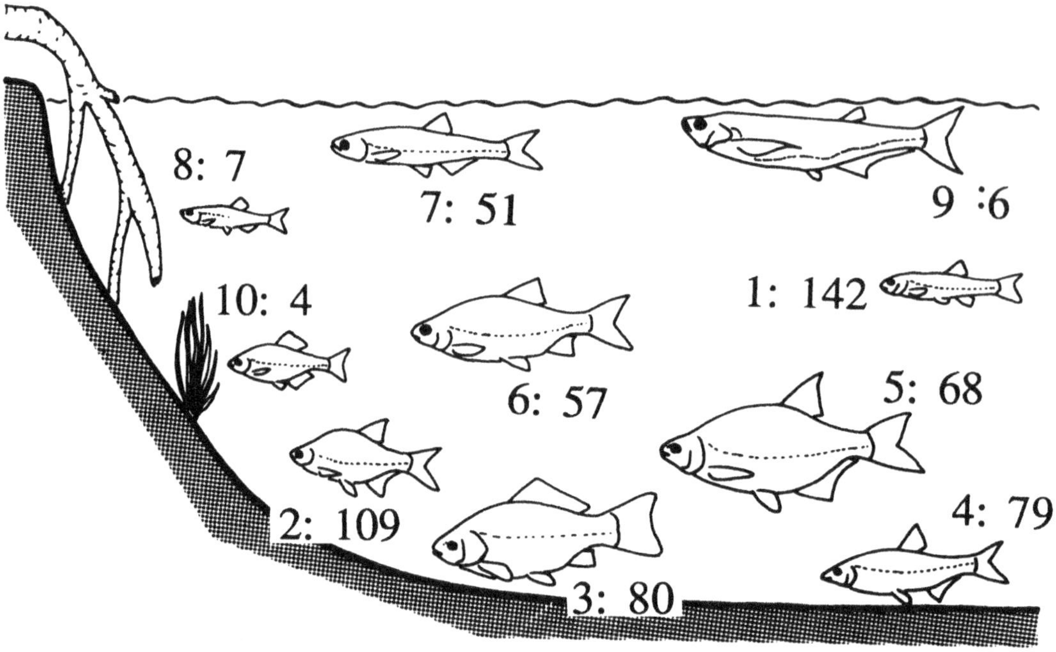

Fig. 4. The species investigated in a typical Danube backwater situation arranged according to habitat preferences and numbered as in Table 1. Values for body mean TB densities per mm^2 shown.

feeding habit. Highest TB densities are found in the benthic and benthivorous species.

TB densities change gradually over the body surface. The steepness of this density gradient does not seem to be the same between different species. Therefore, particular sampling areas (Table 1) were used for interspecific comparison, rather than the density means formed from different areas (Fig. 4). Due to high variance, low numbers of observations and a lack of figures (due to damaged epidermis) from some of the sampling areas, ANOVA found significant TB density differences only between a limited number of species (Table 2). Thus the rank in Table 1 represents a trend from high to low TB density, which is significant only between minnow, white bream, roach and bleak. Remarkably, the TB density on the body seems to be higher in minnow and crucian carp as compared to the other species (Table 1, 2), whereas the highest forehead density was found in white bream.

Intraspecific correlations revealed that variation of TB densities between individuals was significantly higher, in most cases, than between body sides of an individual (opercula, pectoral and pelvic fins were tested bilaterally). Thus methodological flaws, such as insufficient staining of TBs or inconsistent sampling are unlikely to account for high variation between sample areas of different individuals. TB density revealed high intraspecific variability (Fig. 5).

During growth, density of TBs seems to be relatively constant (Fig. 5). No significant regressions could be found between TB density and body surface area at $\alpha \leq 0.05$. This may lead to relatively constant TB densities during late juvenile and adult growth, but (as the body surface increases approxi-

Table 2. Significant interspecific differences in taste bud densities between the listed areas as revealed by ANOVA and the Scheffe-procedure. Level of significance: $\alpha \leq 0.05$. For values and latin names see Table 1, for position of sample areas see Fig. 1.

Area	species significantly different from	
gular	white bream	roach, bleak
forehead	white bream	roach, bleak
belly	minnow	white bream, roach, bleak
lateral trunk	minnow	white bream, roach
anal papilla	vimba	bleak

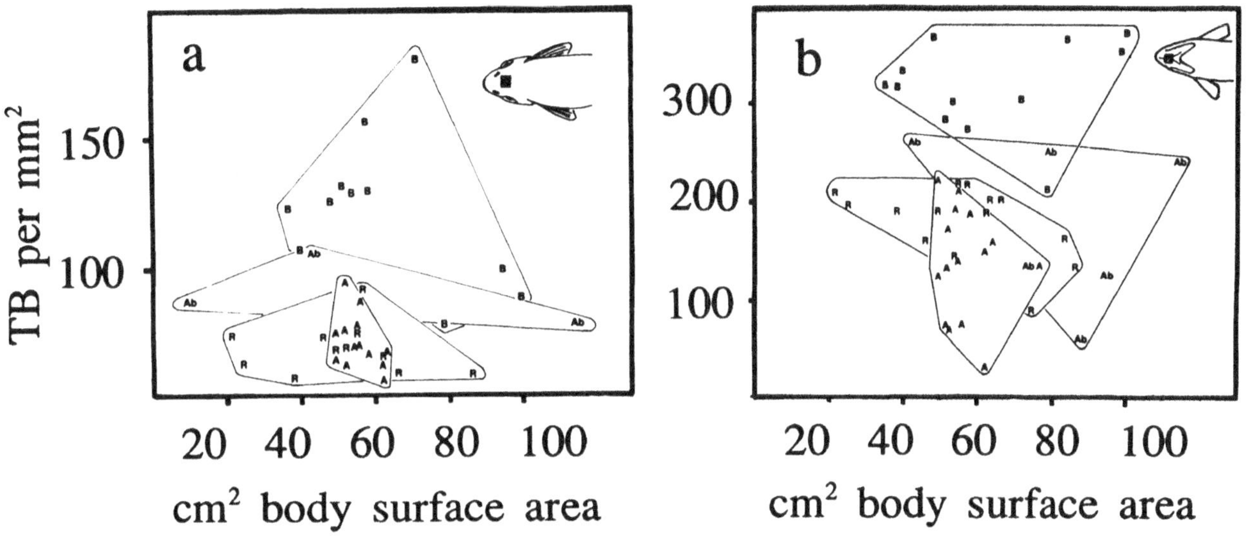

Fig. 5. Taste bud densities versus body surface area at the forehead (a) and the gular (b) of selected species, bleak (A), bream (Ab), white bream (B), and roach (R). Note lack of correlations. Species groups surrounded by minimal polygons.

mately with the second power of body length) exponentially increasing TB numbers. However, caution is necessary, as this study was done with only a limited range of sizes and TB counts showed great intraspecific variation. This does not permit a definitive judgement of how TB numbers and densities develop in growing cyprinids.

Density of external TBs (mean of body areas, Fig. 4) and relative sizes of the brain stem facial lobes of the species in question (brain data from Kotrschal & Palzenberger 1992) were significantly correlated (Spearman's rank test, $\alpha \leq 0.001$), supporting the plausibility of the rank in Table 1. Moreover this rank correlation substantiates that a high density of external TBs is predictive for a relatively large facial lobe, the primary sensory brain stem center for external taste, and vice versa.

In an effort to test the compensatory interaction between vision and taste, no significant negative rank correlation (Spearman's) could be found between external TB density and the relative size of the tectum opticum (the major primary termination area for visual input). This was true for the percentage of total brain volume as well as for residuals on standard length of these percentage (to account for the significant decrease of the relative size of the optic tectum during growth, Brandstätter & Kotrschal 1990). However, in 28 species of cyprinids (Kotrschal & Palzenberger 1992), a negative correlation between the size of the brain stem facial lobe and the size of the optic tectum was highly significant ($\alpha \leq 0.001$).

Discussion

Taste bud distribution

Although the present study shows some relationship between degree of benthivory and density of external TBs (Fig. 4), a purely adaptive interpretation of TB distribution patterns seems questionable when a finer spatial and structural scale is considered. It is known that TBs only survive under the trophic influence of an intact nerve supply (Landacre 1907, Lane 1976, May 1925, Olmsted 1920, Raderman-Little 1979). TB formation is induced by the re-innervating nerve fibers. Thus the information on TB density may mainly reside in the innervating nerves, which are of neural crest origin.

Particularly over unscaled epidermis, TB spacing is fairly irregular, but probably not stochastic, as no major clustering of TBs were observed (Fig.

3a). Whether the TBs themselves exert an inhibitory influence on their surroundings, or whether this is entirely up to the innervating nerves is not known.

Ecomorphology of the external taste system

Bottom dwellers and benthic feeders are plentifully equipped with external TBs, whereas in open water or surface species the density is usually considerably lower (Fig. 4). Generally, body surfaces with a high probability of coming into contact with food sources show a high density of TBs (Schemmel 1967). Still, such functional ecological considerations do not explain why the small, benthos-associated sun bleak and bitterling are low in TB density (Table 1). The relatively high TB densities in the surface dwelling bleak may be related to its occasional benthic feeding excursions (Schiemer 1985). Also, benthic life style may explain the high TB densities in crucian carp, but is unlikely to account for the high densities in minnow, which occurs in mid-water of relatively clear creeks and lakes. The TB densities on the forehead of white bream may be related to its high turbidity habitat, where chemical stimuli may be an important source of information (Bardach & Atema 1971). Thus, at a coarse scale, density and arrangement of the external taste system can be predicted from the species' ecology.

Benthivory and a turbid habitat often occur together. Thus in species and environments where vision is only of limited use (Huber & Rylander 1992) the external taste system may increase in importance for detecting food. External taste may serve for gradient search, as in catfish, but generally elicits ingestion upon contact (Atema 1971, Bardach & Atema 1971, Kapoor et al. 1975, Wunder 1927). Touch seems as important for triggering this feeding reflex as chemo-stimulation (Davenport & Caprio 1982). Indeed, all taste systems investigated are also touch responsive (Kiyohara et al. 1985, Marui 1977, Marui & Funakoshi 1979, Marui & Caprio 1982, Peters et al. 1987, Peterson 1972). TBs on hillocks may even be superior for the perception of tactile stimuli (Reutter et al. 1974, Reutter & Breipohl 1975). TBs thus integrate information on chemical quality with exact spatio-temporal location of prey.

Flexibility of the taste system and taste-vision compensatory interactions

Size of brain taste lobes co-varies with density of external TBs. This was not necessarily expected. Although the inter- and intraspecific variation in numbers of sensory cells per TB seems to be low, the same number of sensory cells may converge on different numbers of nerve fibers. Also, the primary central representation of an identical number of end organs may occupy varying brain volumes. Within the cyprinids considered, this seems not to be the case, supporting the assumption that quantitative comparison of brain structure is a reliable predictor of sensory organization and vice versa (Bullock 1983, Huber & Rylander 1992, Kotrschal & Palzenberger 1992, Northcutt 1988).

Although no negative correlation between TB density and optical tectum size could be found with the present species range, this correlation does exist at the brain level between facial lobe size and optic tectum size when a larger number of species is considered (Kotrschal & Palzenberger 1992). Schemmel's (1967) findings on a spectacular increase of the external taste system in hypogean as compared to epigean *Astyanax* spp. can thus probably be generalized also for cyprinids. Huber & Rylander (1992) found the same correlation in American shiners, where species living in clear water have large optic lobes, but small facial lobes and vice versa in species dwelling in turbid water. This evidence suggests that there is indeed a compensation by taste when vision is impeded by low light levels or high turbidity.

The high intraspecific variability of the external taste system found by the present study may be seen in context with its possible ontogenetic adaptability. The external TB system may be adjustable within wide ranges (Table 1, Fig. 5) according to individual ecological needs. Provided this ontogenetic adaptability exists, we are confronted with a series of questions, the most intriguing of which is, how vision and taste "communicate"? How could a deterministic adjustment of TB density possibly be triggered by deficient or suboptimal visual input? What causes the changes in numbers and/or size of the neural-crest derived sensory ganglion cells, which induce TB formation?

These questions apply also to evolution, where relatively minor genetic changes may lead to significant changes in numbers and sizes of the sensory ganglion cells and to substantial variation in density and distribution of external TBs. We are just beginning to find patterns of interactions between sensory systems. The search for the mechanisms causing these interactions will be a most challenging task of neurobiology and developmental biology. Fish promise to be exciting models.

Acknowledgements

W. Wieser and A. Goldschmid as well as H. Adam provided encouragement and support throughout this project, which was funded by the "Fonds zur Förderung der wissenschaftlichen Forschung in Österreich" (project no. S-35). Special thanks to our fisherman-in-chief W.-D. Krautgartner and also to family Weber in Loimersdorf, who repeatedly let us abuse their neat restaurant facilities as a field laboratory. P. Simonsberger constructed measurement equipment, K. Bernatzky and R. Hametner provided photographic expertise. Revision of an earlier version by M. Whitear substantially improved the manuscript.

References cited

Adam, H. 1964. Die Haut der Kopflosen, Rundmäuler, Fische und Lurche. Studium generale 17: 350–362.

Atema, J. 1971. Structures and functions of the sense of taste in the catfish (*Ictalurus natalis*). Brain Behav. Evol. 4: 273–294.

Bardach, J.E. & J. Atema. 1971. The sense of taste in fishes. pp. 293–336. *In:* L.M. Beidler (ed.) Handbook of sensory physiology, 4 Chemical senses 2. Taste, Springer Verlag, Heidelberg.

Brandstätter, R. & K. Kotrschal. 1990. Brain growth patterns from juveniles to adults in four mid-European cyprinid fishes (Cyprinidae, Teleostei), roach (*Rutilus rutilus*), bream (*Abramis brama*), carp (*Cyprinus carpio*) and sabre carp (*Pelecus cultratus*). Brain Behav. Evol. 35: 195–211.

Bullock, T.H. 1983. Why study fish brains. pp. 361–367. *In:* R.E. Davis & R.G. Northcutt (ed.) Fish Neurobiology, II, The University of Michigan Press, Ann Arbor.

Campos, H. 1969. Die Geschmacksknospen im Vorderdarm von Süßwasserfischen – Zahl, Verteilung und Entwicklung. Z. Wiss. Zool. 179: 253–299.

Caprio, J. 1988. Peripheral filters and chemoreceptor cells in fishes. pp. 313–338. *In:* J. Atema, R.R. Fay, A.N. Popper & W.N. Tavolga (ed.) Sensory Biology of Aquatic Animals, Springer Verlag, Berlin.

Davenport, D.E. & J. Caprio. 1982. Taste and tactile recordings from the ramus recurrents facialis innervating flank taste buds in the catfish. J. Comp. Physiol. (A) 147: 217–229.

Davis, B.J. & R.J. Miller. 1967. Brain patterns in minnows of the genus *Hybopsis* in relation to feeding habits and habitat. Copeia 1967: 1–39.

Evans, H.M. 1931. A comparative study of the brains in British cyprinoids in relation to their habits of feeding, with special reference to the medulla oblongata. Proc. R. Soc. Lond. B 108: 233–257.

Evans, H.M. 1932. Further observation on the medulla oblongata of cyprinoids and a comparative study of the medulla of clupeoids and cyprinoids with special reference to the acoustic tubercles. Proc. R. Soc. Lond. B 111: 247–280.

Evans, H.M. 1935. The brain of *Gadus*, with special reference to the medulla oblongata and its variation according to feeding habits of different Gadidae. Proc. R. Soc. Lond. B 117: 56–68.

Finger, T.E. 1983. The gustatory system in teleost fish. pp. 285–311. *In:* Northcutt & R.E. Davis (ed.) Fish Neurobiology, I, The University of Michigan Press, Ann Arbor.

Finger, T.E. 1988. Organization of chemosensory systems within the brains of bony fishes. pp. 339–363. *In:* J. Atema, R.R. Fay, A.N. Popper & W.N. Tavolga (ed.) Sensory Biology of Aquatic Animals, Springer Verlag, Berlin.

Freihofer, W.C. 1963. Patterns of the ramus lateralis accessorius and their systematic significance in teleostean fish. Stanford Ichthyol. Bull. 8: 81–189.

Gomahr, A., K. Kotrschal & A. Goldschmid. 1988. Die chemosensorischen Zellen der Haut bei den heimischen Karpfenfischen (Teleostei, Cyprinidae): Geschmacksknospen und freie Sinneszellen, Österreichs Fischerei 41: 241–253.

Herrick, C.J. 1901. The cranial nerves and cutaneous sense organs of North American siluroid fishes. J. Comp. Neurol. 11: 177–247.

Herrick, C.J. 1904. The organ and sense of taste in fishes. Bull. U.S. Fish Comm. 22: 237–272.

Huber, R. & M.K. Rylander. 1992. Brain morphology and turbidity preference in *Notropis* and related genera (Cyprinidae, Teleostei) Env. Biol. Fish 33: 153–165.

Jakubowski, M. & M. Whitear. 1990. Comparative morphology and cytology of taste buds in teleosts. Z. mikrosk.-anat. Forsch. (in press).

Kapoor, B.G., H.E. Evans & R.A. Pevzner. 1975. The gustatory system in fish. Adv. Mar. Biol. 13: 53–108.

Kiyohara, S., S. Yamashita & J. Kitoh. 1980. Distribution of taste buds on the lips and inside the mouth in the minnow, *Pseudorasbora parva*. Physiol. Behav. 24: 1143–1148.

Kiyohara, S., I. Hidaka, J. Kitoh & S. Yamashita. 1985. Mechanical sensitivity of the facial nerve fibers innervating the anterior palate of the puffer, *Fugu pardalis*, and their central projection to the primary taste center. J. Comp. Physiol. (A) 175: 705–716.

Kotrschal, K. & H. Junger. 1988. Patterns of brain morphology in mid-European Cyprinidae (Pisces, Teleostei): a quantitative histological study. J. Hirnforsch. 29: 341–352.

Kotrschal, K. & M. Palzenberger. 1992. Neuroecology of cyprinids: comparative, quantitative histology reveals diverse brain patterns. Env. Biol. Fish 33:135–152.

Kotrschal, K., R. Brandstätter, A. Gomahr, H. Junger, M. Palzenberger & M. Zaunreiter. 1991. Brain and sensory systems. pp. 284–331. *In:* I.J. Winfield & J.S. Nelson (ed.) Cyprinid Fishes, Systematics, Biology and Exploitation, Chapman & Hall, London.

Landacre, F.L. 1907. On the place of origin and method of distribution of taste buds in *Ameirus melas*. J. Comp. Neurol. Psychol. 17: 1–67.

Lane, E.B. 1976. Structural aspects of skin sensitivity in the catfish, *Ictalurus*. Ph.D. Thesis, University College London, London. 170 pp.

Leydig, F. 1851. Über die Haut einiger Süßwasserfische. Z. Wiss. Zool. 3: 1–12.

Marui, T. 1977. Taste responses in the facial lobe of the carp, *Cyprinus carpio* (L.). Brain Res. 130: 287–298.

Marui, T. & J. Caprio. 1982. Electrophysiological evidence for the topographical arrangement of taste and tactile neurons in the facial lobe of the channel catfish. Brain Res. 231: 185–190.

Marui, T. & M. Funakoshi. 1979. Tactile input to the facial lobe of the carp, *Cyprinus carpio* (L.). Brain Res. 177: 479–488.

May, R.M. 1925. The relation of nerves to degenerating and regenerating taste buds. J. Exp. Zool. 42: 371–410.

Moore, G.A. 1950. The cutaneous sense organs of barbed minnows adapted to life in the muddy waters of the Great Plains region. Trans. Amer. Microsc. Soc. 69: 69–95.

Northcutt, R.G. 1988. Sensory and other neural traits and the adaptionist program: Mackerels of San Marco. pp. 869–883. *In:* J. Atema, R.R. Fay, A.N. Popper & W.N. Tavolga (ed.) Sensory Biology of Aquatic Animals, Springer Verlag, Berlin.

Olmsted, J.M.D. 1920. The nerve as a formative influence in the development of taste buds. J. Comp. Neurol. 31: 465–468.

Peters, R.C., G.W. van Steenderen & K. Kotrschal. 1987. A chemoreceptive function for the dorsal fin in rocklings (*Gaidropsarus* and *Ciliata*: Teleostei, Gadidae): electrophysiological evidence. J. Mar. Biol. Ass. U.K. 67: 819–823.

Peterson, R.H. 1972. Tactile responses of the goldfish (*Carassius auratus* L.) facial lobe. Copeia 1972: 814–819.

Radermann-Little, R. 1979. The effect of temperature on the turnover of taste bud cells in catfish. Cell Tissue Kinet. 12: 269–280.

Reutter, K. 1986. Chemoreceptors. pp. 587–604. *In:* J. Bereiter-Hahn, A.G. Matoltsy & K.S. Richards (ed.) Biology of the Integument, II, Vertebrates, Springer Verlag, Heidelberg.

Reutter, K. & W. Breipohl. 1975. Rasterelektronenmikroskopische Untersuchungen der Geschmacksknospen von Fischen (*Ameiurus nebulosus* und *Xiphophorus helleri*). Verh. Anat. Ges. 69: 879–884.

Reutter, K., W. Breipohl & G.J. Bijvank. 1974. Taste bud types in fishes. II. Scanning electron microscopical investigations on *Xiphophorus helleri* Heckel (Poeciliidae, Cyprinodontiformes, Teleostei). Cell Tissue Res. 153: 151–166.

Schemmel, C. 1967. Vergleichende Untersuchungen an den Hautsinnesorganen ober- und unterirdisch lebender *Astyanax*-Formen. Z. Morph. Tiere 61: 255–305.

Schiemer, F. 1985. Die Bedeutung der Augewässer als Schutzzone für die Fischfauna. Österreichs Wasserwirtschaft 37: 239–245.

Schiemer, F. 1988. Gefährdete Cypriniden – Indikatoren für die ökologische Intaktheit von Flußsystemen. Natur und Landschaft 63: 370–373.

Sibbing, F.A. 1988. Specializations and limitations of utilization of food by carp, *Cyprinus carpio*: a study of oral food processing. Env. Biol. Fish. 22: 161–178.

Sinha, G.M. & S.K. Moitra. 1978. Studies on the comparative histology of the taste buds in the alimentary tract of a herbivorous fish, *Labeo calbasus* (Ham.), an omnivorous fish, *Catla catla* (Ham.) and a carnivorous fish, *Clarias batrachus* (L.) in relation to food and feeding habits. Zool. Beitr. 24: 43–57.

Weber, H. 1827: Über das Geschmacksorgan der Karpfen und den Ursprung seiner Nerven. Arch. Anat. Physiol. 1827: 309–315.

Wunder, W. 1927. Sinnesphysiologische Untersuchungen über die Nahrungsaufnahme bei verschiedenen Knochenfischarten. Z. vergl. Physiol. 6: 67–98.

Environmental Biology of Fishes **33**: 135–152, 1992.

Neuroecology of cyprinids: comparative, quantitative histology reveals diverse brain patterns

Kurt Kotrschal[1] & Margit Palzenberger
Zoologisches Institut der Universität Salzburg, Hellbrunnerstraße 34, A-5020 Salzburg, Austria
Present address: Konrad-Lorenz-Forschungsstelle für Ethologie, Auingerhof 11, A-4645 Grünau, Österreich

Received 2.2.1990 Accepted 22.10.1990

Key words: Allometry, Comparative morphology, Ecomorphology, Evolutionary patterns, Morphospace, Nervous system, Octavo-lateralis, Sensory diversification, Taste, Vision

Synopsis

Brain patterns are compared by quantitative histology in 28 native and introduced mid-European cyprinid species, considering 17 primary sensory and higher order brain areas. Cluster analysis (CLA) and principal component analysis (PCA) based on relative volumes of these brain areas indicate that cyprinid brains are diversified into four major groups, basic cyprinid, abramine, octavo-lateralis and chemosensory. PCA recognizes the brain of *Phoxinus phoxinus* as a fifth group. Interspecific differences in brain morphology are mainly caused by variability in relative sizes of the brain stem lobes for external and internal taste (lobus facialis and lobus vagus), as well as of octavo-lateralis and visual areas. Higher order brain areas show little interspecific variation in relative size, and were grouped by PCA according to inter- and intraspecific allometries. Hypotheses on brain functions are based on brain area correlations. We propose that the processing of external taste information in the valvula cerebelli may be particularly important for benthivorous cyprinids, whereas the integration of octavo-lateralis input with visual information via the torus longitudinalis – stratum marginale system may play a key role in the planktivores. Brain patterns suggest two major pathways of cyprinid evolutionary and ecological radiation, one leading from the basic cyprinids towards octavo-lateralis dominated midwater and surface planktivores, the second towards taste-dominated benthivores.

Introduction

Quantitative, comparative brain morphology in specious and closely related vertebrate groups reveals within-group trends of sensory diversification, which can be interpreted ecologically as well as evolutionarily (Bullock 1983, Goldschmid & Kotrschal 1989, Northcutt 1988). With these and related goals, quantitative brain morphology has been applied in ecologically diversified taxa, such as birds, bats and other mammals (Stephan 1967, Stephan & Pirlot 1970, Jolicoeur & Baron 1980, Pirlot & Jolicoeur 1982, Pagel & Harvey 1989).

Teleost fish form the largest of all vertebrate groups and frequently have been the subject of comparative, qualitative brain (eco)morphology (Balon 1968, Davis & Miller 1967, Evans 1931, 1932, 1940, 1952, Khanna & Singh 1966, Kirka 1963a, Kishida 1979, Mayser 1881, Miller & Evans 1965, Northcutt & Wullimann 1988, Schnitzlein 1964, Uchihashi 1953, for comprehensive treatments see Ariens Kappers et al. 1936, Davis &

Northcutt 1983, Northcutt & Davis 1983). However, only few investigations were quantitative (Brandstätter & Kotrschal 1990, Bauchot et al. 1977, 1989, Geiger 1956a, b, Kirka 1963b, Kotrschal & Junger 1988, Kotrschal et al. 1991, Ridet et al. 1977, Snow & Rylander 1982). Only recently, multivariate statistics (Huber & Rylander 1992) has been used to extract information from brain data, allowing for a rigid classification and interpretation.

In the present paper we consider 28 species of common mid-European cyprinids. Included are also a few species introduced from eastern Asia, such as silver carp, grass carp and goldfish. Recently (Kotrschal & Junger 1988), brains in 14 of these species have been examined histologically.

Although cyprinids are the most successful freshwater teleost family in the northern hemisphere (Nelson 1984), comparatively little is known on their biology and evolutionary radiation. Therefore we address the following questions:
(1) What is the morphological radiation of cyprinid brains and which brain areas contribute most to interspecific brain variability?
(2) Does a correlation matrix of brain areas lead to novel ideas concerning brain function and brain area interactions?
(3) Can brain morphology be ecologically interpreted?
(4) Can evolutionary trends be recognized?

Table 1. List of species in alphabetical order, common names according to Maitland (1981), n = number of brains measured, SL = range of standard length, origin, basic habitat requirements (i = indifferent towards currents, r = rheophilic) and feeding styles given. The asterisk (*) indicates that specimens originated from the Stopfenreuther Au, a Danube backwater area, Wall. = Wallersee.

Species	Common name	n	SL, cm	Origin	Habitat/feeder type or food
1 *Abramis ballerus*	blue bream	2	19, 20	*	i, midwater/plankton
2 *A. brama*	common bream	7	13–37	*, Wall.	i, benthos/microzoobenthos + plankton
3. *A. sapa*	whiteye bream	2	17, 17	*	i-r, benthos/benthic + plankton
4 *Alburnoides bipunctatus*	schneider	2	9–9.5	Danube tributary	i-r, midwater/omnivorous
5 *Alburnus alburnus*	bleak	2	11.5–13	*	i, surface/surface + omnivorous
6 *Aspius aspius*	asp	2	22.5–23	*	i-r, midwater/piscivorous
7 *Barbus barbus*	barbel	1	32	*	r, benthic/macrozoobenthos
8 *Blicca bjoercna*	white bream	2	14–17.5	*	i, benthic-midwater/macrozoobenthos
9 *Carassius auratus*	goldfish	3	11–13	local pond	i, benthic/benthos-omnivorous
10 *C. carassius*	crucian carp	2	9.5–11	pond-raised	i, benthic/benthos-omnivorous
11 *Chalcalburnus chalcoides*	shemaya	2	16.5–18	Mondsee	i, pelagic, surface/planktivorous
12 *Chondrostoma nasus*	nase	3	22–41	*	r-i, benthic/benthic grazer
13 *Ctenopharyngodon idella*	grass carp	2	13.5–14	fish farm	i, midwater/macrophytes
14 *Cyprinus carpio*	common carp	3	13–24	fish farm	i, benthic/benthic omnivorous
15 *Gobio gobio*	gudgeon	2	11–12	Mur tributary	r-i, benthic/macrozoobenthos
16 *Hypophthalmichthys mollitrix*	silver carp	1	19	fish farm	i, midwater/phyto-zooplankton
17 *Leucaspius delineatus*	sun bleak	2	5–6	pond-raised	i, midwater/zoophageous
18 *Leuciscus cephalus*	chub	2	16–19	*	i-r, midwater/omnivorous
19 *L. idus*	orfe	3	8–33	*, pond-raised	i, benthic-midwater/omnivorous
20 *L. leuciscus*	dace	2	10–11	Danube tributary	i-r, midwater/omnivore-planktivorous
21 *L. souffia*	streamer	2	12, 12	Mur	r, midwater/omnivorous-drift feeder
22 *Pelecus cultratus*	sabre carp	7	17–35	*, Neusiedlersee	i-r, surface/surface + plankton
23 *Phoxinus phoxinus*	minnow	2	6–9	local creek	r-i, benthic-midwater/zoophageous
24 *Rutilus rutilus*	roach	6	10–24	*, Wallersee	i, midwater/omnivorous
25 *Rhodeus sericeus*	bitterling	2	6, 6	pond-raised	i, benthic/zoophageous
26 *Scardinius erythrophthalmus*	rudd	2	14–20	*	i, midwater/omnivorous
27 *Tinca tinca*	tench	2	17–18	fish farm	i, benthic/benthos-omnivorous
28 *Vimba vimba*	vimba	2	14,14	*	i-r, benthic/macrozoobenthos

Materials and methods

Most fish were gill-netted in the Stopfenreuther Au, a Danube-wetland area in the east of Austria. Additional specimens were obtained by electrofishing in Upper Austrian, Styrian and Salzburg creeks and rivers, or were purchased at a local hatcheries and pet shops (Table 1).

Quantitative histology

Specimens were deeply anaesthetized with MS 222 (1 : 10 000) and fixed by perfusion with 10% buffered formaldehyde solution. Brains were removed from the skull, postfixed for several weeks in the same fixative, embedded in gelatine, incubated overnight in 30% sucrose and cryostat-sectioned at 30 or 40 μm, depending on brain size. Cryostat sections were preferred over paraffin in this quantitative study, because of the possibility of heterogenous shrinkage of brain areas during the paraffin embedding procedure. Alternate sections were mounted in series with gelatine-chromalaun, allowed to dry slowly in a moist environment and stained with cresyl-violet.

For purposes of comparative, qualitative histology, brains of selected species were embedded in paraffin and serial sections were stained with the Bodian silver method.

A total of 82 cryo-sectioned brains from 28 cyprinid species were measured, but only 72 cases were considered in the present study. The remaining, juvenile specimens were excluded to avoid intro-

Table 2. List of measured brain areas and their interspecific mean in % of total brain volume, ranked according to decreasing interspecific coefficients of variation (VR). Major source of input given (cf. McCormick & Braford 1988, Northcutt & Davis 1983, Davis & Northcutt 1983). Crosses code for demarcability, x = closed structure, unambiguously demarcable; xx = some boundaries by definition; xxx = major boundaries by definition, a verbal description of defined boundaries is given. Primary sensory areas coded with 1, higher order and/or multimodal centers with 2. The number of cases was 72 in most variables, except for the stratum marginale (n = 59) and the torus semicircularis (n = 71).

Code	Sensory area	Mean brain vol · % ± sd	VR	Input from	Demarcability, borders
1	lobus facialis	1.34 ± 1.48	110.4%	external taste	x
2	nucleus habenularis	0.11 ± 0.09	81.8%	olfactory/multimodal	x
1	lobus vagus	4.71 ± 3.69	78.3%	internal taste/visceral	x including the motor-layer
1	central acoustic area	0.83 ± 0.65	78.3%	?	x
1	crista cerebellaris	2.18 ± 1.06	48.6%	lateral line/cerebellum	x
1	eminentia granularis	1.88 ± 0.8	42.6%	inner ear/lateral line	x
2	valvula cerebelli	6.59 ± 2.44	37.0%	multimodal	xx: underneath tectum opticum
2	stratum marginale	3.08 ± 1.12	36.4%	t. longitudinalis	x
1	bulbus olfactorius	2.47 ± 0.78	31.6%	olfactory mucosa	x
2	mesencephalic tegmentum	7.47 ± 1.4	31.3%	multimodal	xxx: caudoventral of fasciculus retroflexus to nucleus entopeduncularis
2	torus semicircularis	1.35 ± 0.34	25.2%	octavo-lateralis relays	x
2	corpus cerebelli	13.51 ± 3.16	23.6%	multimodal	xx: cerebellar part free from tectal cover
1/2	tectum opticum	17.22 ± 3.92	22.8%	retina/multimodal	x
2	telencephalon	8.45 ± 1.77	21.0%	multimodal	xx: including commissura anterior, excluding recessus praeopticus and pedunculus cerebri
2	diencephalon	0.96 ± 2.18	19.9%	multimodal	xxx: including recessus praeopticus and nucleus habenularis to rostrodorsal fasciculus retroflexus
2	brain stem caudal to mesencephalic brain stem	22.09 ± 3.53	16.0%	multimodal	xxx: obtained by subtraction of sum-volume of all areas measured from total brain volume

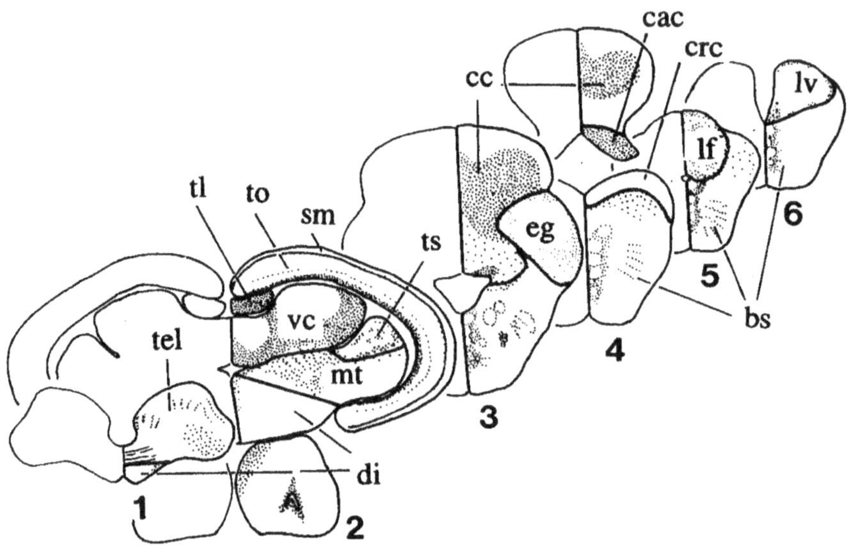

Fig. 1. Demarcation of measured areas exemplified with 6 cross sections through roach brain. 1 = telencephalon at the level of the commissura anterior; 2 = at maximal extension of the tectum opticum; 3 = at maximal cross section areas of the corpus cerebelli and the eminentia granularis; 4 = at the crista cerebellaris; 5 = through the lobus facialis; 6 = through the lobus vagus. Modified after Brandstätter & Kotrschal (1990). Abbreviations: bs = brain stem; cac = central acoustic area (see Materials and methods for comments); cc = corpus cerebelli; crc = crista cerebellaris; di = diencephalon; eg = eminentia granularis; lf = lobus facialis; lv = lobus vagus; mt = mesencephalic tegmentum; sm = stratum marginale; tel = telencephalon; tl = torus longitudinalis; to = tectum opticum; ts = torus semicircularis; vc = valvula cerebelli. Bulbus olfactorius and nucleus habenularis not within section planes 1–6 and therefore not shown.

ducing a major allometric bias (Brandstätter & Kotrschal 1990).

Seventeen areas per brain were measured (Table 2), representing approximately 78% of the total brain volume (of the entire brain from the olfactory bulbs to the caudal end of the lobus vagus). For demarcation of brain areas see Figure 1. Brain areas to be measured were selected according to functional importance and demarcability. The remaining 22% of brain volume consisted mainly of the myelencephalic brain stem, which was not directly measured, but back-calculated by subtracting the sum of all measured brain area volumes from total brain volume (Table 2). Therefore the myelencephalic brain stem was excluded from the multivariate statistical analysis.

By aid of a camera lucida connected to a binocular microscope, cryostat sections were projected to a digitizing tablet and data were fed directly into a computer. Brain areas were measured on 90 (± 15) sections per brain, left and right separately, which resulted in an average of approximately 30 measured planes per brain area. Thus the present study integrates approximately 40 000 individual measurements.

In most brain areas considered the number of planes measured exceeded eight. Area volumes were estimated by multiplication of the mean area of the brain structure measured times its length (left and right separately). As no significant (exceeding the measurement error, see below) and consistent lateralization of brain areas could be found, the sums of equivalent bilateral volumes were used for further comparison.

To make volumes of brain areas comparable despite differences in body size between species, area volumes are expressed as % of total brain volume. Although this creates the problem that % volumes are not independent from each other, we feel that this approach is still more valid than standardizing brain volumes with body length or weight, because the cyprinids considered are relatively heterogenous with respect to body shape (Fig. 2) (cf. Stephan 1960).

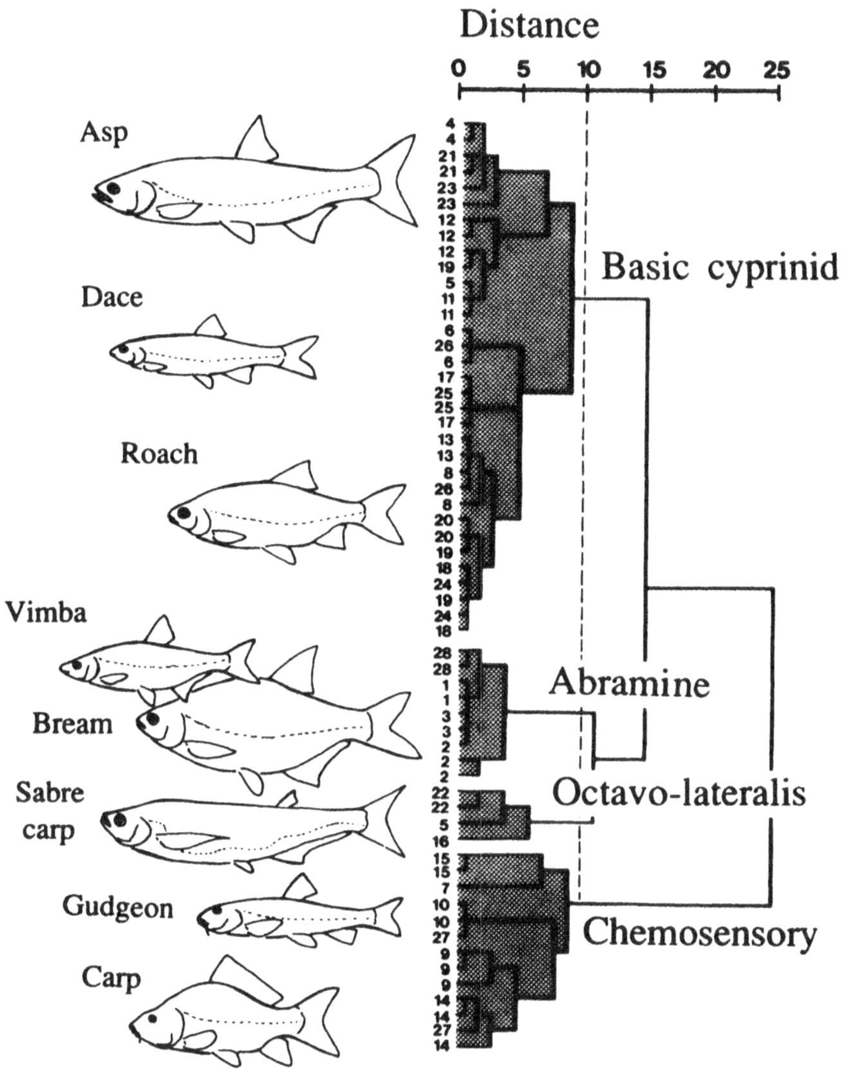

Fig. 2. Dendrogram obtained by cluster analysis (Euclidian distances, Ward's sorting, based on primary sensory brain areas, Table 2) reveals 4 major groups of brain morphologies. Habitus of representative species shown. Numbers at tips of the branches code for species (Table 1).

The measurement error was quantified by tracing 4 differently sized brain structures 5 times successively. The error with respect to area of the structures averaged 4%, the maximal error was 9% in the smallest structure considered, the nucleus habenularis.

Topographic positions of brain areas are given in Kotrschal & Junger (1988) as well as in Figure 1 and Table 2. For information on structure, connectivity and functions of brain areas see Davis & Northcutt (1983), Northcutt & Davis (1983) and Northcutt & Wullimann (1988).

The authors are aware of possible oversimplifications when associating brain areas with sensory functions. The tectum opticum for example, although the primary brain area for most of the visual input, is a multimodal center. The term 'central acoustic area' (Evans 1931, based on the co-occurrence of well developed auxilliary auditory structures with a well developed CAC) lacks experimental confirmation. As this area, including the present study, shows strong correlations with other octavo-lateralis areas, the term central acoustic area is tentatively used as a working hypothesis, although no primary octavo-lateralis terminals were found there (McCormick & Braford 1988).

Multivariate statistics

Data were analyzed by multivariate statistics, cluster analysis (CLA) and principal component analysis (PCA), employing SPSSx (Brosius 1989) and CLUSTAN-software. Each individual brain was treated as a single case. The relative volumetric data matrix (% of areas of total brain volumes) was arc-sine transformed prior to analysis (Sokal & Rohlf 1981) to correct for compressed variances at either ends of the %-scale. Only basic statistical procedures (Table 2) were performed on the original %-matrix.

CLA is an ordination technique which is based on finding similarities or dissimilarities between cases. Squared Euclidian distances in combination with Ward's sorting were employed to generate a dendrogram (Fig. 2). Only the primary sensory brain areas and those showing strong correlations with the latter were used (see PCA for reason, Tables 2, 4), although the classification was found to be relatively robust even when applied with the entire set of variables (not shown).

PCA is a standard ordination and data reduction method, different from CLA. PCA finds a small number of new variables (principal components [PCs], or factors), based on the correlations between original variables. Factor loadings (0–1, Table 4) describe the contribution of the original variables to the newly found PCs. The number of significant PCs (those which explain a high % of the total variance in the data set) was determined by the scree-test (plotting the Eigenvalues in decreasing order). This revealed only 2 PCs in the primary sensory area subset and 3 PCs in the higher order brain area subset (see below, Table 4). The cases factor scores were calculated with the Bartlett-method (for Fig. 3, 4). A reasonable coherence of variables within a data set is a prerequisite for PCA. Communality was estimated by the Kaiser-Meyer-Olkin (KMO) measure for sample adequacy (Brosius 1989, Kaiser 1974). Generally, the KMO was higher in the primary sensory areas as compared to the higher order brain areas. KMO was low (< 0.7) when applied to the entire set of variables. When variables were divided into two groups, primary sensory and higher order, the KMO was within suitable ranges for both data subsets (0.75, 0.72). The tectum opticum is both a primary sensory and higher order brain structure, showing reasonable communalities with both subsets and was therefore included in both. The stratum marginale and the valvula cerebelli were included in the primary sensory subset, because of their strong correlations and high communalities with the latter (Table 3). The correlation matrix was produced with the undivided set of variables.

Results

Brain area variability

The highest interspecific variation (VR) of all brain areas is shown by primary sensory areas, particularly the brain stem taste lobes (Table 2). The nucleus habenularis is also highly variable in size, although part of this variability may be due to a relatively high measurement error in this small structure (see Materials and methods). The octavo-lateralis associated brain areas are next in variability. Still relatively high in interspecific variability are certain higher-order multimodal areas, such as the valvula cerebelli, the stratum marginale of the optic tectum and the torus longitudinalis, all of which show strong correlations with primary sensory areas (Table 3). Two primary sensory areas are relatively low in interspecific variability, the bulbus olfactorius and the tectum opticum. All basal brain areas, such as the myelencephalic, rhombencephalic and mesencephalic tegmentum as well as the diencephalon and telencephalon are at the lower range of interspecific variability.

Brain area correlations

Table 3 presents a matrix of all significant correlations found among the present variables (= brain areas, Table 2), which serves as a base for PCA and functional hypotheses.

External and internal taste lobes are positively correlated (Table 3), as are the octavo-lateralis areas as well as the visual areas among themselves. A highly significant negative correlation exists between the lobus facialis (external taste) and all octavo-lateralis areas, but not between the lobus vagus (internal taste) and the octavo-lateralis areas. There are also highly significant negative correlations between taste and visual areas. Although there is no correlation between the tectum opticum as a whole and octavo-lateralis areas, the stratum marginale of the tectum is significantly correlated with all three, as is the torus longitudinalis, the source of origin of unmyelinated parallel stratum marginale fibers (Vanegas 1983).

There is a highly significant positive correlation of the valvula cerebelli with the lobus facialis. Negative correlations exist between the valvula and one octavo-lateralis area, as well as between valvula and visual areas (Table 3). There is no correlation between the torus semicircularis and octavo-lateralis areas, but the correlation between the torus and the tectum opticum is highly significant.

Aside from being significantly correlated with all octavo-lateralis areas and the stratum marginale, the torus longitudinalis correlates negatively with the lobus facialis.

Positive correlations of the corpus cerebelli were found with two octavo-lateralis areas (Table 3), as well as with the valvula cerebelli and with the torus longitudinalis.

The bulbus olfactorius shows no correlation with any other primary sensory area. The telencephalon is positively correlated with the bulbus olfactorius only, negatively with the valvula cerebelli (Table 3).

Correlations of the bulbus olfactorius, the telencephalon, the diencephalon and mesencephalon (Table 3) seem to a lesser degree be determined by sensory interactions than by intra- and interspecific allometries (see below, Fig. 3).

Cluster analysis

Cluster analysis (CLA) found four major groups of brain morphologies (Fig. 2). Most species show a basic cyprinid brain, which is characterized by fairly well developed visual centers, but only moderately sized octavo-lateralis and brain stem chemo-

Table 3. Matrix of significant, positive and negative correlations between brain areas, based on relative volumes of all specimens considered. *: ≥0.01; **: ≥0.001; ns = not significant. All except for correlations with the myelencephalic brain stem listed. Brain areas associated with 5 functional groups, taste, octavo-lateralis, vision, multimodal areas with strong correlations to sensory areas, and multimodal. For abbreviations see Figure 1.

		lf	lv	eg	cac	crc	to	sm	vc	ts	tl	cc	bo	tel	di	mt
taste	lf	–														
	lv	+**	–													
	eg	–**	ns	–												
oct.-l.	cac	–**	ns	+**	–											
	crc	–**	ns	+**	+**	–										
vision	to	–**	–**	ns	ns	ns	–									
	sm	–**	n.s.	+*	+*	+**	+**	–								
sens.-	vc	+**	ns	–*	ns	ns	–**	–*	–							
multimod.	ts	ns	ns	ns	ns	ns	+**	ns	ns	–						
	tl	–**	ns	+**	+**	+**	ns	+**	ns	ns	–					
	cc	ns	ns	ns	+*	+**	–*	ns	+*	ns	+**	–				
multimodal	bo	ns	ns	ns	ns	ns	ns	ns	ns	–*	ns	ns	–			
	tel	ns	ns	ns	ns	ns	ns	ns	–*	ns	ns	ns	+*	–		
	di	ns	ns	–*	–*	–**	+*	ns	–**	ns	–*	–**	ns	+*	–	
	mt	–*	–**	ns	ns	ns	+**	ns	–**	+*	ns	–**	ns	ns	+*	–

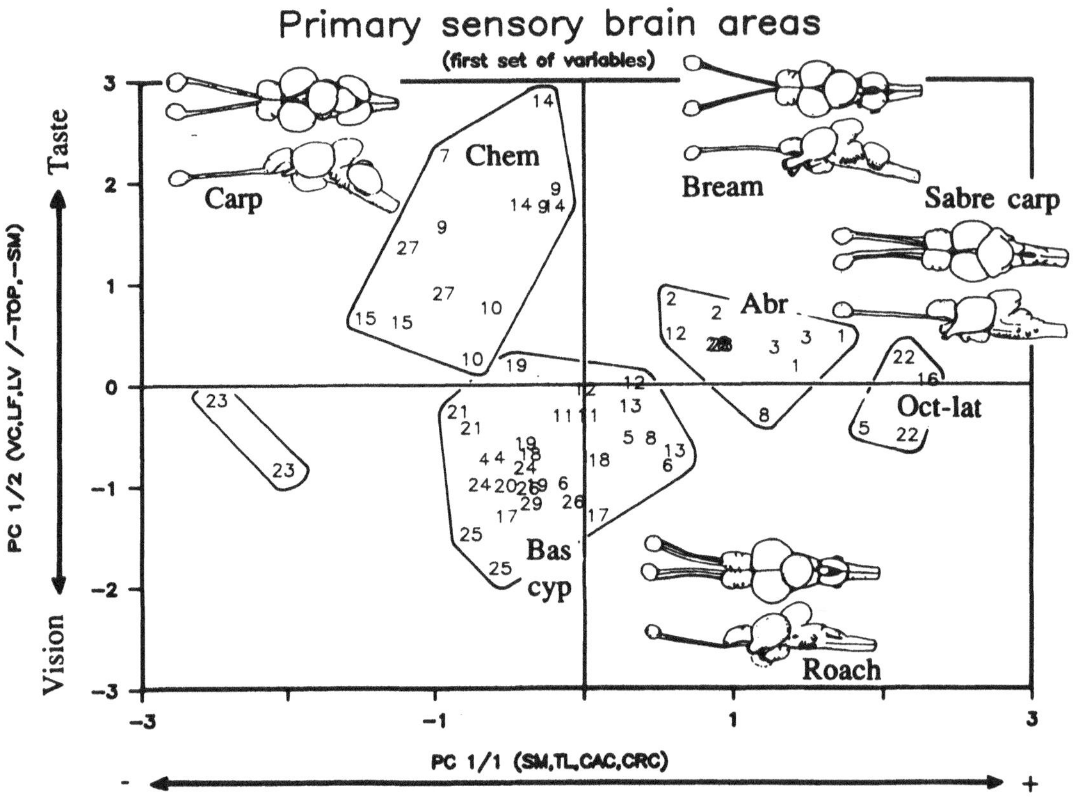

Fig. 3. Principal component (PC) plot of PC scores of individual brains, based on primary sensory brain areas (Table 2). Numbers code for species (Table 1). PC1/1 represents an octavo-lateralis and torus longitudinalis-stratum marginale axis. PC1/2 represents a visual (−) versus brain stem taste (+) axis. For factor loadings see Table 4a. Groups, basic cyprinid (Bas cyp), abramine (Abr), octavo-lateralis (Oct-lat), chemosensory (Chem) and *Phoxinus phoxinus* (23) surrounded by polygons. External morphologies of representative brains shown in dorsal and lateral view.

centers (e.g. roach brain; for representative examples of external brain morphologies from the different groups see Fig. 3). Although appropriate from a purely morphological viewpoint, the potentially misleading term 'generalized' was avoided for this brain group, as it contains also ecologically specialized species, such as the piscivorous *Aspius aspius.*

Relatively close to the basic cyprinid brains are the abramine brains, including *Vimba vimba* and *Blicca bjoerkna* being intermediate between the basic cyprinid and abramine brains. These breams are characterized by a relatively distinct development of brain areas representing all three major sensory faculties, vision, octavo-lateralis and brain stem chemosense (e.g. bream, Fig. 3).

The octavo-lateralis brains of sabre carp, silver carp and bleak show relatively large octavo-lateralis and visual centers, but only very small brain stem chemosensory lobes (e.g. sabre carp, Fig. 3).

The chemosensory brains were most distantly placed by CLA from the basic cyprinid brains. The latter contain mainly benthivorous, often barbed murky water species, such as the barbel, common carp, crucian carp, tench and gudgeon (Fig. 2). Chemosensory brains are characterized by large brain stem taste lobes and a large valvula cerebelli, but small octavo-lateralis and visual centers (e.g. common carp, Fig. 3).

Within the dendrogram the majority of individuals within species are grouped together. Interspecific brain variation is greater than intraspecific variation and brain patterns are indeed species-specific.

Principal component analysis

Principal component analysis (PCA) with the primary sensory subset of variables (see Materials and methods) confirms the robustness of the classification found by CLA (Fig. 2). Two significant principal components (PCs) were extracted. As indicated by factor loadings, PC1/1 is an octavo-lateralis axis, whereas PC1/2 is a taste-visual axis (Table 4a). By plotting the cases factor scores, the groups found by PCA are essentially the same as those found by CLA (Fig. 2, 3).

Due to its outlier position, *Phoxinus phoxinus* has to be recognized as a fifth group. *P. phoxinus* brains are characterized by a relatively large tectum opticum, a small vagal, but well developed facial lobe and small octavo-lateralis centers. A large tectum opticum, but only moderate brain stem taste lobes are shared by all small-sized species present, thus body size influences the PC1/2 axis (Fig. 5). On the other hand, small sized species are distributed over the entire range of PC1/1.

Table 4. Factor loadings of principal components (PC) for the 2 subsets of data: (a) primary sensory including stratum marginale, torus longitudinalis and valvula cerebelli, (b) higher order, multimodal areas including the bulbus olfactorius and the tectum opticum. Areas ranked according to decreasing coefficients of variation. Only loadings > 0.5 shown.

(a)	PC1/1	PC1/2	
lobus facialis		+ 0.72	
lobus vagus		+ 0.79	
central acoustic area	+ 0.80		
crista cerebellaris	+ 0.92		
eminentia granularis	+ 0.55		
valvula cerebelli		+ 0.76	
stratum marginale	+ 0.65	− 0.56	
torus longitudinalis	+ 0.78		
tectum opticum		− 0.92	
(b)	**PC2/1**	**PC2/2**	**PC2/3**
nucleus habenularis			+ 0.94
bulbus olfactorius		+ 0.70	
mesencephalic tegmentum			+ 0.80
torus semicircularis	+ 0.52		
corpus cerebelli	− 0.73		
tectum opticum	+ 0.75		
telencephalon		+ 0.79	
diencephalon	+ 0.65	+ 0.51	

Distributions of two tribes, the leuciscines and the abramines, are restricted to small areas of the PC1/1-PC1/2 morphospace (Fig. 3).

No particular grouping was found by PCA within the second subset of variables, representing mainly the multimodal brain areas (Fig. 4, Tables 2, 4b). In this subset, PCA extracted 3 significant PCs. PC2/1 represents an axis of tegmentum versus cerebellum. PC2/2 may be interpreted as an olfactory axis (Table 4b, Fig. 4) and PC2/3 only loads with the negatively allometric nucleus habenularis.

Within the PC2/1-PC2/2 morphospace cases are separated according to size. Thus, the vertical position of species polygons (Fig. 4) indicates that PC2/2 is an axis determined by intraspecific allometry, with small individuals at the + range of PC2/2. Also, it seemed that PC2/1 represents a grade from large (negative range of PC2/1) to small (positive range of PC2/2) species. To test this hypothesis, the PC scores of all four PCs (1/1-2/2) were plotted versus standard body length (SL, from tip of the snout to tail fin base). Only the plot of PC1/2 versus SL (Fig. 5) resulted in a significant negative correlation, which confirmed that PC2/1 is indeed mainly determined by interspecific allometry. Large fish (within and between species) have a relatively large cerebellum, but small tectum opticum and vice versa in small fish (Fig. 4, Table 4b, Brandstätter & Kotrschal 1990).

Discussion

Most studies on functional brain morphology focus on fiber connections and input-output characteristics of areas (Northcutt & Davis 1983, Davis & Northcutt 1983), but offer little explanation of the patterns of morphological variation and their biological basis. In the following we attempt to scratch the surface of these issues.

Two assumptions have to be met to be able to draw meaningful conclusions from a quantitative interspecific comparison of brain areas, when relative size is the only parameter considered:

1. The species in question need to be sufficiently

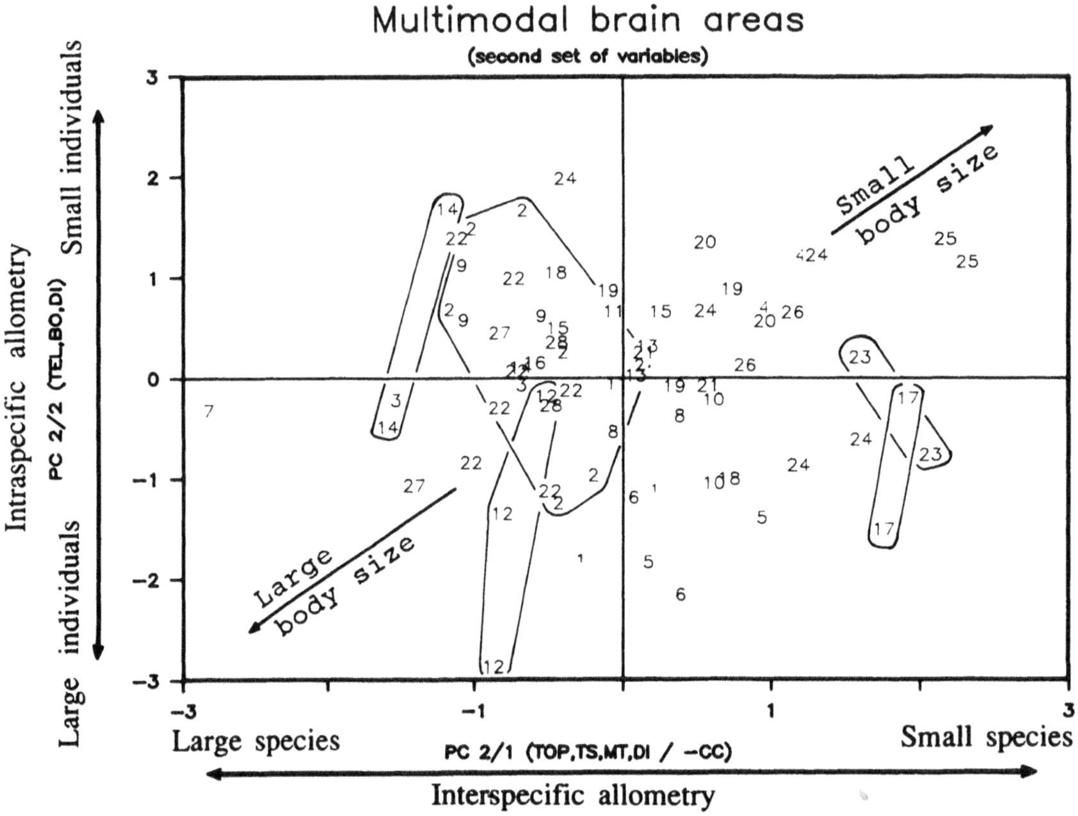

Fig. 4. Plot of PC scores of individuals based on multimodal brain areas (Table 2). Numbers code for species (Table 1). PC2/1 represents an axis of corpus cerebelli (−) versus mesencephalic and diencephalic areas (+). PC2/2 represents an olfactory-telencephalic-diencephalic axis. Examples of vertical orientation of species groups highlighted by polygons. Body size distribution indicates that PC2/1 sorts specimens according to interspecific allometry (cf. Fig. 4), PC2/2 separates according to intraspecific allometry.

closely related to ensure a high probability that the compared brain areas fulfill comparable functional roles (Goldschmid & Kotrschal 1989, Northcutt 1988). If this is the case, then relative sizes of major sensory brain lobes may indicate the 'relative importance of sensory faculties within and between species'. Clearly, questions as to whether this means different spatio-temporal fidelity, sensitivity or differences in filter properties (e.g. degree of color vision, Blaxter 1988, Lythgoe 1988), demand a closer look at the sensory systems in question (Gomahr et al. 1992, Junger & Kotrschal 1989, Kotrschal et al. 1991, Zaunreiter et al. 1991) and cannot be addressed by quantitative brain morphology.

2. As a further prerequisite, contributions of brain-internal allometries to inter- and intraspecific brain variability should be minor and detectable, if the goal of the investigation is to identify those brain areas which account for the sensory interspecific variation. This is evidently important in vertebrates with non-terminal growth such as fish (Geiger 1956a, Brandstätter & Kotrschal 1989). In an ontogenetic study on four major cyprinid brain morphologies (Kotrschal & Junger 1988), Brandstätter & Kotrschal (1990) found substantial brain-internal allometries, which decreased in dynamics with growth. For this reason, only adults or large juveniles were considered in the present study. Still, size effects remained and were recognized by principal component analysis (see below).

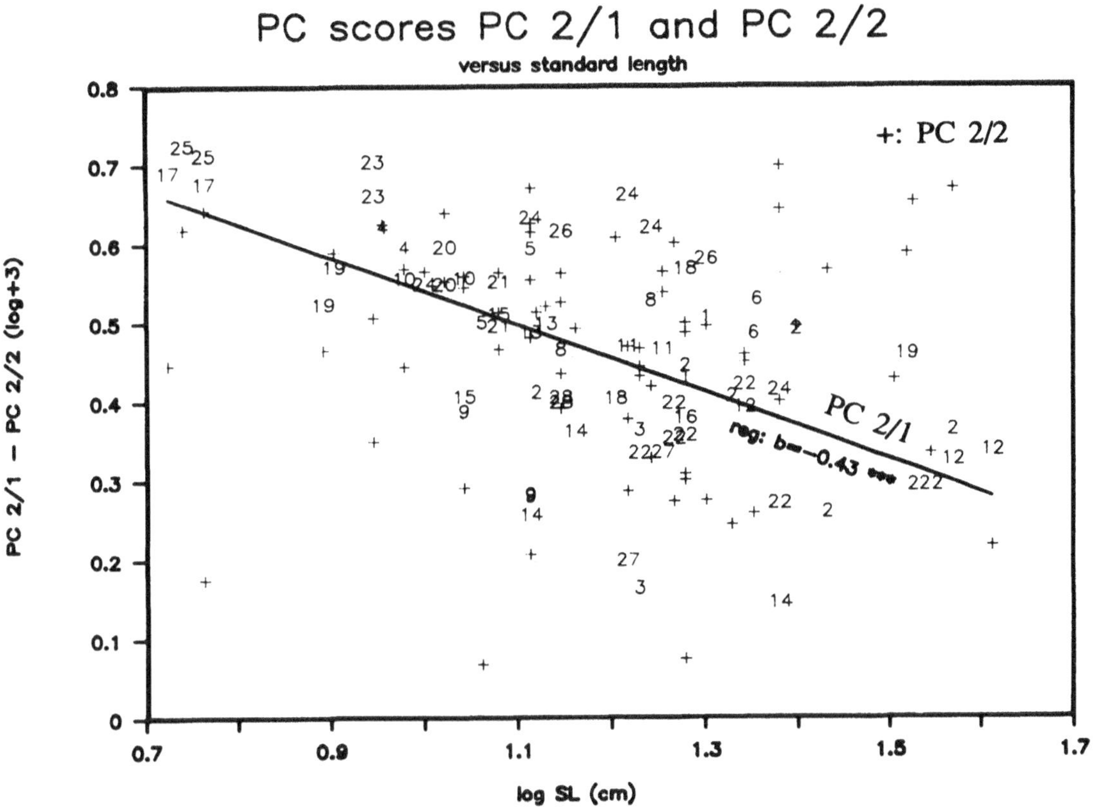

Fig. 5. Regression (log/log scale) of the PC scores PC2/1 (data points coded by species numbers, Table 1) and PC2/2 (crosses) versus standard body length (SL) of individuals. Only the regression of PC2/1 with SL is significant ($\alpha < 0.001$, $b = -0.43$).

Brain area correlations as a source for functional hypotheses

Correlations between brain areas may or may not indicate causal relationships and if they do, it is appropriate to invoke a nested set of explanations from different causal levels, allometric, evolutionary-adaptive and functional (the lack of correlations, however, does not indicate the lack of functional relationships).

One category of correlations is caused by allometries and thus growth, rather than by functional interactions between brain areas. This is the case with the negative correlation between the tectum opticum and the cerebellum and between other multimodal areas.

The evolutionary-adaptive history of cyprinids may generate the 'construction rules' for their brains as discussed below, whereas functional interactions between brain areas are defined by the nature of present information exchange. Secondary and higher order brain areas may or may not covary with primary sensory areas from which they receive information. To preserve and process spatio-temporal information at higher system levels, synaptic connectivity has to be elaborated, causing higher order centers of a labelled pathway to covary in size with their primary centers, as is the case between brain stem taste lobes and the secondary gustatory nucleus (Finger 1988).

Sensory information may diverge widely within the brain, or travel along modality-specific pathways through a series of brain areas, as is the case with lemniscal pathways for taste (Finger 1983a, 1987, 1988, Herrick 1905, Morita & Masai 1980) and octavo-lateralis (Echteler 1985, McCormick & Braford 1988). Input, processing and output may even be mainly localized within a single brain lobe, as is the case with the vagal oropharyngeal reflex pathways (Morita & Finger 1985a, b, 1987). Striking differences in relative sizes and structural organization of the cyprinid vagal lobes (Kotrschal & Junger 1988, Mayser 1881) may therefore, indicate interspecific differences in oropharyngeal sorting capabilities (Sibbing 1988).

Multimodal convergence onto higher order areas may make it difficult to define their major tasks merely on the basis of their fiber connections, as obtained in tracing experiments. One of the strengths of comparative quantitative morphology is to generate functional hypotheses based on covariation of brain areas, even if fiber connections between the areas involved are still insufficiently known. Within the present correlation matrix (Table 3) surprising examples for such relationships emerged: the clear positive correlation between the facial lobe and the valvula cerebelli suggests that the valvula may play an important role in the processing of external taste information. The function of the valvula is still unclear (Finger 1983b); the valvula seems to receive no direct, afferent fibers from the brain stem taste centers (Wullimann & Northcutt 1989).

The second, even more striking example is provided by correlations between the octavo-lateralis associated lobes and the torus longitudinalis-stratum marginale system (Table 3). This suggests integration of lateral line (highly significant correlation with the crista cerebellaris) with visual information within the stratum marginale of the tectum opticum. This hypothesis is backed by physiological and ecological evidence. Torus longitudinalis unit activity is related to eye movements (Northmore et al. 1983, Northmore 1984) and superficial stimulation of the frontal tectum opticum elicits eye convergence and food search behavior (Vanegas 1983). Plankton feeders show the greatest development of the octavo-lateralis areas as well as the stratum marginale system within the cyprinids (Kotrschal & Junger 1988) and within the teleosts in general (Kishida 1979, Winkelmann & Winkelmann 1968). It was shown, that lateral line input may be used for localizing plankton prey (Bleckmann 1988, Montgomery & Macdonald 1987). We therefore suggest that lateral line information may be used to attract visual attention, may prime and guide vision for focussing at, and discriminating small-sized prey.

Although the major source of input into the torus longitudinalis in common carp was found to be the valvula cerebelli, the torus longitudinalis also has afferent connections with the torus semicircularis and other areas (Ito & Kishida 1978). It seems reasonable to assume that the quantity of afferents from different sources varies according to relative importance of sensory systems in different species and that the outcome of Ito & Kishida's (1978) study, had it been conducted with breams or sabre carps, would have been different than in the taste-oriented carp.

Brain diversification in cyprinids

Virtually identical associations of species with brain morphology groups were found by both CLA and PCA. If the major cause of interspecific size variation of primary sensory areas is quantitative, variation in afferent nerve endings, as well as degree of intra-lobe processing (Gomahr et al. 1992, Kotrschal & Junger 1988, Kotrschal et al. 1990), the present four major groups indicate sensory diversification.

Brain area correlations and patterns of cyprinid brain diversification (the PC1/1-PC1/2 morphospace, Fig. 2, 3) allow us to formulate a set of 'construction rules' for cyprinid brains. Some of these rules apply for all cyprinids investigated, irrespective of body size: brains with relatively large visual centers may or may not show large octavo-lateralis centers, but brains with large octavo-lateralis centers always show a relatively large tectum opticum and a thick stratum marginale in particular. Taste lobe-determined brains always show relatively small visual and octavo-lateralis centers. Similar brain area relationships in cyprinids were recognized by earlier workers and sight feeders, mouth and skin tasters were distinguished (Evans 1940, Evans 1952).

The cause for these rules may be some sort of constructional constraint (Maynard Smith et al. 1985). More likely these sensory patterns are due to adaptation, because certain sensory combinations are more useful than others to cope with certain sets of habitat variables and thus were selected for. Therefore, the brain groups found may represent occupied peaks in the adaptive landscape (Bock 1980, Simpson 1944). Under relatively clear midwater conditions a combination of lateral line and vision may be more adaptive than an elaborate chemosensory apparatus, whereas for benthic and benthivorous species, often dwelling in turbid water, the focus of sensory selection is evidently on external taste (cf. Huber & Rylander 1992). Also in the blind cave fish (*Astyanax* spp.), the loss of vision was evidently compensated by hypertrophy of the external taste, but not the lateral line (Schemmel 1967).

Adaptation is both, a state of being and a process (Bock 1980). Therefore the present adaptive interpretation of cyprinid brain diversification also implies our present evolutionary hypothesis. The scenario is summarized by two arrows in the sensory PCA morphospace. After attaining the cyprinid brain synapomorphies, such as a distinctive brain stem (Fig. 3), both, the octavo-lateralis as well as the chemosensory brains evolved out of the pool of basic cyprinid brain morphologies. The two arrows are not intended as a statement concerning the number of evolutionary pathways, but simply indicate the direction of morphological change. We have to assume multiple and parallel evolution has occurred, as not all species (but some, see position of abramines and leuciscines, Fig. 3, cf. Pagel & Harvey 1989) within the chemosensory and octavolateralis groups are closely related.

This evolutionary hypothesis is backed by ontogeny. Abramine and chemosensory brains at least, diversify lifelong into their specific morphologies during juvenile and adult growth (Brandstätter & Kotrschal 1990) and brains of small juveniles are relatively similar to the basic cyprinid brains.

Some construction rules are size related. This applies to a changing relationship in relative sizes of the tectum opticum and the cerebellum with growth and in differently sized species (Fig. 4, 5; Brandstätter & Kotrschal 1989, 1990). In the case of the tectum opticum this allometry probably reflects release from size constraints at the level of the retina during postlarval growth (Kotrschal et al. 1990). The reason for the positive allometry of the corpus cerebelli is less clear (Brandstätter & Kotrschal 1990), but may be related to its late development in vertebrates (Finger personal communication).

Within the second set of variables, including mainly the multimodal brain areas, only a few correlations, such as between the bulbus olfactorius and the telencephalon merit functional interpretation (Table 3). The majority of correlations within this group seems to result from within brain allometries (F2/1-F2/2, Fig. 4, 5). Allometry is therefore a major source of intra- and interspecific fish brain variability.

A recent study (Brandstätter & Kotrschal 1989) on adult brain growth patterns in representatives of the four major cyprinid brain morphologies (the same as shown in Fig. 3) revealed two types of brain-internal allometries: Type 1 allometries, either positive or negative, are common to all species and affect the cerebellum and tectum opticum (Geiger 1956a), but also the telencephalon, the bulbus olfactorius and the nucleus habenularis. Type 2 allometries are species specific and affect different primary sensory areas. In roach for example, the facial lobe decreases in relative size during juvenile to adult growth, whereas the facial and vagal lobes in carp increase steadily in size.

In the present study, type 1 allometry is not only found within specific size ranges of particular species, but is also valid in terms of absolute size: as a rule of thumb, small species show a relatively large tectum opticum, but a small cerebellum and vice versa in the large species. In this respect, the brains of small species match the juvenile brains of large species.

To a lesser extent, absolute body size may also affect taste areas and thus type 2 allometries, as indicated by the restricted position of small species (bitterling, bleak, dace, minnow, schneider, streamer, sun bleak) in the sensory PCA morphospace (PC1/1-PC1/2, Fig. 3, 6). Within the chemosensory group, the small gudgeon occupies a position close to the basic cyprinid brain type (Fig. 2, 3). This

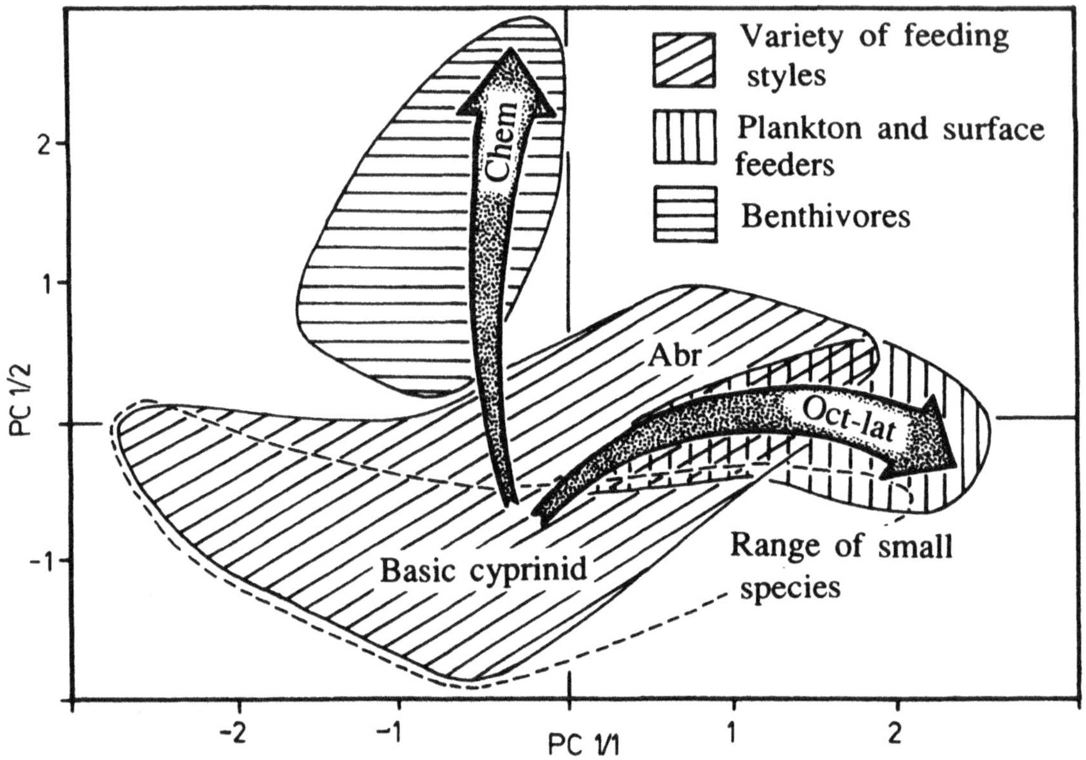

Fig. 6. Ecological (habitat and feeding style) and evolutionary interpretation superimposed over the sensory brain area PCA plot (identical with Fig. 3). Distribution of small species shown by broken lines. Arrows indicate proposed directions of ecological shifts with changing morphology from basic cyprinid brains to chemosensory (Chem) and octavo-lateralis (Oct-lat) brains. Therefore these arrows also represent our hypothesis on cyprinid neuroecological evolutionary diversification. Finally, arrows indicate the direction of brain differentiation during growth (Brandstätter & Kotrschal 1990).

suggests, that absolute body size is a major constraint for potential taste specialization.

This lack of small, taste-orientated species in temperate European freshwater may be caused by brain internal or size constraints as is the case with the retina (Kotrschal et al. 1990). However, the latter hypothesis is rendered implausible by the existence of small, taste-orientated cyprinids, such as the genus *Barbus* or *Labeo* in the Old World tropics. Geological and evolutionary histories as well as differences in the distribution of ecological factors (such as temperature, food, habitat complexity, etc.) may account for the scarcity of small, taste-oriented species in central Europe.

Ecomorphology: is brain morphology related to the life style?

Distinct relationships between brain morphologies and life styles have been found (Bauchot et al. 1989, Davis & Miller 1967, Evans 1931, 1932, 1935, 1940, Evans 1952, Kirka 1963b, Miller & Evans 1965, Peter 1979, Uchihashi 1953). However, in these qualitative studies interpretations mellowed at coarse levels and it was unclear how reliable brain structures were as predictors of ecology.

When superimposing cyprinid life styles (Balon et al. 1986, Ladiges & Vogt 1965, Maitland 1981, Schiemer 1985, 1988) on the sensory morphospace (Fig. 3, 6) consistent trends can be recognized only at the more specialized ranges of brain morphology. Octavo-lateralis brains are associated with plankton or surface feeding, whereas chemosensory brains are found in benthic and benthivorous species. Particularly the octavo-lateralis brain stem areas in combination with very small chemosensory brain stem lobes as well as the relative thickness of the stratum marginale seem excellent predictors for planktivorous or surface-feeding life styles. In the benthivores, particularly diagnostic brain areas are large brain stem taste centers and a large valvula cerebelli.

No definite correlation between life style and brain morphology is evident within the bulk of species with a basic cyprinid brain. This group contains very different ecotypes, such as the generalized roach and the piscivorous asp. Sensory specializations, e.g. the elaborate visual system in asp only become apparent at a closer inspection of sensory systems (Gomahr et al. 1992, Junger & Kotrschal 1989, Kotrschal et al. 1990, Zaunreiter & Kotrschal 1990).

Species with basic cyprinid brains, such as roach and chub may be prone to a high degree of ecological flexibility (Schiemer 1985, 1988). Generalized sensory and motor patterns may allow for flexible reactions to environmental change (Brabrand 1985, Lammens et al. 1987) and may be a major component of their success.

Two distinct tribes, the abramines including the closely related white bream and vimba and the leuciscines occupy relatively restricted areas of morphospace (Fig. 3). High within-group similarity in brain morphology may indicate relatively recent speciation. Still, lifestyles vary within these groups, spanning the range from different types of benthivory to planktivory. Thus, in contradiction to most of the recent work on fish brain ecomorphology quoted above, it has to be concluded that brain structure is not necessarily a close predictor of ecology. As is the case with fish jaws (Kotrschal 1989), and in other animal taxa (Haslett 1989), it seems that morphology at least at the present level of consideration is more conservative than behavior and ecology. It is necessary to recognize the patterns of diversification within a taxon to deduct ecological predictions at least in the morphologically more 'specialized' ranges.

In conclusion, quantitative comparative brain morphology is a powerful tool to generate novel hypotheses on brain area interactions and to investigate sensory diversification within groups of closely related fish such as the cyprinids.

Acknowledgements

This study is part of a research program of the 'Fonds zur Förderung der wissenschaftlichen Forschung in Österreich' (grant no S-35, Wieser 1986). H. Adam, A. Goldschmid, A. Lametschwandtner and W. Wieser provided logistic and technical support. Without the histological assistance by H. Junger this project would not have been possible. The 'Rechenzentrum der Universität Salzburg' provided the VAX-machine including software to analyze some of our data. K. Bernatzky and R. Hametner provided photographic expertise. The supply of specimens by A. Herzig, A. Jagsch and E. Kainz is gratefully acknowledged. Special thanks to our fisherman-in-chief, W.-D. Krautgartner, and his junior helpers. We praise the patience and generosity of Fam. Weber in Loimersdorf, who repeatedly hosted our smelly field laboratory in their neat restaurant facilities. The manuscript improved by comments of M. Braford, T. Finger, J. Kanwal and J. Motta.

References cited

Ariens Kappers, C.U., G.C. Huber & E.C. Crosby. 1936. The comparative anatomy of the nervous system of vertebrates, including man. Vol. 1–3. Macmillan, New York. 1350 pp.

Balon, E. 1968. Beitrag zur Systematik, Ökologie, Morphologie, Alter, Wachstum und Eizahl der Eier der Ziege [*Pelecus cultratus* (L.)] aus der Donau bei Medevedov. Biologické práce (Bratislava) 13: 63–88.

Balon, E.K., S.S. Crawford & A. Lelek. 1986. Fish communities of the upper Danube River (Germany, Austria) prior to the new Rhein-Main-Donau connection. Env. Biol. Fish. 15: 243–271.

Bauchot, R., M.L. Bauchot, R. Platel & J.M. Ridet. 1977. Brains of Hawaiian tropical fishes: brain size and evolution. Copeia 1977: 42–46.

Bauchot, R., J.M. Ridet & M.L. Bauchot. 1989. The brain organization of butterflyfishes. Env. Biol. Fish. 25: 205–219.

Blaxter, J.H.S. 1988. Sensory performance, behavior and ecology of fish. pp. 203–232. *In:* J. Atema, R.R. Fay, A.N.

Popper & W.N. Tavolga (ed.) Sensory Biology of Aquatic Animals, Springer Verlag, New York.

Bleckmann, H. 1988. Prey identification and prey localization in surface-feeding fish and fishing spiders. pp. 619–641. *In:* J. Atema, R.R. Fay, A.N. Popper & W.N. Tavolga (ed.) Sensory Biology of Aquatic Animals, Springer Verlag, New York.

Bock, W.J. 1980. The definition and recognition of biological adaptation. Amer. Zool. 20: 217–227.

Brabrand, A. 1985. Food of roach (*Rutilus rutilus*) and ide (*Leuciscus idus*): significance of diet shifts for interspecific competition in omnivorous fishes. Oecologia 66: 461–467.

Brandstätter, R. & K. Kotrschal. 1989. Life history of roach, *Rutilus rutilus* (Cyprinidae, Teleostei): a qualitative and quantitative study on the development of sensory brain areas. Brain, Behav. Evol. 34: 35–42.

Brandstätter, R. & K. Kotrschal. 1990. Brain growth patterns in four European cyprinid fish species (Cyprinidae, Teleostei): roach (*Rutilus rutilus*), bream (*Abramis brama*), common carp (*Cyprinus carpio*) and sabre carp (*Pelecus cultratus*). Brain, Behav. Evol. 35: 195–211.

Brosius, G. 1989. Spss/pc. Advanced statistics and tables. McGraw Hill, Hamburg. 350 pp.

Bullock, T.H. 1983. Why study fish brains? pp. 361–368. *In:* R.E. Davis & R.G. Northcutt (ed.) Fish Neurobiology, The University of Michigan Press, Ann Arbor.

Davis, B.J. & R.J. Miller. 1967. Brain patterns in minnows of the genus *Hybopsis* in relation to their feeding habits. Copeia 1987: 1–39.

Davis, R.E. & R.G. Northcutt (ed.). 1983. Fish neurobiology. Vol. 2. The University of Michigan Press, Ann Arbor. 370 pp.

Echteler, S. 1985. Organization of central auditory pathways in a teleost fish, *Cyprinus carpio*. J. Comp. Physiol. A. 156: 267–280.

Evans, H.M. 1931. A comparative study of the brains in British cyprinoids in relation to their habits of feeding, with special reference to the anatomy of the medulla oblongata. Proc. R. Soc. Lond. B 108: 233–257.

Evans, H.M. 1932. Further observations on the medulla oblongata in cyprinoids; and a comparative study of the medulla of clupeoids and cyprinoids with special reference to the acoustic tubercles. Proc. R. Soc. Lond. B 111: 247–280.

Evans, H.M. 1935. The brain of *Gadus,* with special reference to the medulla oblongata and its variation according to feeding habits of different Gadidae-I. Proc. R. Soc. B. 117: 56–68.

Evans, H.M. 1940. Brain and body of fish. A study of brain pattern in relation to hunting and feeding in fish. Technical Press, London. 224 pp.

Evans, H.E. 1952. The correlation of brain patterns and feeding in four species of cyprinid fishes. J. Comp. Neurol. 97: 133–142.

Finger, T.E. 1978. Gustatory pathways in the bullhead catfish. II. Facial lobe connections. J. Comp. Neurol. 180: 691–706.

Finger, T.E. 1983a. The gustatory system in teleost fish. pp. 285–311. *In:* R.G. Northcutt & R.E. Davis (ed.) Fish Neurobiology, Vol. 1, The University of Michigan Press, Ann Arbor.

Finger, T.E. 1983b. Organization of the teleost cerebellum. pp. 261–284. *In:* R.G. Northcutt & R.E. Davis (ed.) Fish Neurobiology, Vol. 1, The University of Michigan Press, Ann Arbor.

Finger, T.E. 1987. Gustatory nuclei and pathways in the central nervous system. pp. 331–354. *In:* T.E. Finger & W.L. Silver (ed.) Neurobiology of Taste and Smell, J. Wiley Publishers, New York.

Finger, T.E. 1988. Organization of the chemosensory systems within the brains of bony fishes. pp. 339–363. *In:* J. Atema, R.R. Fay, A.N. Popper & W.N. Tavolga (ed.) Sensory Biology of Aquatic Animals, Springer Verlag, New York.

Geiger, W. 1956a. Quantitative Untersuchungen über das Gehirn der Knochenfische mit besonderer Berücksichtigung seines relativen Wachstums. I. Acta anat. 26: 121–163.

Geiger, W. 1956b. Quantitative Untersuchungen über das Gehirn der Knochenfische mit besonderer Berücksichtigung seines relativen Wachstums. II. Acta anat. 27: 324–350.

Goldschmid, A. & K. Kotrschal. 1989. Ecomorphology: developments and concepts. Progr. Zool. 35: 501–512.

Gomahr, A., M. Palzenberger & K. Kotrschal. 1992. Density and distribution of external taste buds in cyprinids. Env. Biol. Fish. 33: 125–134.

Haslett, J.R. 1989. Adult feeding by holometabolous insects: pollen and nectar as complementary nutrient sources for *Rhingia campestris* (Diptera: Syrphidae). Oecologia 81: 361–363.

Huber, R. & M.K. Rylander. 1992. Brain morphology and turbidity preference in *Notropis* and related genera (Cyprinidae, Teleostei). Env. Biol. Fish. 33: 153–165.

Herrick, C.J. 1905. The central gustatory paths in the brains of bony fishes. J. Comp. Neurol. Psychol. 15: 375–456.

Ito, H. & R. Kishida. 1978. Afferent and efferent fiber connections of the carp torus longitudinalis. J. Comp. Neurol. 181: 465–476.

Jolicoeur, P. & G. Baron. 1980. Brain center correlations among Chiroptera. Brain, Behav. Evol. 17: 419–431.

Junger, H. & K. Kotrschal. 1989. Developmental changes in the optic tracts of cyprinids (Cyprinidae, Teleostei). p. 191. *In:* N. Elsner & W. Singer (ed.) Dynamics and Plasticity in Neuronal Systems, Proceedings of the 17th Göttingen Neurobiology Conference, Georg Thieme Verlag, Stuttgart.

Kaiser, H.F. 1974. An index of factorial simplicity. Psychometrica 39: 31–36.

Khanna, S.S. & H.R. Singh. 1966. Morphology of the teleostean brain in relation to feeding habits. Proc. Nat. Acad. Sci. India 336: 306–316.

Kirka, A. 1963a. Brain structures in the Danube representatives of perciformes and the role of brain study for the systematic and ecology of fishes. Zool. Zhurnal 42: 400–407 (in Russian).

Kirka, A. 1963b. Die äußere Gehirnform des Donaunerflings (*Rutilus pigus virgo* He.) und ein Vergleich seines Gehirns mit der Donauplötze (*Rutilus rutilus carpathorossicus* Vladykov) und der Elbeplötze (*Rutilus rutilus frici* Misik). Zeitschrift für Fischerei und deren Hilfswissenschaften 11: 129–141.

Kishida, R. 1979. Comparative study on the teleostean optic tectum. J. Hirnforsch. 20: 57–67.

Kotrschal, K. 1989. Trophic ecomorphology in eastern Pacific blennioid fishes: character transformation of oral jaws and associated change of their biological roles. Env. Biol. Fish. 24: 199–218.

Kotrschal, K. & H. Junger. 1988. Patterns of brain morphology in mid-European Cyprinidae (Pisces, Teleostei): a quantitative histological study. J. Hirnforsch. 29: 341–352.

Kotrschal, K., H. Adam, R. Brandstätter, H. Junger, M. Zaunreiter & A. Goldschmid. 1990. Larval size constraints determine directional ontogenetic shifts in the visual system of teleosts. A mini-review. Z. Zool. Syst. Evolutionsforsch. 28: 166–182.

Kotrschal, K., R. Brandstätter, A. Gomahr, H. Junger, M. Palzenberger & M. Zaunreiter. 1991. Brain and sensory systems. *In:* I.J. Winfield & J.S. Nelson (ed.) Cyprinid Fishes, Systematics, Biology and Exploitation, Chapman and Hall, London.

Ladiges, W. & D. Vogt. 1965. Die Süßwasserfische Europas. Paul Parey, Hamburg. 299 pp.

Lammens, E.H.R.R., J. Geursen & P.J. McGillavry. 1987. Diet shifts, feeding efficiency and coexistence of bream (*Abramis brama*), roach (*Rutilus rutilus*) and white bream (*Blicca bjoercna*) in hypertrophic lakes. pp. 153–162. *In:* S. Kullander & B. Fernholm (ed.) Proceedings of the V Congress of European Ichthyologists, Stockholm.

Lythgoe, J.N. 1988. Light and vision in the aquatic environment. pp. 57–82. *In:* J. Atema, R.R. Fay, A.N. Popper & W.N. Tavolga (ed.) Sensory Biology of Aquatic Animals, Springer Verlag, New York.

Maitland, P.S. 1981. Freshwater fishes of Britain and Europe. Hamlyn, London. 256 pp.

Maynard Smith, J., R. Burian, S. Kauffman, P. Alberch, J. Campbell, B. Goodwin, R. Lande, D. Raup & L. Wolpert. 1985. Developmental constraints and evolution. Quart. Rev. Biol. 60: 260–287.

Mayser, P. 1881. Vergleichend anatomische Studien über das Gehirn der Knochenfische mit besonderer Berücksichtigung der Cyprinoiden. Arch. wiss. Zool. 36: 259–366.

McCormick, C.A. & M.R. Braford, Jr. 1988. Central connections of the octavolateralis system: evolutionary considerations. pp. 733–756. *In:* J. Atema, R.R. Fay, A.N. Popper & W.N. Tavolga (ed.) Sensory Biology of Aquatic Animals, Springer Verlag, New York.

Miller, R.J. & H.E. Evans. 1965. External morphology of the brains and lips in catastomid fishes. Copeia 1965: 467–487.

Montgomery, J.C. & J.A. Macdonald. 1987. Sensory tuning of lateral line receptors in Antarctic fish to the movement of planktonic prey. Science 235: 195–196.

Morita, Y. & T.E. Finger. 1985a. Topography and laminar organization of the vagal gustatory system in the goldfish, *Carassius auratus*. J. Comp. Neurol. 238: 187–201.

Morita, Y. & T.E. Finger. 1985b. Reflex connections of the facial and vagal gustatory systems in the brain stem of the bullhead catfish, *Ictalurus nebulosus*. J. Comp. Neurol. 231: 547–558.

Morita, Y. & T.E. Finger. 1987. Topographic representation of the sensory and motor roots of the vagus nerve in the medulla of goldfish, *Carassius auratus*. J. Comp. Neurol. 264: 231–249.

Morita, Y. & H. Masai. 1980. Central gustatory paths in the crucian carp, *Carassius carassius*. J. Comp. Neurol. 191: 119–132.

Nelson, J.S. 1984. Fishes of the world. J. Wiley & Sons, New York. 523 pp.

Northcutt, R.G. 1988. Sensory and other neural traits and the adaptationist program: mackerels of San Marco? pp. 869–883. *In:* J. Atema, R.R. Fay, A.N. Popper & W.N. Tavolga (ed.) Sensory Biology of Aquatic Animals, Springer Verlag, New York.

Northcutt, R.G. & R.E. Davis (ed.). 1983. Fish neurobiology, Vol. 1. The Michigan University Press, Ann Arbor. 345 pp.

Northcutt, R.G. & M.F. Wullimann. 1988. The visual system in teleost fishes: morphological patterns and trends. pp. 515–552. *In:* J. Atema, R.R. Fay, A.N. Popper & W.N. Tavolga (ed.) Sensory Biology of Aquatic Animals, Springer Verlag, New York.

Northmore, D.P.M. 1984. Visual and saccadic activity in the goldfish torus longitudinalis. J. Comp. Physiol. A 155: 333–340.

Northmore, D.P.M., B. Williams & H. Vanegas. 1983. The teleostean torus longitudinalis: responses related to eye movements, visuotopic mapping, and functional relations with the optic tectum. J. Comp. Physiol. A 150: 39–50.

Pagel, M.D. & P.H. Harvey. 1989. Taxonomic differences in the scaling of brain on body weight among mammals. Science 244: 1589–1593.

Peter, R.E. 1979. The brain and feeding behavior. pp. 121–159. *In:* Hoar & Randall (ed.) Fish Physiology, Vol. 8, Academic Press, New York.

Pirlot, P. & P. Jolicoeur. 1982. Correlations between major brain regions in Chiroptera. Brain. Behav. Evol. 20: 172–181.

Ridet, J.M., R. Bauchot, M. Diagne & R. Platel. 1977. Croissance ontogenetique et phylogenetique de l'encephale des Teleosteens. Cahiers Biol. Marine 18: 163–176.

Schemmel, C. 1967. Vergleichende Untersuchungen an den Hautsinnesorganen ober- und unterirdisch lebender *Astyanax*-Formen. Z. Morph. Tiere 61: 255–305.

Schiemer, F. 1985. Die Bedeutung der Augewässer als Schutzzonen für die Fischfauna. Österr. Wasserwirtschaft 37: 239–245.

Schiemer, F. 1988. Gefährdete Cypriniden – Indikatoren für die ökologische Intaktheit von Flußsystemen. Natur und Landschaft 63: 370–373.

Schnitzlein, H.N. 1964. Correlation of habit and structure in the fish brain. Amer. Zool. 4: 21–32.

Sibbing, F.A. 1988. Specializations and limitations in the utilization of food resources by the carp, *Cyprinus carpio:* a study of oral food processing. Env. Biol. Fish. 22: 161–178.

Simpson, G.G. 1944. Tempo and mode in evolution. Columbia University Press, New York. 420 pp.
Snow, J.L. & M.K. Rylander. 1982. A quantitative study of the optic system of butterflyfishes (family Chaetodontidae). J. Hirnforsch. 23: 121–125.
Sokal, R.R. & F.J. Rohlf. 1981. Biometry, 2nd ed. Freeman, San Francisco. 859 pp.
Stephan, H. 1960. Methodische Studien über den quantitativen Vergleich architektonischer Struktureinheiten des Gehirns. A. wiss. Zool. 164: 143–172.
Stephan, H. 1967. Quantitative Vergleiche zur phylogenetischen Entwicklung des Gehirns der Primaten mit Hilfe der Progressionsindices. Mitt. Max-Planck Ges. 2: 63–68.
Stephan, H. & P. Pirlot. 1970. Volumetric comparison of brain structures in bats (an attempt to a phylogenetic interpretation). Bijdr. Dierkunde 40: 95–98.
Uchihashi, K. 1953. Ecological study of Japanese teleosts in relation to brain morphology. Bull. Jap. Sea Regional Fish. Res. Lab. 2: 1–166.
Vanegas, H. 1983. Organization and physiology of the teleostean optic tectum. pp. 43–87. *In:* R.E. Davis & R.G. Northcutt (ed.) Fish Neurobiology, The University of Michigan Press, Ann Arbor.
Werner, E.E. 1984. The mechanisms of species interactions and community organization in fishes. pp. 360–383. *In:* D.R. Strong, D. Simberloff, L.G. Abele & A.B. Thistle (ed.) Ecological Communities, Conceptual Issues and Evidence, Princeton University Press, Princeton.
Wieser, W. 1986. Die Ökophysiologie der Cyprinidenfauna österreichischer Gewässer. Ein Forschungsschwerpunkt des Fonds zur Förderung der wissenschaftlichen Forschung. Österreichs Fischerei 36: 88–93.
Winkelmann, E. & L. Winkelmann. 1968. Vergleichend histologische Untersuchungen zur funktionellen Morphologie des Tectum opticum verschiedener Teleostier. J. Hirnforsch. 10: 1–16.
Wullimann, M.F. & R.G. Northcutt. 1989. Afferent connections of the valvula cerebelli in two teleosts, the common goldfish and the green sunfish. J. Comp. Neurol. 289: 554–567.
Zaunreiter, M. & K. Kotrschal. 1989. Shifting retinal parameters during growth in roach (*Rutilus rutilus*, Cyprinidae, Teleostei). p. 192. *In:* N. Elsner & W. Singer (ed.) Dynamics and Plasticity in Neuronal Systems, Proceedings of the 17th Göttingen Neurobiology Conference, Georg Thieme Verlag, Stuttgart.
Zaunreiter, M., H. Junger & K. Kotrschal. 1991. Ecomorphology of the cyprinid retina: a quantitative histological study on ontogenetic shifts and interspecific variation. Vision. Res. 31: 383–394.

Environmental Biology of Fishes **33**: 153–165, 1992.

Brain morphology and turbidity preference in *Notropis* and related genera (Cyprinidae, Teleostei)

Robert Huber[1, 2] & Michael K. Rylander[1]
[1] *Department of Biological Sciences, Texas Tech University, Lubbock, TX 79409, U.S.A.*
[2] *Department of Neurobiology, Harvard Medical School, 220 Longwood Avenue, Boston, MA 02115, U.S.A.*

Received 21.1.1990 Accepted 6.7.1990

Key words: Neuroanatomy, Ecology, Vision, Olfaction, Gustation, Plasticity, Adaptation, Evolution

Synopsis

The size of seven neural structures was compared in 51 species of *Notropis, Pteronotropis, Cyprinella, Luxilus, Lythrurus,* and *Hybopsis,* and related to the turbidity of the species' habitat. This last parameter was assessed for each species by personal communication with 42 ichthyologists. To control for size differences among species, all analyses were performed on the residuals from a regression of each character on standard length. Principal components analysis (PCA) of the residuals produced four significant PC-axes that together explained 65% of the total variation represented in the original variables. The size of brain structures concerned with vision, olfaction, and gustation was correlated with habitat turbidity. Two-way Analyses of Covariance (ANCOVAs) revealed significant differences between species in the size of all structures. Sexual dimorphism was found in the size of the olfactory bulb and the cerebellum, and significant two-way interactions (species vs. sex) were detected for the telencephalon, optic lobes, cerebellum, vagal lobe, and the eye. Cluster analysis indicated that neither similar turbidity preference nor shared phylogeny is alone sufficient to explain the observed differences in brain morphology.

Introduction

Parallel changes in the size of specific brain parts and ecological adaptations have been demonstrated in many vertebrate taxa, ranging from fish (e.g. Kishida 1979, Kotrschal & Junger 1988) to bats (Baron & Jolicoeur 1980) and primates (see Jolicoeur et al. 1983). Fish are particularly useful for comparative studies because the primary targets of sensory modalities are distinct brain divisions which can be measured easily in the intact brain (e.g. the facial and vagal lobe for taste, the optic lobe for vision, and the olfactory bulb for smell). Minnows of the genus *Notropis* and related genera are especially useful for such studies because this group comprises a large number of closely related species that occupy a variety of diverse habitats.

When investigating the relationship between brain morphology and ecological parameters one presumes that the size of a specific neural structure is related to its functional potential. Such a relationship is evident, for example, in zebra finches where differences in song performance are paralleled by changes in the size and neuronal characteristics of brain nuclei controlling song (Nottebohm & Arnold 1976, Arnold 1980, Nottebohm 1980). To identify correlations between the size of specific brain parts and ecological parameters the primary structures for the senses should be most useful, as they generally have fewer functions than those as-

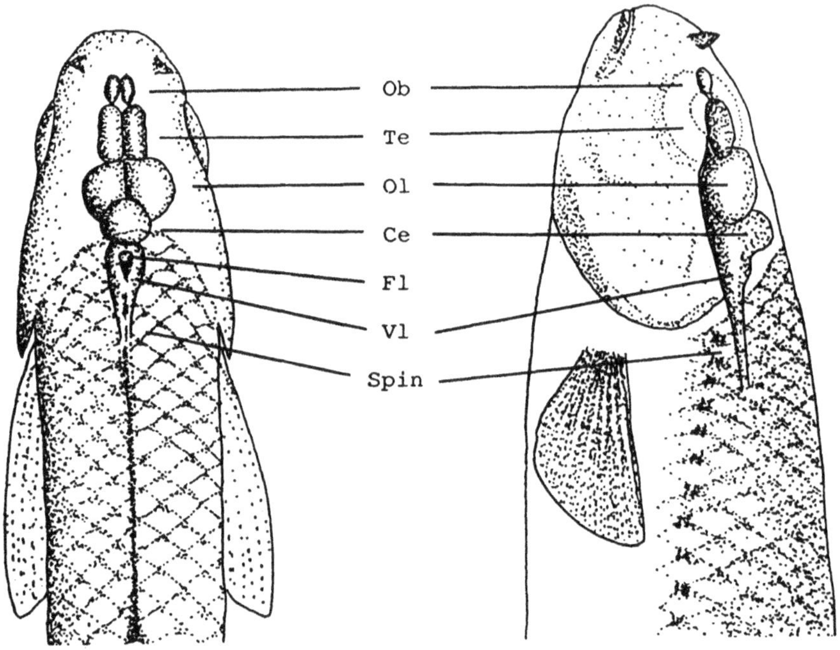

Fig. 1. Dorsal (left) and lateral (right) view of the brain of *Notropis stramineus*. Abbreviations are olfactory bulb (Ob), telencephalon (Te), optic lobes (Ol), cerebellum (Ce), facial lobe (Fl), vagal lobe (Vl), and spinal chord (Spin).

sociated with complex behaviors, such as learning or courtship. We have hypothesized that, in clear water, vision is superior to all other sensory modalities for evaluating the environment and that with increasing turbidity species rely more on other senses, such as taste, hearing, smell or touch. Therefore, we predict that species adapted to turbid environments have relatively smaller structures associated with vision, and relatively larger brain parts associated with taste, hearing, olfaction or taste.

Teleost olfaction, which is routed to the olfactory bulb (Fig. 1), is used primarily for social communication, and in most species is of little or no use in foraging (Finger 1988). The telencephalon appears to be important in the integration of olfactory as well as visual and gustatory stimuli (Friedlander 1983, Davis & Kassel 1983), as well as for many more complex tasks such as learning, and agonistic or social behaviors (Demski 1983). Eyes and optic lobes are intimately related to vision, the latter being the primary target of retinal fibers. The cerebellum plays a central role in proprioception, motor coordination, and eye movement (Demski 1983). Taste in fish is used both in locating food and in evaluating it once found. Stimuli are relayed primarily to structures in the medulla oblongata, specifically sensory nuclei of the nervus glossopharyngeus (IX), n. facialis (VII), and n. vagus (X). The last two form prominent lobes in minnows. The main neural structure involved in locating food is the facial lobe, innervating taste buds on the general body surface and the front portion of the mouth cavity. Information about the quality of a food item is derived from the glossopharyngeal lobe, which innervates taste buds in the back of the mouth and the first gill slit, and the vagal lobe, which supplies taste buds in the remaining gill slits, as well as the palatal and sublingual organs (Kanwal & Caprio 1987, Gomahr et al. 1988).

This study investigates whether the relative size of the eye, olfactory bulb, telencephalon, optic tectum, cerebellum, facial lobe, and vagal lobe in 51 species of minnows is related to the level of turbidity in the preferred habitat of each species. Next, the size of these brain structures is compared between sexes and among species. Finally, species are clustered according to brain morphology to determine whether such an arrangement more closely approximates an ecological or phylogenetic classification.

Materials and methods

Although the genus *Notropis* was once thought to contain just under one half of all nearctic cyprinids (Coburn 1982), it recently has been revised extensively by Mayden (1985, 1990) whose nomenclature is followed herein.

The validity of this study depends on accurately classifying the species according to their habitat. The determination of the physical parameters of the preferred habitat is a difficult task, as these parameters vary in accordance with the season and the weather. This necessitates a large number of fish collections and measurements of water clarity throughout the range of the species which must span all seasons and include several years. At best, such data are available for only a few of the 51 species included in this study (cf. Felley & Hill 1983) and it would be unrealistic to empirically determine the habitat turbidity of the remainder. As a more economic approach, we asked 42 ichthyologists who have extensive field experience with this group of fish, to identify the turbidity levels in which 90% of individuals of each species might be expected to occur. The categories were very clear (1), clear (2), average (3), turbid (4), and very turbid (5). The responses for a particular species were averaged to obtain an estimate of the mean turbidity (MT) of that species.

Only adult specimens, a total of 476 (4–24 per species), were examined to avoid the potential effects of ontogenetic changes (e.g. developmental constraints, allometry, and changes in the functional importance of sensory modalities during development). Total and standard length (tip of the snout to base of caudal fin) were measured to the nearest 0.5 mm, and weight to the nearest 0.001 g. Sex was determined by examining the gonads. Under a dissecting microscopc, the skull was opened dorsally and the brain exposed from olfactory bulbs to the rostral end of the spinal cord. The length and width of olfactory bulb, telencephalon, optic tectum, cerebellum, facial lobe, and vagal lobe were measured to the nearest 0.05 mm, using 10 × magnification and an ocular grid. The diameter of the eye was measured along the nasal-temporal axis to the nearest 0.05 mm. In the case of paired structures, measurements were made on the right side except in a few cases where this side was damaged. Although procedures for the preservation of specimens, such as fixation in formalin or storage in ethanol, result in tissue shrinkage, shrinkage was assumed to be uniform across brain areas. As the general shape of the brain was very similar in all species, the size of each structure was estimated as the product of length and width.

Comparative neuroanatomical studies have commonly suffered from the difficulty of comparing portions of the brain in specimens of different size (Geiger 1956a, b, Stephan 1960, Kotrschal & Junger 1988). Considerable controversy has been raised in the literature concerning the effects of body size and methods for an adequate removal (see Sneath & Sokal 1973). As all species included in this study are similar in size (ranging in length from 4–9 cm) and shape, such problems are reduced. However, as a precaution, all subsequent analyses were performed on the residuals of a linear regression on standard length, either directly or by the use of Analyses of Covariance (ANCOVAs). Standard length, rather than body weight, was chosen to represent body size, because it is less influenced by physiological and nutritional factors.

A principal components analysis (PCA) was performed on the log-transformed residuals (Procedure FACTOR, SPSSX Inc. 1988) to determine the main factors subsuming the variation in brain morphology. These residuals represent 'size-free' neuroanatomical variation. Pearson's correlation coefficients (Procedure CORRELATION, SPSSX Inc. 1988) were calculated between the scores on the significant PC-axes with Eigenvalues ≥ 1.00, and the mean turbidity. For each species, means for the first three PC-axes were plotted (Proc G3D, SAS Institute Inc. 1985) to display the dispersion of different sensory modalities among species with different turbidity preferences. Two-way ANCOVAs with species and sex as treatment effects, and standard length as covariate (Procedure MANO-

VA, SPSSX Inc. 1988), were performed on the size of each structure measured, permitting the evaluation of differences in brain morphology among species, between sexes, or as a result of an interaction between species and sex. Sufficient sample sizes of each sex were obtained for 28 species (cf. Table 2). To achieve homogeneity of variances all variables (unless otherwise stated) were subjected to the Box-Cox transformation (Sokal & Rohlf 1981). A sine transformation was applied to telencephalon, and cosine transformations were applied to the optic lobe and eye. Finally, cluster analysis (Procedure CLUSTER, SPSSX Inc. 1988) was performed on the species averages of the transformed data. Similarities were based upon the squared Euclidian distance derived from the standardized variables, and clusters were assembled by the unweighted pair-group method using arithmetic averages (UPGMA).

Material examined

For each species the museum acronym and number is listed first, followed by the number of specimens in parenthesis, and the location and the year of collection. Institutional acronyms are: University of Kansas (KU); Fort Hayes State University (FHSU); William J. Matthews (WJM); Jeffrey R. Bek and Eric M. Surat (BS); A.J. Gatz (AJG); own collection (RH).

Genus *Pteronotropis: P. euryzonus* KU 21905 (5), unknown location AL, 1987.

Genus *Cyprinella: C. analostana* KU 15367 (5), Alamance Creek NC, 1967; AJG, (4), Maho Creek DE, year unknown. *C. callistia* KU 20483 (5), Crooked Cr. AL, 1982. *C. camura* KU 15821 (5), Spring River KS, 1974; FHSU 150 (5), Shoal Creek KS, 1966. *C. galactura* KU 7866 (5), James River MO, 1963; WJM 1285 (5), Piney Creek AR, 1982. *C. lepida* WJM 1661 (5), Nueces River TX, 1984. *C. lutrensis* RH (10), Brazos River TX, 1987. *C. ornata* KU 4446 (5), Rio Papigochic Mex, 1958. *C. proserpina* WJM 362 (5), Devils River TX, 1978. *C. spiloptera* KU 8628 (5), Wapsipinicon River IA, 1964. *C. venusta* RH (5), Main Llano River TX, 1987; RH (5), Pennington Creek OK, 1987. *C. whipplei* KU 18088 (5), Big River MO, 1975; WJM 835 (3), Kiamichi River OK, 1981; WJM 836 (2), Kiamichi River OK, 1981.

Genus *Luxilus: L. albeolus* WJM 258 (10), Dixie Caverns Channel VA, 1978. *L. cardinalis* KU 21284 (5), Spring Creek OK, 1980. *L. cerasinus* BS 76 (10), Roanoke River VA, 1979. *L. chrysocephalus* RH (10), Blue Creek OK, 1987. *L. cornutus* KU 3758 (5), Mill Creek kS, 1957; FHSU 1852 (5), Byram River CN, 1966. *L. pilsbryi* KU 8023 (5), Mill Creek AR, 1964; WJM 1281 (5), Mill Creek AR, 1982. *L. zonatus* KU 19091 (5), Roubideaux Creek MO, 1981.

Genus *Lythrurus: Ly. ardens* KU 11614 (5), Pitman Creek KY, 1966; AJG (6), Maho Creek DE, year unknown. *Ly. fumeus* WJM 562 (10), Old River Slough AR, 1980. *Ly. roseipinnis* KU 16857 (7), Bayou Bacon MS, 1976. *Ly. umbratilis* KU 15562 (5), Diamond Creek KS, 1984; FHSU 126 (5), Shoal Creek KS, 1966.

Genus *Notropis: N. amabilis* KU 2144 (5), Guadalupe River TX, 1951. *N. atherinoides* WJM 2300 (10), North Canadian River OK, 1987. *N. baileyi* KU 14537 (5), Mill Creek AL, 1969. *N. bairdi* KU 3400 (10), Wichita River TX, 1955. *N. boops* RH (10), Pennington Creek OK, 1987. *N. buchanani* KU 18449 (2), Spring River KS, 1980. *N. chihuahua* WJM 1339 (5), Terlingua Creek TX, 1983. *N. chiliticus* WJM (4), Dan River VA, 1979. *N. girardi* KU 2330 (5), Canadian River OK, 1952. *N. greenei* KU 6632 (5), Illinois River AR, 1960. *N. heterodon* KU 14255 (5), Mississippi River MN, 1968. *N. hudsonius* FHSU 1849 (10), Shetucket River CN, 1969. *N. nubilis* KU 20818 (5), Flat Creek MO, 1983; WJM 1284 (5), Piney Creek AR, 1982. *N. ozarcanus* KU 12656 (5), White River MO, 1967. *N. petersoni* KU 17161 (5), Alexander Springs FL, 1976. *N. potteri* WJM 474 (5). Lake Texoma OK, 1980; WJM 485 (3), Lake Texoma OK, 1980. *N. procne* BS 14 (1), Bottom Creek VA, 1978; BS 59 (5), Bottom Creek Pool VA, 1978. *N. rubellus* KU 18441 (5), Shoal Creek KS, 1980; FHSU 132 (5), Spring River KS, 1966. *N. snelsoni* WJM (7), Cucumber Creek OK, ?. *N. stramineus* WJM 2268 (10), Crutcho Creek OK, 1987. *N. telescopus* KU 5621 (5), Current River MO, 1960; WJM 1285 (5), Piney Creek AK, 1982. *N. texanus* WJM 562 (10),

Old River Slough AR, 1980. *N. topeka* FHSU 293 (10), Mill Creek KS, 1968. *N. volucellus* KU 10584 (5), Lower Cullen Lake MN, 1965; WJM 562 (5), Old River Slough AR, 1980. *N. xaenocephalus* KU 20268 (5), Hillabee Creek AL, 1982.

Genus *Hybopsis: H. dorsalis* KU 4839 (5), North Platte River NA, 1959; WJM (5), Big Creek MO, 1974. *H. longirostris* KU 8799 (5), Upatoi Creek GA, 1964. *H. sabinae* WJM 1281 (10), Mill Creek AR, 1982.

undescr. sp. WJM 2345 (5), Blue River OK, 1988.

Results

The MTs and coefficients of variation for each species are listed along with the phylogenetic classification in Table 1. According to this survey, species of the genera *Notropis, Pteronotropis, Cyprinella, Luxilus,* and *Lythrurus* generally occupy habitats ranging from very clear to moderately turbid.

Species means for all measured morphological variables are listed in Tables 2 and 3. PCA of the regression residuals produced four PC-axes, each with an Eigenvalue ≥ 1, cumulatively accounting for 64.8% of the total variation among individuals. Different sensory structures showed high loadings (Table 4) on different PC-axes. Visual structures and the cerebellum loaded most highly on PC1, the olfactory bulb on PC2, the facial lobe on PC3, and the vagal lobe on PC4. PCA separates the species according to the relative importance of vision, smell, taste on general body surface, and taste in the back of the mouth, respectively.

A significant correlation existed between MT and each of the first three PC-axes, indicating that brain morphology is intimately related to the turbidity level where the species is found. The correlation between the turbidity of the preferred habitat and PC1 was negative ($r = -0.372$, $N = 48$, $0.001 \leq p < 0.01$), demonstrating that the size of structures with high loadings on this axis, specifically the primary optic structures and the cerebellum, are larger in clear water species. A positive association between PC2 ($r = 0.439$, $N = 48$, $0.001 \leq p < 0.01$) and PC3 ($r = 0.285$, $N = 48$, $0.01 \leq p < 0.05$) and the preferred turbidity suggests the olfactory bulb and the facial lobe are larger in turbid water species. No significant correlation was found between PC4 and mean turbidity ($r = 0.033$; $N = 48$, $p > 0.05$). Species means of the scores on the first three PCA-axes are plotted in three dimensions in Figure 2.

The results of two-way ANCOVAs of the size of the olfactory bulb, telencephalon, optic lobe, cerebellum, facial lobe, vagal lobe, and eye indicate that the size of these structures differed significantly ($p < 0.001$) among species (Table 5). The olfactory bulb was larger in males ($0.01 \leq p < 0.05$) and the cerebellum was larger in females ($0.001 \leq p < 0.01$). Significant interactions ($p < 0.001$) were found in the size of the telencephalon, optic lobes, cerebellum, and eye, as well as the vagal lobe ($0.01 \leq p < 0.05$).

The dendrogram produced by the UPGMA cluster analysis indicated that differences in brain size cannot be explained sufficiently by either MT or phylogenetic relatedness alone. The existence of several clusters (Fig. 3) provides evidence for similarities in brain morphology based on common descent. The brain of *C. lepida* proved very similar to its close relative *C. lutrensis* (A), and most species of *Luxilus* (B) and *Lythrurus* (C) clustered together, regardless of MT. On the other hand, brains from several less closely related species with similar MTs clustered together, such as in the case of *N. boops, L. chrysocephalus* and *C. venusta* (D). In addition, species of *Cyprinella* are spread across the entire dendrogram. *L. zonatus* and *L. pilsbryi,* as well as *L. cornutus* and *L. chrysocephalus,* clustered far apart, even though they are closely related phylogenetically and are found in similar habitat turbidity.

Discussion

The correlations between brain structure and habitat turbidity reported here are in accord with other studies that relate brain morphology and ecological parameters in fish (e.g. Miller & Evans 1965, Davis & Miller 1967, Rao 1967, Kishida 1979, Kotrschal

Table 1. Results of a survey of 42 ichthyologists to estimate habitat turbidity of 100 species of *Notropis* and related genera. Averages (mean turbidity, MT) range from 1 to 5, with 1 representing very clear water and 5 very turbid water. N indicates the number of responses for a particular species, and CV variation in opinions expressed.

Phylogenetic classification	N	MT	CV	Phylogenetic classification	N	MT	CV
Genus Pteronotropis:				N. stramineus	23	3.13	0.23
P. hypselopterus	2	2.00	–	N. atrocaudalis	8	2.75	0.38
P. signipinnis	3	2.17	0.35	N. mekistocholas	3	2.17	0.35
P. welaka	5	2.40	0.52	N. procne	8	2.94	0.14
P. hubbsi	6	2.42	0.38	N. braytoni	2	2.50	–
Genus Cyprinella:				N. chihuahua	4	2.75	0.32
C. lutrensis	27	3.52	0.16	subgenus Hydrophlox:			
C. formosa	2	2.25	–	N. rubellus	27	2.22	0.33
C. lepida	2	1.75	–	N. rubricroceus	5	1.60	0.26
C. proserpina	4	1.63	0.30	N. chiliticus	8	2.44	0.23
C. spiloptera	17	2.79	0.24	N. baileyi	6	2.08	0.28
C. camura	13	2.46	0.26	N. chlorocephalus	5	1.90	0.29
C. whipplei	21	2.52	0.28	N. lutipinnis	8	2.00	0.33
C. analostana	9	2.72	0.19	N. leuciodus	10	2.10	0.29
C. chloristia	3	3.33	0.09	N. nubilus	10	1.65	0.25
C. venusta	19	2.82	0.25	N. chrosomus	4	1.63	0.39
C. galactura	15	2.00	0.30	texanus species group:			
C. pyrrhomelas	4	1.50	0.27	N. petersoni	7	2.93	0.18
C. xaenura	1	3.00	–	N. heterodon	5	1.50	0.33
C. caerulea	3	1.33	0.22	N. anogenus	2	1.75	–
C. trichroistia	4	2.13	0.30	N. topeka	2	2.50	–
C. gibbsi	1	2.00	–	N. ortenburgeri	7	2.36	0.34
C. callistia	5	2.40	0.27	N. altipinnis	7	2.71	0.18
C. nivea	6	2.75	0.22	N. cummingsae	8	2.19	0.24
C. leedsi	2	3.25	–	N. texanus	10	2.75	0.29
C. callitaenia	1	3.00	–	N. chalybaeus	17	2.77	0.26
Genus Luxilus:				N. boops	24	1.94	0.38
L. zonatus	8	1.63	0.22	N. xaenocephalus	3	1.67	0.17
L. pilsbryi and				subgenus Notropsis:			
L. cardinalis	16	1.69	0.26	N. candidus			
L. cornutus	19	2.47	0.27	N. shumardi	10	3.50	0.19
L. chrysocephal.	22	2.41	0.23	N. oxyrhynchus	3	3.67	0.16
L. albeolus	9	2.33	0.24	N. jemezanus	5	3.10	0.29
L. coccogenis	8	2.13	0.27	N. atherinoides	24	3.29	0.17
L. cerasinus	7	2.21	0.18	N. perpallidus	8	2.44	0.20
Genus Lythrurus:				N. amabilis	5	1.90	0.66
L. fumeus	14	2.71	0.30	N. amoenus	5	2.90	0.23
L. lirus	3	2.17	0.35	N. stilbius	6	2.17	0.12
L. ardens	14	2.43	0.17	N. photogenis	11	2.46	0.28
L. umbratilis	24	2.85	0.22	N. telescopus	13	1.89	0.27
L. roseipinnis	3	2.83	0.20	N. ariommus	5	2.30	0.33
L. bellus	4	3.25	0.15	N. scepticus	4	2.63	0.28
L. atrapiculus	2	3.00	–	N. semperasper	3	2.17	0.13
Genus Notropis:				unclear position within Notropis:			
subgenus Alburnops:				N. scabriceps	3	2.50	0.00
N. edwardreneyi	2	3.75	–	N. asperifrons	1	1.00	–
N. blennius	14	3.61	0.15	N. hypsilepis	2	3.25	–
N. simus	3	2.50	0.20	N. hudsonius	13	2.73	0.26
N. girardi	12	3.67	0.16	Genus Hybopsis:			
N. potteri	10	3.60	0.23	H. amnis	10	3.10	0.23
N. buccula	2	3.25	–	H. alborus	6	2.83	0.18
N. bairdi	11	3.77	0.22	H. dorsalis	11	3.05	0.09
volucellus species group:				H. sabinae	7	3.07	0.33
N. heterolepis	9	1.78	0.25	H. longirostris	6	3.33	0.18
N. ozarcanus	5	1.60	0.26	H. bifrenatus	3	3.00	0.33
N. spectrunculus	8	1.81	0.33	unclear phylogenetic position:			
N. emiliae	23	2.96	0.23	N. greenei	11	1.73	0.20
N. maculatus	12	2.92	0.23	N. harperi	1	2.00	–
N. buchanani	15	3.67	0.22	N. rupestris	1	2.50	–
N. volucellus	24	2.73	0.20	N. snelsoni	2	3.25	–

Table 2. Summary statistics for all examined species. The statistics for each species are sample size (N), ratio of males to females (M : F), weight (W) in grams, total length (TL), standard length (SL), brain length (BL), olfactory bulb length (OBL), olfactory bulb width (OBW), telencephalon length (TEL), telencephalon width (TEW), and optic lobe length (OLL) in mm.

Species	N	M:F	W	TL	SL	BL	OBL	OBW	TEL	TEW	OLL
L. albeolus	10	5:2*	2.02	71.55	57.87	7.73	0.97	0.67	1.93	0.89	2.47
N. amabillis	16	13:3	1.54	58.19	46.91	6.43	0.91	0.56	1.69	0.94	2.40
C. analostana	9	5:4	1.77	60.22	48.58	6.53	0.94	0.59	1.72	0.84	2.04
Ly. ardens	11	2:9	2.35	70.00	56.60	7.55	0.98	0.69	1.93	0.87	2.53
N. atherinoides	20	15:5	1.72	61.35	49.51	6.75	1.15	0.74	1.80	0.91	2.00
N. baileyi	5	1:4	1.48	55.00	44.30	6.96	0.85	0.63	1.74	0.57	2.17
N. bairdi	20	13:7	1.69	55.90	45.04	6.66	1.01	0.71	1.83	0.86	1.82
N. boops	21	13:8	1.94	62.00	50.04	6.71	1.08	0.71	1.86	0.98	2.43
N. buchanani	2	0:2	0.52	41.00	32.82	5.10	0.60	0.50	1.40	0.50	1.55
C. callistia	5	3:2	2.49	71.20	57.58	8.06	1.05	0.80	2.08	0.77	2.64
C. camura	10	3:7	3.76	72.20	58.40	7.63	1.08	0.83	1.70	0.88	2.35
L. cerasinus	10	3:7	4.25	77.15	64.46	8.41	1.07	0.77	2.08	0.98	2.63
N. chihuahua	5	2:3	0.90	49.00	38.38	6.49	0.75	0.51	1.80	0.78	1.93
N. chiliticus	4	3:1	1.61	57.63	46.45	6.73	0.76	0.56	1.81	0.65	2.14
L. chrysoceph.	10	5:5	7.94	87.50	70.94	8.67	1.44	1.03	2.23	1.23	2.99
L. cornutus	10	6:4	8.38	91.90	74.55	8.99	1.22	0.84	2.20	0.98	2.83
N. dorsalis	10	0:0*	1.80	61.10	49.30	6.74	1.04	0.75	1.81	0.82	1.90
N. euryzonus	5	1:4	0.65	43.20	34.63	5.28	0.54	0.41	1.11	0.63	1.87
Ly. fumeus	10	1:9	0.601	49.10	39.46	5.50	0.68	0.53	1.43	0.76	1.88
C. galactura	10	3:7	5.52	86.60	70.21	8.06	1.14	0.79	1.91	0.94	2.74
N. girardi	9	8:1	1.23	50.77	40.84	5.74	0.80	0.61	1.52	0.71	1.71
N. greenei	5	0:5	1.46	62.20	50.20	6.76	0.89	0.54	1.82	0.58	2.02
N. heterodon	5	1:4	1.07	55.00	44.30	5.76	0.65	0.39	1.53	0.76	1.96
N. hudsonius	10	0:10	3.41	74.20	60.04	8.03	1.29	0.90	2.02	0.88	2.35
C. lepida	5	2:3	1.60	54.20	43.64	7.01	0.82	0.66	1.81	0.93	2.39
H. longirostris	5	3:2	1.04	51.60	41.51	6.36	0.80	0.63	1.74	0.67	1.84
C. lutrensis	24	14:10	3.34	66.42	53.66	7.16	0.93	0.69	1.94	1.02	2.37
N. nubilus	10	3:7	2.72	71.00	57.42	7.34	1.06	0.68	1.87	0.79	2.35
N. ornatus	5	2:3	1.90	54.80	44.14	6.44	0.78	0.62	1.42	0.82	2.08
N. ozarkanus	5	2:3	1.40	61.00	49.22	6.14	0.89	0.60	1.96	0.82	1.94
N. petersoni	5	2:3	0.75	47.80	38.40	6.13	0.71	0.54	1.70	0.56	1.86
L. pilsbryi	15	4:11	5.05	86.87	70.42	9.07	1.19	0.72	2.18	0.88	3.11
N. potteri	8	7:1	1.01	48.31	38.82	6.16	0.78	0.62	1.66	0.71	1.91
N. procne	8	4:4	1.36	60.13	48.50	6.39	0.84	0.57	1.79	0.59	1.80
C. proserpina	5	2:3	0.80	46.30	37.17	5.76	0.75	0.49	1.52	0.83	1.79
Ly. roseipinnis	7	5:2	1.05	55.43	44.65	6.46	0.79	0.60	1.67	0.75	1.99
N. rubellus	10	6:4	1.67	58.80	47.42	7.11	1.07	0.69	1.74	0.71	2.19
H. sabinae	10	8:2	1.61	57.45	46.31	6.76	0.74	0.59	1.98	0.77	2.03
N. snelsoni	7	4:3	0.31	38.79	31.01	5.51	0.46	0.36	1.64	0.53	1.70
C. spiloptera	5	2:3	1.54	60.60	48.89	6.33	0.82	0.56	1.63	0.83	2.03
N. stramineus	10	1:9	1.76	58.30	47.01	6.89	1.10	0.65	1.88	0.76	2.03
N. telescopus	10	4:6	2.11	69.90	56.52	8.18	0.90	0.62	2.33	0.65	2.67
N. texanus	10	5:5	1.06	53.95	43.44	6.69	0.71	0.53	1.91	0.85	2.14
N. topeka	10	1:9	1.48	53.50	43.07	6.22	0.94	0.58	1.47	0.71	1.92
Ly. umbratilis	10	4:6	1.55	58.40	47.09	6.42	0.91	0.62	1.56	0.66	2.10
undescr. sp.	5	3:2	1.94	57.10	46.02	6.66	0.82	0.61	1.62	0.86	2.11
C. venusta	24	20:4	2.92	68.17	55.09	7.25	1.05	0.79	1.95	1.05	2.65
N. volucellus	10	1:9	1.15	55.75	44.92	6.28	0.80	0.52	1.72	0.76	1.82
C. whipplei	10	2:8	3.59	69.50	56.19	7.17	1.03	0.68	1.68	0.77	2.33
N. xaenoceph.	5	5:0	1.65	60.40	48.73	6.72	0.66	0.57	1.88	0.70	3.06
L. zonatus	5	0:5	5.45	88.80	72.01	8.96	1.23	0.85	2.03	0.89	2.96

* The sex of the remaining specimens could not be determined, because the gonads had been removed previously.

Table 3. Summary statistics for all examined species. The statistics for each species are optic lobe width (OLW), cerebellum length (CBL) and width (CBW), facial lobe length (FLL) and width (FLW), vagal lobe length (VLL) and width (VLW), and eye diameter (ED). All measurements are in millimeters.

Species	OLW	CBL	CBW	FLL	FLW	VLL	VLW	ED
L. albeolus	2.02	1.38	1.56	0.58	0.58	1.39	0.60	4.63
N. amabilis	1.87	1.62	2.08	0.31	0.33	1.38	0.49	4.29
C. analostana	1.66	1.28	1.27	0.46	0.44	1.38	0.44	3.37
Ly. ardens	1.91	1.57	1.71	0.40	0.41	1.32	0.53	4.07
N. atherinoides	1.54	1.59	1.92	0.46	0.43	1.46	0.44	3.47
N. baileyi	1.55	1.24	1.38	0.70	0.51	1.33	0.50	3.24
N. bairdi	1.52	1.54	1.94	0.65	0.66	1.74	0.49	3.07
N. boops	1.98	1.62	2.09	0.45	0.47	1.46	0.54	4.48
N. buchanani	1.18	0.95	1.10	0.35	0.28	0.95	0.40	3.05
C. callistia	2.12	1.90	2.00	0.45	0.56	1.58	0.59	3.92
C. camura	1.98	1.65	1.62	0.49	0.50	1.59	0.54	3.80
L. cerasinus	2.17	1.61	1.70	0.61	0.67	1.58	0.61	4.73
N. chihuahua	1.61	1.38	1.66	0.49	0.51	1.15	0.45	3.45
N. chiliticus	1.61	1.28	1.28	0.50	0.53	1.18	0.45	3.88
L. chrysoceph.	2.36	2.12	2.32	0.58	0.66	1.86	0.67	5.16
L. cornutus	2.26	1.87	2.13	0.62	0.64	2.28	0.68	5.36
N. dorsalis	1.62	1.43	1.69	0.58	0.61	1.58	0.55	3.68
N. euryzomus	1.47	1.25	1.14	0.28	0.36	1.12	0.40	2.78
Ly. fumeus	1.41	1.18	1.30	0.25	0.24	0.92	0.48	3.11
C. galactura	2.13	1.67	1.74	0.53	0.59	1.52	0.58	4.17
N. girardi	1.44	1.37	1.78	0.50	0.52	1.60	0.46	2.84
N. greenei	1.55	1.30	1.41	0.39	0.44	1.44	0.44	4.32
N. heterodon	1.71	1.24	1.41	0.37	0.31	1.26	0.43	3.72
N. hudsonius	2.04	1.67	1.86	0.58	0.56	1.79	0.59	4.54
C. lepida	1.83	1.64	1.75	0.54	0.55	1.24	0.48	3.46
H. longirostris	1.57	1.44	1.62	0.54	0.55	1.41	0.46	3.04
C. lutrensis	1.88	1.84	2.06	0.52	0.55	1.74	0.51	3.58
N. nubilus	1.95	1.47	1.85	0.56	0.57	1.54	0.58	4.34
N. ornatus	1.67	1.51	1.59	0.49	0.55	1.46	0.44	2.96
N. ozarkanus	1.77	1.40	1.61	0.40	0.37	1.32	0.48	3.58
N. petersoni	1.48	1.26	1.48	0.33	0.29	1.25	0.47	3.36
L. pilsbryi	2.53	2.16	2.19	0.68	0.69	1.79	0.59	6.14
N. potteri	1.44	1.16	1.49	0.43	0.49	1.08	0.46	2.66
N. procne	1.49	1.21	1.36	0.48	0.44	0.98	0.40	3.60
C. proserpina	1.54	1.28	1.21	0.46	0.42	0.95	0.42	2.91
Ly. roseipinnis	1.70	1.53	1.73	0.27	0.22	1.39	0.48	4.00
N. rubellus	1.48	1.48	1.57	0.53	0.53	1.64	0.53	3.65
H. sabinae	1.62	1.46	1.66	0.53	0.60	1.18	0.51	2.98
N. snelsoni	1.20	1.08	1.20	0.29	0.31	0.83	0.42	2.73
C. spiloptera	1.68	1.47	1.52	0.50	0.45	1.52	0.50	3.49
N. stramineus	1.63	1.48	1.66	0.57	0.63	1.64	0.56	3.43
N. telescopus	2.11	1.61	1.82	0.45	0.46	1.40	0.50	5.47
N. texanus	1.71	1.42	1.70	0.43	0.42	1.05	0.43	3.51
N. topeka	1.54	1.28	1.31	0.32	0.36	1.31	0.46	3.13
Ly. umbratilis	1.54	1.32	1.31	0.27	0.26	1.41	0.48	3.39
undescr. sp.	1.64	1.43	1.50	0.54	0.53	1.20	0.50	3.49
C. venusta	1.96	1.86	2.11	0.46	0.48	1.42	0.55	3.90
N. volucellus	1.59	1.19	1.43	0.41	0.39	1.23	0.45	3.86
C. whipplei	1.68	1.38	1.37	0.44	0.50	1.33	0.49	3.67
N. xaenoceph.	1.80	1.22	1.34	0.45	0.37	1.62	0.48	4.26
L. zonatus	2.11	1.86	1.72	0.68	0.65	1.92	0.64	5.72

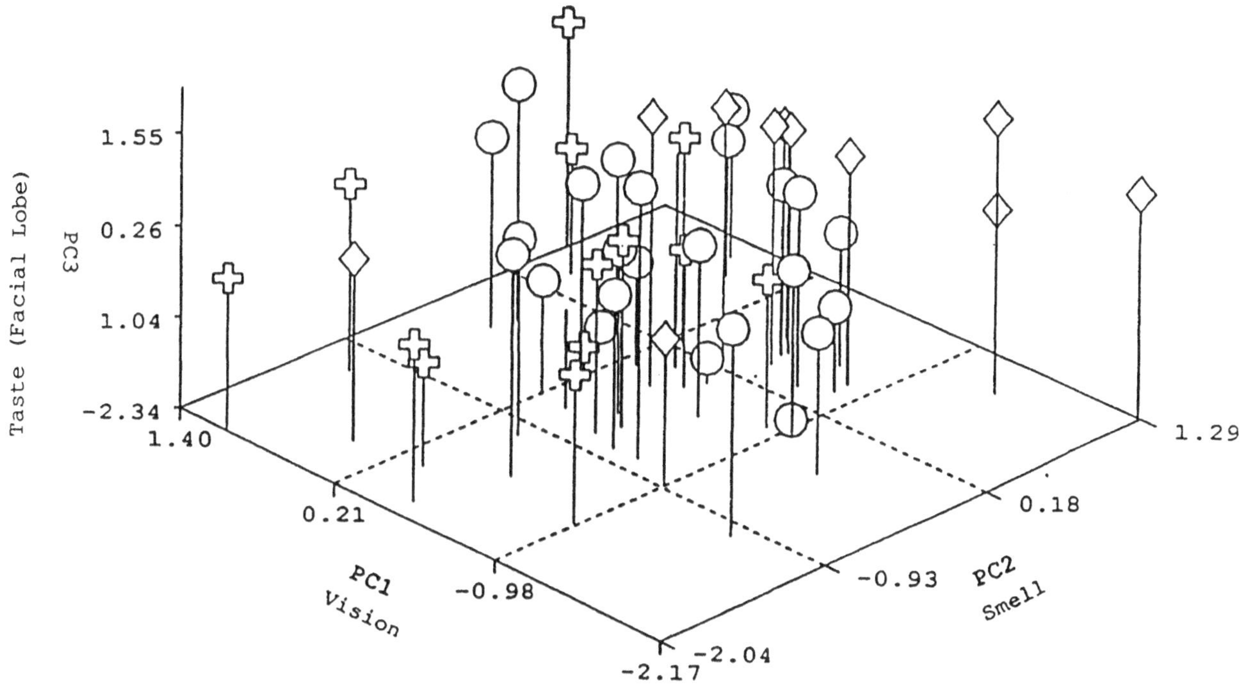

Fig. 2. 3-dimensional plot of species means of the scores on PC1, PC2, and PC3. Well developed sensory structures are characterized by high scores on PC1 for vision, PC2 for olfaction, and PC3 for taste. Plotting symbols represent species preferring turbid ◇, average ○, and clear ✣ habitats.

Table 4. Factor loadings of residuals of each variable on PC-axes following varimax rotation and variance explained by each axis (V). Abbreviations are Eigenvalue (EV), length (L), and width (W).

Structure	PC1	PC2	PC3	PC4
Olfactory bulb L.	− 0.107	0.678	0.029	0.329
Olfactory bulb W.	− 0.050	0.742	0.138	0.260
Telencephalon L.	0.508	− 0.063	0.389	− 0.042
Telencephalon W.	0.414	0.614	0.001	− 0.311
Optic lobe L.	0.783	− 0.072	− 0.073	0.013
Optic lobe W.	0.783	− 0.082	0.009	0.071
Cerebellum L.	0.675	0.465	0.083	0.136
Cerebellum W.	0.675	0.426	0.204	0.158
Facial lobe L.	− 0.019	0.064	0.913	0.142
Facial lobe W.	0.043	0.164	0.890	0.122
Vagal lobe L.	− 0.024	0.235	0.124	0.805
Vagal lobe W.	0.247	0.052	0.106	0.660
Eyediameter	0.532	− 0.467	0.204	0.172
EV	3.443	2.405	1.442	1.140
V	26.5%	18.5%	11.1%	8.8%

& Junger 1988). However, previous studies have suffered from one or more of the following problems: comparisons of distantly related taxa, comparisons of members of polyphyletic groups, comparisons of individuals of different size, and small sample size. Having reduced or avoided many of these difficulties, the results of this study strongly support the following two hypotheses: (1) species of minnows rely on different sensory modalities, which correlate with the physical parameters in their preferred habitat, and (2) the importance of a particular modality is reflected in the size of the corresponding neural structures.

The existence of two gustatory subsystems was demonstrated in ictalurid catfish species: a facial system, with an importance for the localization of a food source in the environment, and a glossopharyngeal-vagal system, which is critical for the ingestion of food placed into the mouth (Atema 1971). The present study supports this functional division in cyprinids, as habitat turbidity is correlated with the facial lobe but not with the vagal system.

The high concordance of the responses on the survey suggests that this approach provided an accurate estimate of preferred turbidity. PCA accounted for 65% of the variation among individuals. Further studies may help to account for the remainder by evaluating: (1) microhabitat use; (2) feeding habits; (3) territorial and non-territorial species; (4) schooling and solitary species; and (5) day and night feeders. A high correlation between size of cerebellum and visual structures (high loadings on PC1) suggests a functional association between these structures. Although this result may represent an effect of body size that remained after the regression on standard length, several alternative hypotheses should be considered. A visually oriented species, which can pursue faster moving prey, may require a more highly developed sense of motor coordination than a species relying on taste. In visually oriented species, more elaborate behavioral patterns may necessitate increased proprioceptive performance. Alternatively, Gatz (1979) and Felley (1984) suggested that 'swimming ability' is especially prominent in species inhabiting fast currents. Fast flowing habitats are often clear (Felley 1984), and swimming ability and vision may represent independent adaptations to this environment. Retinal fibers have been found that project to the cerebellum in at least some teleosts (Uchiyama et al. 1988) but it is unlikely that these would appreciably affect the size of the cerebellum.

Intersexual differences in the size of the olfactory bulb and cerebellum, and significant interaction

Table 5. Results of 2-way ANCOVAs of olfactory bulb, telencephalon, optic lobe, cerebellum, facial lobe, vagal lobe, with species and sex as treatment effects (standard length was used as covariate). Abbreviations are significance (p), and degrees of freedom (df), ns non-significant, * $0.01 \leq p < 0.05$, ** $0.001 \leq p < 0.01$, *** $p < 0.001$.

Source	df	F	p
Olfactory bulb			
Covariate: standard length	1	90.908	***
Species	27	6.806	***
Sex	1	6.376	*
Species × sex	27	0.719	ns
Telencephalon			
Covariate: standard length	1	231.887	***
Species	27	13.833	**
Sex	1	0.246	ns
Species × sex	27	2.369	***
Optic lobe			
Covariate: standard length	1	324.602	***
Species	27	26.026	***
Sex	1	2.560	ns
Species × sex	27	3.560	***
Cerebellum			
Covariate: standard length	1	128.037	***
Species	27	15.772	***
Sex	1	7.477	**
Species × sex	27	3.429	***
Facial lobe			
Covariate: standard length	1	63.557	***
Species	27	22.295	***
Sex	1	0.920	ns
Species × sex	27	1.231	ns
Vagal lobe			
Covariate: standard length	1	54.528	***
Species	27	6.693	***
Sex	1	0.004	ns
Species × sex	27	1.772	*
Eye			
Covariate: standard length	1	420.409	***
Species	27	48.106	***
Sex	1	0.883	ns
Species × sex	27	2.904	***

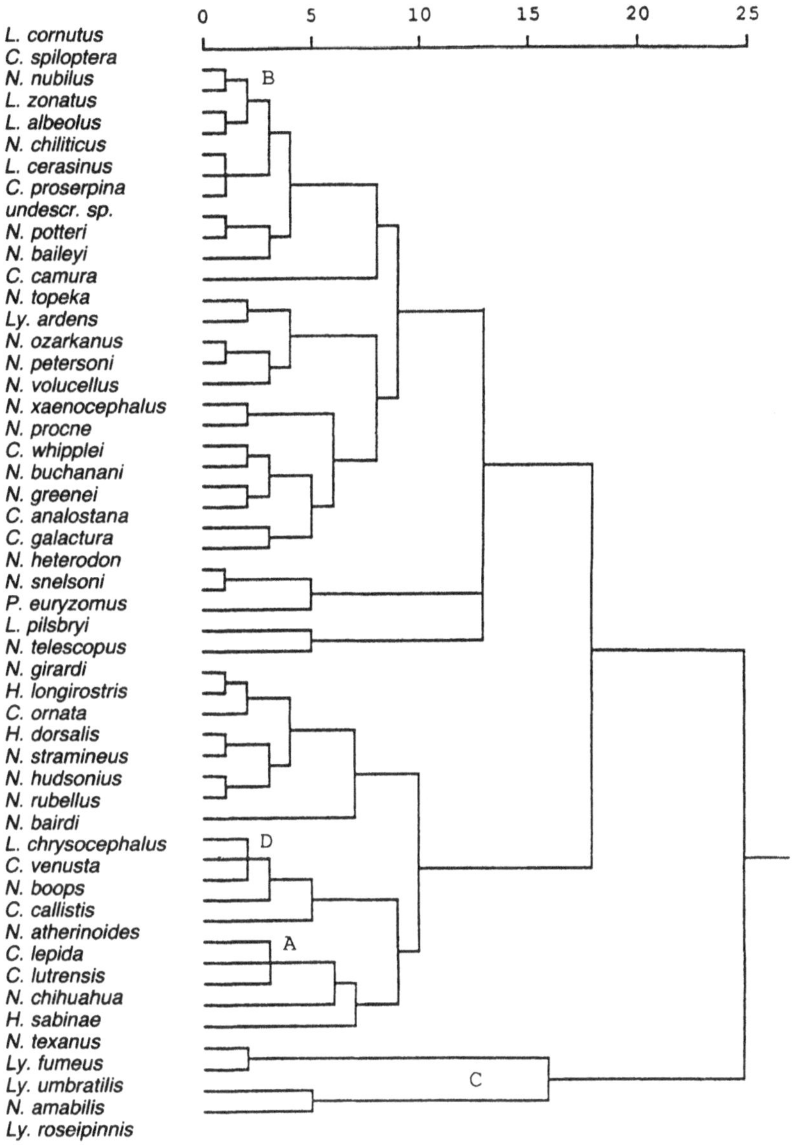

Fig. 3. Dendrogram of the size of olfactory bulb, telencephalon, optic lobe, cerebellum, facial lobe, vagal lobe, and eye in 51 species of the genera *Notropis, Pteronotropis, Cyprinella, Luxilus, Lythrurus,* and *Hybopsis*. Scale represents measures of phenetic dissimilarity.

terms between species and sex in the size of many structures suggests the existence of intersexual differences in ecology and behavior in this group of fishes.

Aknowledgements

We are grateful to Frances P. Gelwick, Richard W. Manning, William J. Matthews, Joel M. McKinney, Douglas G. Sheeley, and Moira J. van Staaden who provided assistance in the field, and to Frank B. Cross, A. John Gatz, Thomas L. Wenke, and William J. Matthews for the loan or donation of specimens. We thank Penelope W. Coates, Michael R. Gannon, H. Bernard Hartman, Robert R. Hollander, Kurt Kotrschal, Richard W. Manning, William J. Matthews, Ernst S. Reese, Moira van Staaden and Michael R. Willig for providing us with many helpful suggestions throughout the study, and for critically reviewing the manuscript. We wish to thank all respondents (Appendix 1) for their efforts in completing the questionnaire, without which this study would have no ecological basis. Support for this research was provided by The Graduate School and the Department of Biological Sciences at Texas Tech University, and the Austrian-American Educational Commission.

References cited

Arnold, A.P. 1980. Sexual differences in the brain. Amer. Sci. 68: 165–173.

Atema, J. 1971. Structures and functions of the sense of taste in the catfish (*Ictalurus natalis*). Brain Behav. Evol. 4: 273–294.

Baron, G. & P. Jolicoeur. 1980. Brain structure in Chiroptera: some multivariate trends. Evolution 34: 386–393.

Coburn, M.M. 1982. Anatomy and relationships of *Notropis atherinoides*. Ph.D. Dissertation, Ohio State University, Columbus. 384 pp.

Davis, B.J. & R.J. Miller. 1967. Brain patterns in minnows of the genus *Hybopsis* in relation to feeding habits and habitat. Copeia 1967: 1–39.

Davis, R.E. & J. Kassel. 1983. Behavioral functions of the teleostean telencephalon. pp. 237–263. *In:* R.E. Davis & R.G. Northcutt (ed.) Fish Neurobiology, Vol. 2, University of Michigan Press, Ann Arbor.

Demski, L.S. 1983. Behavioral effects of electrical stimulation of the brain. pp. 317–358. *In:* R.E. Davis & R.G. Northcutt (ed.) Fish Neurobiology, Vol. 2, University of Michigan Press, Ann Arbor.

Felley, J.D. 1984. Multivariate identification of morphological-environmental relationships within the Cyprinidae (Pisces). Copeia 1984: 442–455.

Felley, J.D. & L.G. Hill. 1983. Multivariate assessment of environmental preferences of cyprinid fishes of the Illinois River, Oklahoma. Amer. Midl. Nat. 109: 209–221.

Finger, T.E. 1988. Organization of chemosensory systems within the brains of bony fishes. pp. 339–363. *In:* J. Atema, R.R. Fay, A.N. Popper & W.N. Tavolga (ed.) Sensory Biology of Aquatic Animals, Springer Verlag, New York.

Friedlander, M.J. 1983. The visual prosencephalon of teleosts. pp. 91–115. *In:* R.E. Davis & R.G. Northcutt (ed.) Fish Neurobiology, Vol. 2, University of Michigan Press, Ann Arbor.

Gatz, A.J. 1979. Ecological morphology of freshwater stream fishes. Tulane Stud. Zool. Bot. 21: 91–124.

Geiger, W. 1956a. Quantitative Untersuchungen über das Gehirn der Knochenfische, mit besonderer Berücksichtigung seines relativen Wachstums. Acta anat. 26: 121–163.

Geiger, W. 1956b. Quantitative Untersuchungen über das Gehirn der Knochenfische, mit besonderer Berücksichtigung seines relativen Wachstums. Acta anat. 27: 324–350.

Gomahr, A., K. Kotrschal & A. Goldschmid. 1988. Die chemosensorischen Zellen der Haut bei den heimischen Karpfenfischen (Teleostei, Cyprinidae): Geschmacksknospen und freie Sinneszellen. Österreichs Fischerei 41: 241–253.

Jolicoeur, P., P. Pirlot, G. Baron & H. Stephan. 1983. Brain structure and correlation patterns in Insectivora, Chiroptera, and Primates. System. Zool. 33: 1892–1908.

Kanwal, J.S. & J. Caprio. 1987. Central projections of the glossopharyngeal and vagal nerves in the channel catfish, *Ictalurus punctatus:* clues to differential processing of visceral inputs. J. Comp. Neur. 264: 216–230.

Kishida, R. 1979. Comparative study on the teleostean optic tectum. Lamination and cytoarchitecture. J. Hirnforsch. 20: 57–67.

Kotrschal, K. & H. Junger. 1988. Patterns of brain morphology in Mid-European Cyprinidae (Pisces, Teleostei): a quantitative histological study. J. Hirnforsch. 29: 341–352.

Mayden, R.L. 1985. Phylogenetic studies of North American minnows, with emphasis on the genus *Cyprinella* (Teleostei: Cypriniformes). Ph.D. Dissertation, University of Kansas, Lawrence. 566 pp.

Mayden, R.L. 1990. Phylogenetic studies of North American minnows, with emphasis on the genus *Cyprinella* (Teleostei: Cypriniformes). Misc. Pub. University of Kansas, Mus. Nat. Hist. (in press).

Miller, R.J. & H.E. Evans. 1965. External morphology of the brain and lips in catostomid fishes. Copeia 1965: 467–487.

Nottebohm, F. 1980. Testosterone triggers growth of brain vocal control nuclei in adult female canaries. Brain Res. 189: 429–436.

Nottebohm, F. & A.P. Arnold. 1976. Sexual dimorphism in vocal control areas of the song bird brain. Science 194: 211–213.

Rao, P.D. 1967. Studies on the structural variations in the brain of teleosts and their significance. Acta anat. 68: 379–399.

SAS Institute Inc. 1985. SAS User's Guide. Carry.

Sneath, P.H.A. & R.R. Sokal. 1973. Numerical taxonomy. W.H. Freeman and Co., San Francisco. 573 pp.

Sokal, R.R. & F.J. Rohlf. 1981. Biometry. W.H. Freeman and Co., New York. 859 pp.

SPSSX Inc. 1988. SPSS-X User's guide. Chicago.

Stephan, H. 1960. Methodische Studien über den quantitativen Vergleich architektonischer Struktureinheiten des Gehirns. Z. wiss. Zool. 164: 143–172.

Uchiyama, H., S. Matsutani & H. Ito. 1988. Pretectum and accessory optic system in the filefish *Navodon modestus* (Balistidae, Teleostei) with special reference to visual projections to the cerebellum and oculomotor nuclei. Brain Behav. Evol. 31: 170–180.

Appendix 1

Names and institutional affiliation of respondents.

John M. Aho - Savannah River Ecology Laboratory; Chris T. Amemiya - Showa Univ. Research Institute; Paul Angermaier - Virginia Polytechnic Institute; John A. Baker - Univ. Southern Mississippi; Hank Bart - Univ. Illinois; Kevin Bestgen - Univ. New Mexico; Albert Blair - Univ. Tulsa; Branley Branson - Eastern Kentucky Univ.; Noel Burkhead - Roanoke College; Miles M. Coburn - John Carrol Univ.; Frank B. Cross - Univ. Kansas Museum of Natural History; Don Distler - Wichita State University; Anthony A. Echelle - Oklahoma State Univ.; David Edds - Oklahoma State Univ.; David Etnier - Univ. Tennessee; William L. Fisher - Oklahoma Fish and Wildlife Service; Jeffrey W. Foltz - Clemson Univ.; A. John Gatz - Ohio Wesleyan Univ.; Gene Helfman - Univ. Georgia; Loren G. Hill - Univ. Oklahoma; Richard J. Horwitz - Acad. Natural Sciences, Philadelphia; Clark Hubbs - Univ. Texas at Austin; Donald C. Jackson - Mississippi State Univ.; Robert E. Jenkins - Roanoke College; K. Jack Kilgore - USCE Waterways Experiment Station; Flavius Killebrew - West Texas State Univ.; Edie Marsh - Angelo State Univ.; William J. Matthews - Univ. Oklahoma Biological Station; Richard L. Mayden - Univ. Alabama; Paul M. McKee - Beak Consultants Limited; David McNeely - Morehead State Univ.; Gary Meffe - Savannah River Ecology Laboratory; Edward F. Menhinick - Univ. North Carolina at Charlotte; William D. Pearson - Univ. Louisville; William J. Pflieger - Missouri Dept. Conservation; Jimmie Pigg - Oklahoma State Dept. Health; Gerald B. Pottern - North Carolina State Univ.; David L. Propst - New Mexico Dept. of Game and Fish; C. Richard Robins - Univ. Miami; Fred C. Rohde - North Carolina Div. Marine Fisheries; Isaac Schlosser - Univ. North Dakota; Richard H. Stasiak - Univ. Nebraska at Omaha.

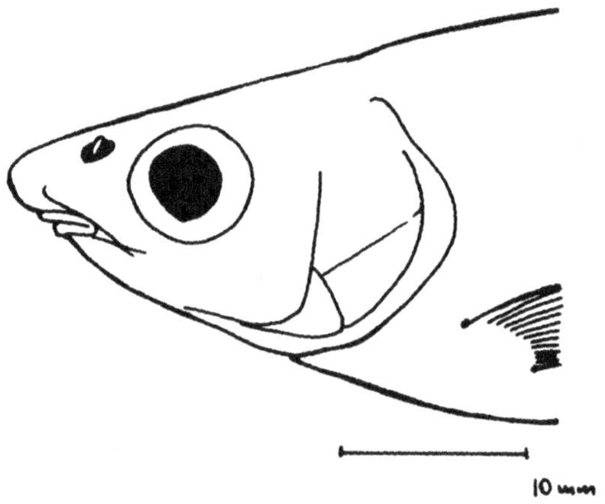
10 mm

Environmental Biology of Fishes **33**: 167–180, 1992.

How dams on the River Danube might have caused hybridization and influenced the appearance of a new cyprinid taxon

Eugene K. Balon
Institute of Ichthyology, Department of Zoology, University of Guelph, Guelph, Ontario N1G 2W1, Canada

Received 29.9.1989 Accepted 14.8.1990

Key words: Gobio spp., *Gobio albipinnatus, Vimba vimba, Blicca bjoerkna,* Hybrid, Backcross, Spawning grounds, Allochrony, Sympatric speciation, Altricial, Precocial, Alprehost, Epigenesis, Evolution

Synopsis

When a new species is found, or a known one collected outside its recognized distribution, most people either assume that the species has been overlooked in the past or that it has recently invaded the area in question from elsewhere. While other, more evolutionary explanations are possible, these are nearly always ignored, in spite of the common perception that animals are constantly in states of change. Recent findings in the Danube River are used as examples for such alternative interpretations. The case of two forms of *Vimba* – short and long snouted – which live in sympatry in the upper Danube is dispelled, as one of them is identified as a hybrid between female *V. vimba* and male *Blicca bjoerkna*. Frequent occurrences of this hybrid (and a few of its backcrosses) in recent times are explained by man-induced alteration of reproductive isolation between their parent species, caused by the impoundment of the river by dams built as part of the Rhein-Main-Donau Kanal. These impoundments, however, might be responsible not only for the creation of hybrids but also for direct speciation, as the case of *Gobio albipinnatus* may represent. The ability for epigenetic creation of two alternative life-history states (as part of the theory of **alprehost**) and the survival of an alternative state when environmental conditions change might be ultimately responsible for the appearance of a new taxon.

Introduction

Science has its dogmas which are as persistent as dogmas in other fields. When new records of a species are found the explanations given most frequently are that they were earlier overlooked or that they are new invaders from elsewhere. Almost never is the abundant and accepted evidence that animal forms are constanly changing used to explain the appearance of a taxon in a new area.

Fishes are no exception. For example, the only explanation given by Wanzenböck et al. (1989) for the discovery of new gudgeons (*Gobio* spp.) in the Austrian Danube drainage is lack of attention given in the past to such coarse fishes. What I would like to emphasize in this paper is that other interpretations are possible – evolutionary interpretations which incorporate historical changes.

Wanzenböck's et al. (1989) obvious defence would be that perusal of museum material revealed that the same species already occurred in the area hundred years ago. This assertion can be challenged by the following arguments: (a) Even if the recent reidentifications are trusted, the presence of *Gobio albipinnatus,* for example, in the old museum collections is very infrequent in comparison to *G. gobio* and *G. uranoscopus*. This may indicate that formation of the *G. albipinnatus* taxon increased markedly in recent times, a phenomenon expected by the alternative explanation given in this paper. (b) The characters used to distinguish among the species are inadequate without comparative material from other parts of the Danube (e.g. from museums in Bratislava and Bucarest), and possibly Dnieper, Don and Volga, which Wanzenböck et al. (1989) failed to obtain.

While I was familiar with the species in question from elsewhere, precise separation in our studies was accomplished only after the use of comparative material and detailed analysis (cf. Balon et al. 1986, 1988).

Of course, I am not claiming that every new distribution is a consequence of new speciation or invasion caused by changed ecological conditions (e.g. Balon 1974). Presence of a taxon missed in the past or caused by recent invasion is real enough; but the other possible explanation of its sudden occurrence, less dogmatic, should not be neglected.

The case of cichlid fishes in the African Great Lakes can serve well as an example. While the speciation in these fishes is considered relatively recent and many new species are still being described, the species diversity and their trophic specializations are known to exist only at a given species composition (e.g. Greenwood 1981, Hoogerhoud et al. 1983, McKaye et al. 1982, 1984, Witte 1984, Ribbink 1986). Changes caused, for example, by trawling or by the introduction of the Nile perch in Lake Victoria, were sufficient to eliminate some species (Barel et al. 1985, Balon & Bruton 1986, Coulter et al. 1986, Ribbink 1987, Ogutu-Ohwayo 1990, Bruton 1990) and may yet cause the appearance of some new species.

The relative time of species formation is not easy to measure and is often obscured by continuing processes of speciation. The charr (*Salvelinus alpinus*) of Thingvallavatn in Iceland is one of the better studied examples (e.g. Sandlund et al. 1987, Magnússon & Ferguson 1987, Jonsson et al. 1988, Noakes et al. 1989). The four distinct forms found there are a result of recent sympatric speciation in the making, not recognized as yet to be distinct species but as real as some of the cichlid 'species' in Lake Malawi (e.g. Ribbink et al. 1983).

For the above reasons I decided to present this paper, using data already published (Balon et al. 1986, 1987, 1988, 1991) but this time placing emphasis solely on the epigenetic interpretation. It goes without saying that more conventional interpretations are still considered plausible but ignored here to accentuate the other possibility.

The case of the two forms of *Vimba*

Bănărescu et al. (1970) believed that two forms of *Vimba vimba,* one deep and the other low bodied, occur in the upper Danube River sometimes in sympatry with *V. elongata.* In our collection of over 23 000 specimens, representing 42 species and 6 cyprinid hybrids throughout the upper Danube (Balon et al. 1986, 1991), 51 adult and 62 juvenile *V. vimba* were identified in samples from Kelheim to Wallsee. The professional fishermen in the area claimed that two forms are known to them – short and long snouted – and among our samples near Vilshofen they identified both forms (Fig. 1). We eventually distinguished at this locality 8 specimens of the second form among the 57 *Vimba* spp. collected in 1976 and 2 among the 10 collected in 1984.

Our initial hope of having documented both *V. vimba* and *V. elongata* in the Danube near the future junction of the Rhein-Main-Donau Kanal or at least the two forms professed by Bănărescu et al. (1963) was dispelled as soon as we took a closer look (Balon et al. 1987). The less frequent form turned out to be a hybrid between *V. vimba* and *Blicca bjoerkna,* the third most common cyprinid in the Danube between Riedlingen and Wallsee, represented by 2486 specimens. Most of them (2290 specimens) were collected at Neuburg, just above the future canal junction, 87 at Kelheim and 19 at Vilshofen. Detailed principal component analyses later enabled us to distinguish even 5 individual backcrosses between the hybrid and its maternal species *V. vimba* (Fig. 2). Both variation coefficients and hybrid indices indicated higher similarity of the hybrids to *V. vimba* (Balon et al. 1987). Consequently, the hybrids and their backcrosses

Fig. 1. Complete view (a) and head view (b) of *Vimba vimba* (below) and the hybrid of *V. vimba* and *Blicca bjoerkna* considered a short snouted form of *Vimba*. From the Danube below Vilshofen in 1976 by the author.

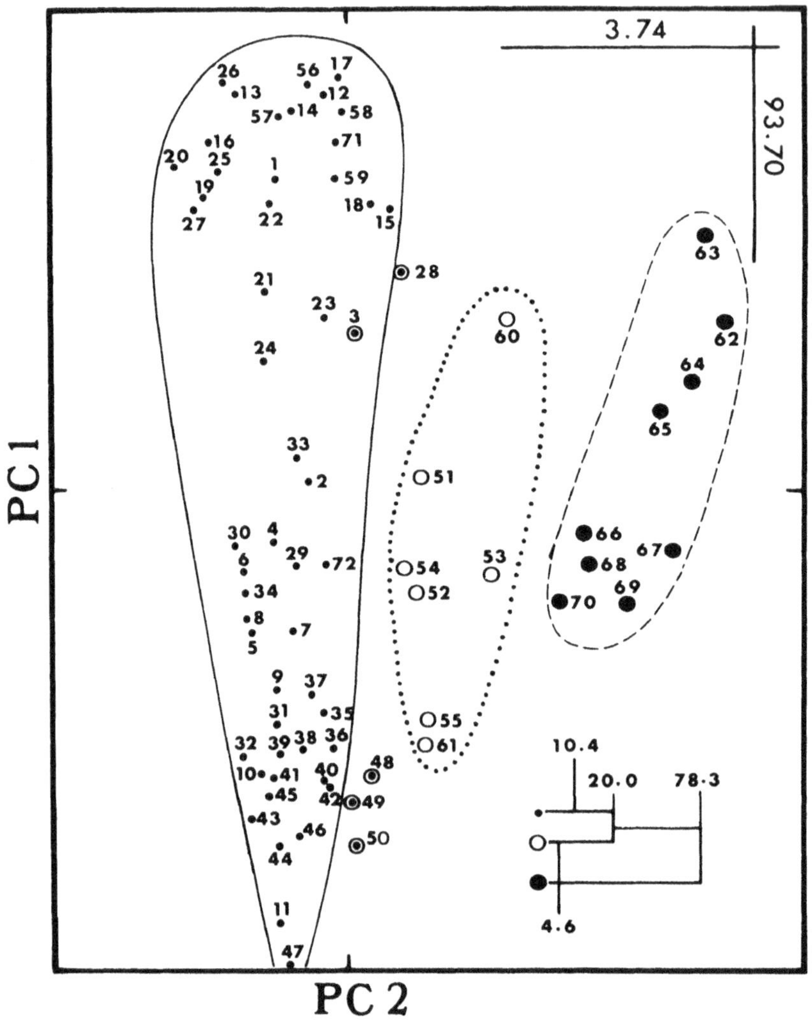

Fig. 2. Projection of the first two principal components for *Vimba vimba* and *Blicca bjoerkna*. From an analysis of 6 mensural characters four clusters were recognized: • = *V. vimba*, ○ = *V. vimba* × *B. bjoerkna* hybrids, ● = *B. bjoerkna*, and ⦿= (*V. vimba* × *B. bjoerkna* hybrid) × *V. vimba* backcrosses. Inset dendrogram indicates the coefficients of dissimilarity within and between 'significantly' separated groups of specimens. From Balon et al. (1987).

were assumed to be a progeny of *V. vimba* females and *B. bjoerkna* or hybrid males.

The reproductive style and early ontogeny are different for *V. vimba* and *B. bjoerkna*. Under normal circumstances their spawning grounds are separated and hybridization is unlikely. *V. vimba* spawns on gravel in running water, *B. bjoerkna* on soft green vegetation in shallow stagnant waters. The first species deposits adhesive eggs 2 mm in diameter which sink into rock crevices, the latter scatters its eggs (1.5 mm in diameter) over submersed plants to which they adhere. *V. vimba* embryos hatch when 6–8 mm long, are photophobic and hide deep in gravel crevices, whereas embryos of *B. bjoerkna* hatch at a smaller size (5 mm), are not photophobic and immediately attach themselves with the help of cement glands on the head (which are missing in *V. vimba*) to plants above the soft bottom (cf. Kryzhanovsky 1949, Smirnova 1957, Koblickaya 1966, Smirnova et al. 1970, and personal unpublished observations). First oral ingestion and intestinal digestion (exogenous feeding) coincide in *V. vimba* with the disappearance of photophobia and emergence from among the grav-

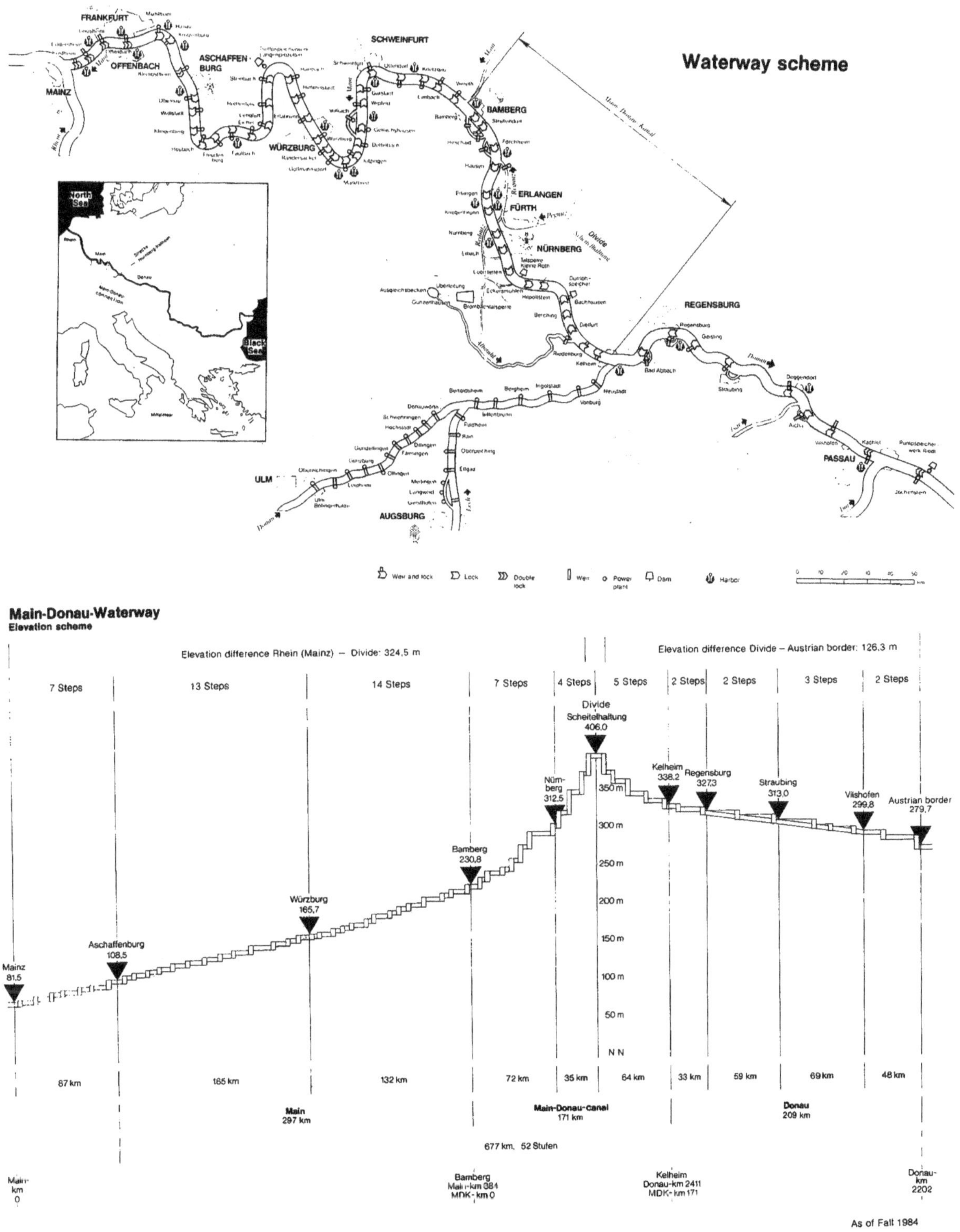

Fig. 3. The 1984 scheme of the Rhein-Main-Donau Kanal (above) and its elevation profile. Insert shows the entire scenario of the European Waterway. From Balon et al. (1986).

Fig. 4. Danube above Vilshofen: a – the river bed with alternating lentic and lotic habitats; b – rock-gravel and submerged vegetation shore. Both by Ivan Sazima in 1984.

el at about 8 mm length, in *B. bjoerkna* with cement gland detachment and free swimming at 5.5 mm length. In both species the yolk is nearly absorbed at that time – the active larva period has started (Balon 1984a, 1985, Flegler-Balon 1989). *V. vimba* belongs to the reproductive guild of 'rock and gravel spawners' (lithophils) while *B. bjoerkna* at best is a 'non-obligatory plant spawner' (phytolithophil) (Balon 1975, 1981). *V. vimba* also spawns somewhat earlier in the season than *B. bjoerkna* and so spawning allochrony further safeguards their reproductive isolation.

For the *V. vimba* × *B. bjoerkna* hybrids to occur the seasonal allochrony as well as spawning ground separation must be removed. The main river bed of the upper Danube, in preparation for the opening of the Rhein-Main-Donau connection, is interrupted by a dam-powerplant-lock combination so frequently that the water is impounded stepwise from one to the next (Fig. 3). Above these dams, stagnant water masses enable submerged vegetation to grow in warmer water due to heat retention, than below the dams, where fluvial conditions over gravel and sand bars are predominating (Fig. 4). At the interface of these two man-made habitats, conditions have been created for the spawning grounds of *V. vimba* and *B. bjoerkna* to overlap in topography as well as temperature regimes.

Hybridization has been known to create new taxa (e.g. Buth 1984, Echelle & Kornfield 1984), but man-made changes can have an even more direct influence on speciation!

The case of the whitefin gudgeon

The whitefin gudgeon, *Gobio albipinnatus,* nearly always occurs together with the common gudgeon, *Gobio gobio.* The whitefin has long been known from the lower part of the Danube; since about 40

Table 1. Localities on the Danube River and numbers of gudgeons collected and analysed in the initial study (Balon et al. 1988).

Locality	Name and description	Date of collection	Distance from source (in km)	Elevation (in m)	Gradient (in m per (10 km)	Specimen numbers	*Gobio gobio* N	*G. albipinnatus* N	Spec. #	Locality total[1]
1	*Schollach* Creek, tributary to Breg at hatchery (Langenfeld's)	22. 7.1976	11	747	61	—	—	—	—	—
2	*Breg* below Braunlingen	22. 7.1976	25	693	19	1—8	8	—	—	8
3	Danube in *Pforen*, at the road bridge to Hufingen	23. 7.1976	36	678	10	9-40	32	*	—	32
4	Danube at *Immendingen*, between foot and railroad bridge	23. 7.1976	56	657	12	41	1	—	—	1
5	Danube below *Fridingen* at Bronnenmuhle	23. 7.1976	91	611	12	42-50	9	—	—	9
6	Danube at *Beuron*, above the wooden bridge	23. 7.1976	96	604	11	51-65	15	—	—	15
7	Danube below *Sigmaringen* 2 km downstream from the last bridge	25. 6.1984	129	554	11	66-110	45	*	—	408 (363)
8	Danube at *Hundersingen*	25. 6. 1984	146	541	7	111-154	44	—	—	44
9	Danube below *Riedlingen* at Daugent	26. 6.1984	162	522	11	155-159	5	—	—	5
10	Danube at *Munderkingen*, river below and impoundment above	26. 6.1984	185	503	7	160-177	18	—	—	85 (67)
11	Danube above *Ulm*, in arms Lichternsee and Gronne	26. 6.1984	216	471	7	178, 179	2	—	—	2
12	Danube below *Neuburg*	2. 8.1976	336	375	6	180-191	6	6	180, 182-186	24 (12)
13	(a) Danube from Herrensal to *Kelheim*	3. 8.1976	390	335	4	—	—	—	—	—
	(b) Altmuhl above Kelheim					192-196	3	2	193, 194	5
14	Danube in *Frengkofen*, between river km 2360-2362	22. 6.1984	437	326	3	197, 198	2	—	—	2
15	Danube from *Straubing* to river km 2314	4. 8.1976	482	324	<1	—	*	*	—	—
16	Danube below *Deggendorf*, from Isar entry to km 2281	19. 6.1984	518	310	2	199	1	—	—	1
17	(a) Danube below (5 km) *Vilshofen*	4. 8.1976				—	—	—	—	—
	(b) above (4 km) Vilshofen	5. 8.1976	549	298	3	200, 201	1	1	200	2
	(c) above (4 km) Vilshofen	20. 6.1984				202-211	6	4	204,206,210	10
	(d) old canal from Vils River and					—	*	—	211	
	(e) Vils River at Mettenhausen	21. 6.1984				212-200	9	—	—	9
18	Danube below *Passau*, to km 2222 on the old-water side	20. 6.1984	576	285	3	221, 222	2	—	—	2
19	Danube at *Wallsee*, in the old riverbed and below the dam	9. 8.1976	707	231	8	223-225	3	*	—	27 (24)
	Upper Danube mixed	1976				226-238	8	5	230,231, 236-238	1167 (1154)
	Danube arm Vajas near Baka	29. 9.1976					44	1	272	45
	(below Bratislava), SNM	17. 10.1977	1088	110	—	239-330		18	284-301	18
	specimens	12. 10.1977						29	302-330	29

[1]Numbers in parenthesis indicate numbers of specimens which were not measured.

Table 2. Character loadings onto the first three principal components of variation for the 13 mensural characters sets for all specimens of *Gobio* spp. analyzed (n = 330).

Variable	Principal components 1	2	3
1. Standard length	0.284	0.025	−0.111
2. Head length	0.283	−0.115	−0.079
3. Orbit diameter	0.269	0.366*	−0.521
4. Postorbital distance	0.280	−0.219	0.024
5. Interorbital distance	0.269	−0.319*	0.648
6. Barbel length	0.243	0.771*	0.486
7. Pelvic fin length	0.282	0.085	0.053
8. Predorsal distance	0.284	−0.078	−0.036
9. Prepectoral distance	0.282	−0.143	0.010
10. Prepelvic distance	0.284	−0.076	−0.047
11. Prevent distance	0.284	−0.089	−0.073
12. Pelvic fin — anal fin	0.281	0.116	−0.168
13. Pelvic fin — vent	0.278	−0.218	−0.116
Percent total variance	94.65	3.05	1.06
Percent cumulative variance	94.65	97.70	98.76

years ago from the middle part of the river, and only recently from the upper part (Balon et al. 1988). With time, it seems the proportion of whitefin has increased in the gudgeon populations in the middle Danube (e.g. Balon 1962, 1967, Holčík & Bastl 1976, Holčík et al. 1981). No data exist for the other parts of the river.

The whitefin is rather difficult to separate from the common gudgeon, especially as small individuals. Nevertheless, using comparative material from the middle Danube and principal component analysis, separation from *G. gobio* was done with high confidence (Balon et al. 1988). Among samples from the upper Danube *G. gobio* occurred throughout the entire section (Table 1) while *G. albipinnatus* was identified only in samples below Kelheim, i.e. downstream from the future connection of the Rhein-Main-Donau Kanal. In all of the upper Danube the whitefin occurred in fewer numbers than the common gudgeon, contrary to the present situation in the middle Danube where *G. albipinnatus* nearly always prevails.

Statistical evaluation of the character data sets from both gudgeon species and from both the upper and middle Danube revealed that the upper-Danube whitefin is much closer to the common gudgeon than its counterpart in the middle Danube (Table 2, Fig. 5), except for some coded and meristic characters (Table 3). Had this species been invading the upper Danube from downstream, this similarity would not be expected even if the parochial explanation of introgression is invoked. A better alternative explanation is demanded for the first time.

As details on the method, data and interpretation are already given in the original study (Balon et al. 1988) only selected variables will be presented

Table 3. Coded and meristic characters for *G. gobio* and *G. albipinnatus* from both the middle and upper Danube. Values in parentheses represent less than 10% of the group, and mean values are given below the ranges. From Balon et al. (1988).

Specimens	Relative length (Fig. 7) of Barbels	Pelvic fins	Melanophore patterns on Dorsal fin	Caudal fin	Lateral body
Middle Danube					
G. albipinnatus (N=48)	(2) 3 (4) 2.98	0 0.00	(0) 1,2 (3) 1.21	1,2 (3) 1.40	7-10 (11) 8.63
G. gobio (N=44)	1,2 (3) 1.80	1 1.00	2-4 (5) 3.23	3-5 (6) 4.30	(7) 8-11 9.16
Upper Danube					
G. albipinnatus					
(a) medium (N=4)[1]	2,3	(0) 1	0,2,3	0-2	7,8
(b) small (N=14)	2,3 (4) 2.86	0 (1) 0.07	0,1 0.36	0 0.00	7-9 8.00
(c) combined (N=18)	2,3 (4) 2.83	0 (1) 0.06	0,1 (2,3) 0.56	0 (1,2) 0.28	7-9 8.00
G. gobio					
(a) large (N=56)	(0) 1,2 (3) 1.64	(0) 1 (2) 1.00	3-6 (7,8) 4.70	(4) 5-8 (9) 6.20	(5-7) 8,9 (10,11) 8.46
(b) medium (N=81)	1,2 (3) 1.83	(0) 1 (2) 1.02	(0,1) 2-5 (6) 3.63	(1,2) 3-6 (7,8) 4.68	(5,6) 7-10 (11) 8.26
(c) small (N=83)	1,2 (3) 1.70	1 1.00	(0,1) 2,3 (4) 2.30	(0,1) 2-4 (5) 2.73	(6) 7-9 7.89
(d) combined (N=220)	(0) 1,2 (3) 1.73	(0) 1,2 1.01	(0,1) 2-5 (6,8) 3.40	(0-2) 3-6 (7-9) 4.33	(5-6) 7-9 (10,11) 8.17

[1]Mean values were not calculated where N<10.

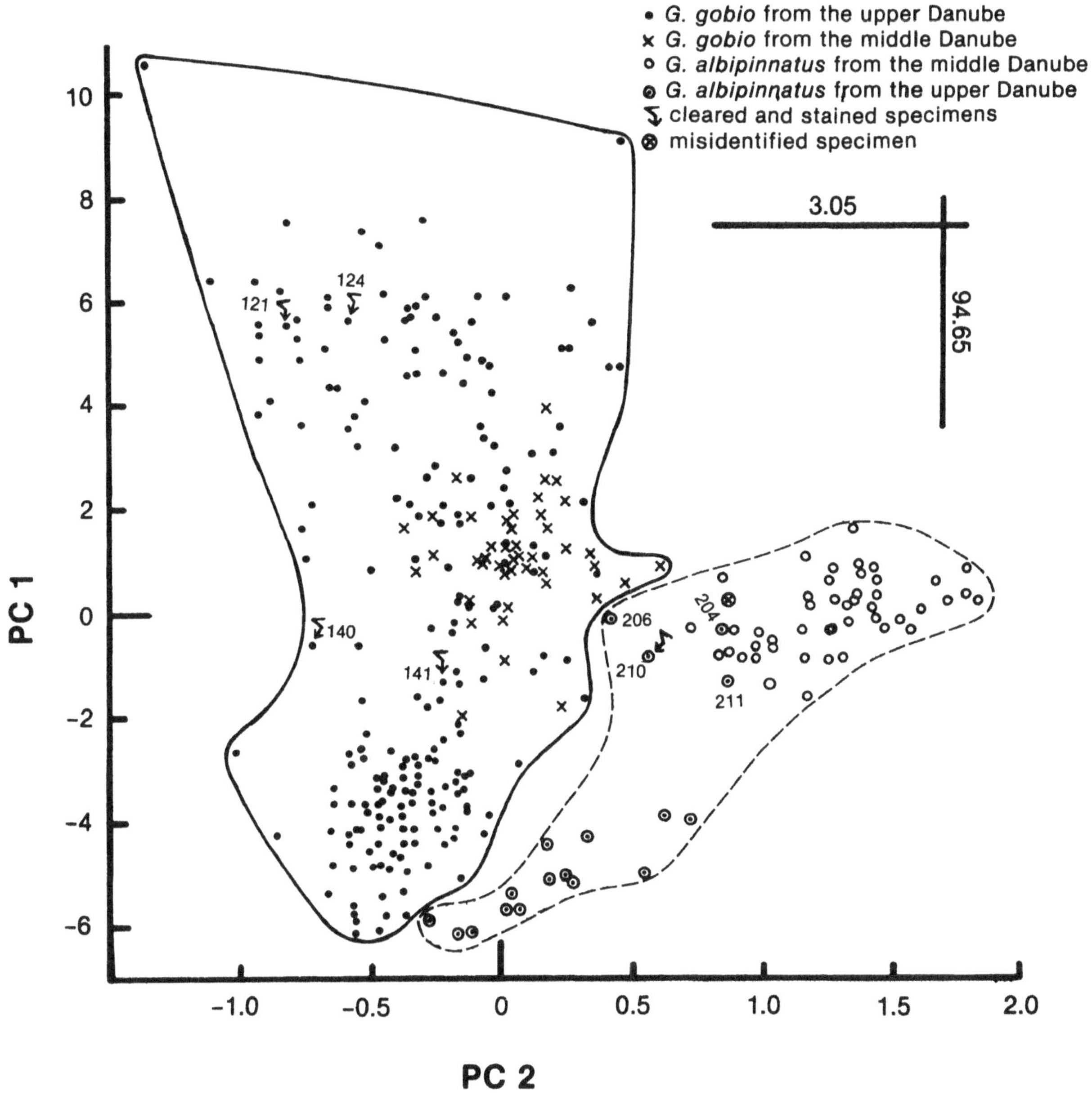

Fig. 5. Projection of the first two principal components of variation for the 13 character data sets of all gudgeons studied from the upper and middle Danube. From Balon et al. (1986).

here. From cleared and stained as well as radiographed specimens the following picture of skeletal differences emerged: Some bones of the axial skeleton appear more robust and more calcified in the whitefin (Fig. 6); especially the supra- or interneurals, located between the dorsal tips of predorsal neural spines, are larger and more calcified in *G. albipinnatus*. The largest of these interneural bones is between the Weberian and the first neural spine, followed by 4 increasingly smaller interneurals among the first 5 neural spines. In *G. gobio* the fifth interneural is missing. The proximal pterygiophores are longer in whitefin and the first more robust with its longer spine inserted over the split-

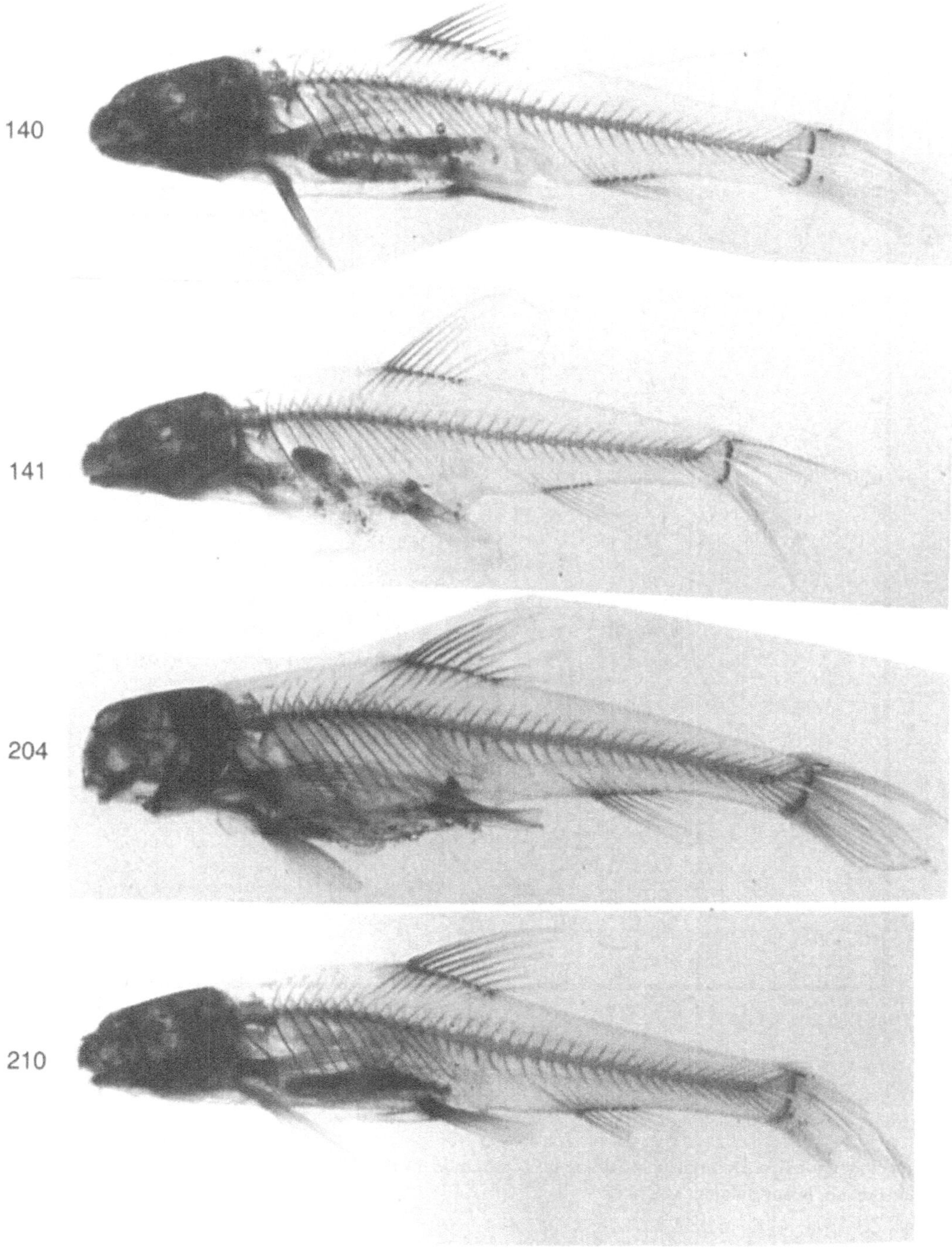

Fig. 6. Skeletons of gudgeons cleared and stained for cartilage and bone: specimens 140 and 141 are *Gobio gobio* 67 and 60 mm SL, specimens 204 and 210 are *G. albipinnatus* 68 and 67 mm SL, respectively.

140

204

Fig. 7. Anterior parts of *Gobio gobio* (above, 140) and *G. albipinnatus* (204) to illustrate the differences in the predorsal interneurals and neural spines.

end neural spine which belongs to the 11th vertebra. In *G. gobio* the proximal pterygiophores are shorter and less calcified, and the longer spine of the first proximal pterygiophore is inserted between the neural spines of the 11th and 12th or 12th and 13th vertebrae (Fig. 7).

Other interpretations given in the original study aside, the above skeletal differences very much resemble similar bone developments of altricial and precocial charrs (Balon 1984b). The more generalistic altricial charr often delay calcification of many skeletal elements or retain them in a cartilaginous state as in *G. gobio,* while the more precocial charr form early a robust, fully calcified skeleton as in *G. albipinnatus.* Both forms of charr have been experimentally created from zygotes of the same parent pair (Balon 1980).

The increasing numbers of both lotic and lentic habitats in the middle Danube may facilitate survival of both altricial *G. gobio* and its precocial alternate *G. albipinnatus* (Fig. 8). Increased numbers of dams on the upper Danube have created similar lotic and lentic habitat frequencies and increased the survival of the precocial form. The abundance of the precocial form represented by *G. albipinnatus,* relative to the ancestral and altricial *G. gobio,* will increase with time if demands for further specialization persist. The questions which remain and which I tried to answer elsewhere (Balon 1989a, b) are:

When do the altricial and precocial forms separate enough to each achieve the status of a species?

When do the phenotypic intraspecific clusters become genetic entities?

Fig. 8. A typical (a) altricial *Gobio gobio* from Vils River at Mettenhausen in 1984 by Ivan Sazima, and the (b) precocial *Gobio albipinnatus* from the Danube above Vilshofen in 1984 by the author.

Epilogue

As I said in the introduction, the interpretation stressed here is only one of many possible. I am far from insisting that it is the only correct one. All I am trying to point out is that this alternative interpretation is worthy of consideration at these times (e.g. Løvtrup 1987, Hoffman 1989). The data, especially in the second case devoted to gudgeons, are insufficient for convincing conclusions; but a comparative study of the early ontogeny of both gudgeons and of some decisive variables of their life histories may resolve the deadlock and support or falsify my interpretation. This is what it is all about (e.g. Popper 1968).

The interpretation of the two case histories was based on an epigenetic theory of altricial-precocial homeorhetic states named 'alprehost' (Balon 1988). Long in honing, this theory combines the concept of saltatory ontogeny and life-history models with bifurcatory processing of information in development and its resultant dichotomous elaboration in phylogeny. It is a theory of epigenetic mechanisms linked to self-organization of living systems (e.g. Riedl 1975, Jantsch 1980, Prigogine 1980, Maturana & Varela 1988). I feel helpless when faced with a task to explain it briefly and direct the reader to the longer essays devoted to the topic (e.g. Balon 1983, 1989a, b).

Specifically, the altricial intraspecific form or already a distinct species should have more numerous but less yolky ova in comparison with the precocial form or taxon. Endowed with more yolk per egg the precocial form should start the exogenous feeding at a larger and more differentiated state. Its larval period could be shorter, as it needs less external nutrients in order to metamorphose into a definitive phenotype. [Acquisition of external nutrients is the main purpose of a larva (Balon 1986).] The definitive phenotype, a juvenile at first, will be more efficiently achieved by the precocial form with its double source of nutrients: an overlap in time of endogenous yolk supply and exogenous oral feeding. This form also shows, among other advantageous features, the earlier calcification of its skeleton. Altricial forms like the less specialized *G. gobio* are often the sole survivors in an ancestral environment and are dominant over harsh but biotically unsaturated conditions, hence, the efficient invaders into new habitats. More specialized precocial forms survive better in a more saturated and competitive environment, much like the *G. albipinnatus* (e.g. Bruton 1989), however, it is not an efficient invader.

Acknowledgements

First, the co-authors of the initial papers, Anton Lelek and Steve Crawford, have to be thanked for all of their contributions without which the present interpretation would not be possible. After all, it is only an amplification of former ideas compiled in haste, which Christine Flegler-Balon and Steve Crawford had to bring into an acceptable shape over my resistance but with my gratitude. The organizers of the 'cyprinid' Workshop in Salzburg deserve my thanks for inviting me and for allowing to present the cases. Most of the other support came from NSERC of Canada operating grants.

References cited

Balon, E.K. 1962. Ökologische Bemerkungen über die Standorte der Donaufische mit einer Beschreibung des Fundes des *Carassius auratus gibelio* (Bloch, 1783) und *Alburnoides bipunctatus* (Bloch, 1782). Věst. Čs. spol. zool. 26: 333–351.

Balon, E.K. 1967. Vývoj ichtyofauny Dunaja, jej súčasný stav a pokus o prognózu ďalších zmien po výstavbe vodných diel (Evolution of the Danube River ichthyofauna, its recent state and an attempt to predict changes after the construction of the hydro-electric power plants). Biologické práce 13: 3–121. (In Slovak).

Balon, E.K. 1974. Fishes from the edge of Victoria Falls, Africa: demise of a physical barrier for downstream invasions. Copeia 1974: 643–660.

Balon, E.K. 1975. Reproductive guilds of fishes: a proposal and definition. J. Fish. Res. Board Can. 32: 821–864.

Balon, E.K. 1980. Early ontogeny of the lake charr, *Salvelinus (Cristivomer) namaycush*. pp. 485–562. *In:* E.K. Balon (ed.) Charrs: Salmonid Fishes of the genus *Salvelinus*, Dr W. Junk Publishers, The Hague.

Balon, E.K. 1981. Additions and amendments to the classification of reproductive styles in fishes. Env. Biol. Fish. 6: 377–390.

Balon, E.K. 1983. Epigenetic mechanisms: reflections on evolutionary processes. Can. J. Fish. Aquat. Sci. 40: 2045–2058.

Balon, E.K. 1984a. Reflections on some decisive events in the early life of fishes. Trans. Amer. Fish. Soc. 113: 178–185.

Balon, E.K. 1984b. Life histories of Arctic charrs: an epigenetic explanation of their invading ability and evolution. pp. 109–141. *In:* L. Johnson & B. Burns (ed.) Biology of the Arctic Charr, University of Manitoba Press, Winnipeg.

Balon, E.K. 1985. The theory of saltatory ontogeny and life history models revisited. pp. 13–30. *In:* E.K. Balon (ed.) Early Life Histories of Fishes: New Developmental, Ecological and Evolutionary Perspectives, Dr W. Junk Publishers, Dordrecht.

Balon, E.K. 1986. Types of feeding in the ontogeny of fishes and the life-history model. Env. Biol. Fish. 16: 11–24.

Balon, E.K. 1988. Tao of life: universality of dichotomy in biology. 2. The epigenetic mechanisms. Rivista di Biologia/ Biology Forum 81: 339–380.

Balon, E.K. 1989a. The Tao of life: from the dynamic unity of polar opposites to self-organization. pp. 7–40. *In:* M.N. Bruton (ed.) Alternative Life-History Styles of Aminals, Kluwer Academic Publishers, Dordrecht.

Balon, E.K. 1989b. The epigenetic mechanisms of bifurcation and alternative life-history styles. pp. 467–501. *In:* M.N. Bruton (ed.) Alternative Life-History Styles of Animals, Kluwer Academic Publishers, Dordrecht.

Balon, E.K. & M.N. Bruton. 1986. Introduction of alien species or why scientific advice is not heeded. Env. Biol. Fish. 16: 225–230.

Balon, E.K., S.S. Crawford & A. Lelek. 1986. Fish communities of the upper Danube River (Germany, Austria) prior to the new Rhein-Main-Donau connection. Env. Biol. Fish. 15: 243–271.

Balon, E.K., S.S. Crawford & A. Lelek. 1987. Are there sympatric forms of *Vimba* Fitzinger 1873 in the Danube near the future connection to the Main River? Senckenbergiana biol. 67: 321–248.

Balon, E.K., S.S. Crawford & A. Lelek. 1988. Is the occurrence of *Gobio albipinnatus* Lukasch 1933 in the upper Danube a result of upriver invasion or sympatric speciation? Senckenbergiana biol. 68: 275–299.

Balon, E.K., S.S. Crawford & A. Lelek. 1991. Die Fische der oberen Donau: taxonomische Merkmale, Grössen- und Gewichtsvariablen vor der Vollendung des Rhein-Main-Donau-Kanals. Cour. Forsch.-Inst. Senckenberg (in press).

Bănărescu, P., M. Papadopol & G. Müller. 1963. Le genre *Vimba* (Piscus, Cyprinidae) dans le basin du Danube. Trav. Mus. Hist. nat. 'Gr. Antipa' 4: 381–400.

Bănărescu, P., M. Papadopol & L. Mikhailova. 1970. Systematics. pp. 23–70. *In:* P. Zayanchkauskas (ed.) Biology and Fisheries of *Vimba* in Europe, Mintis Press, Vilnius. (In Russian).

Barel, C.D.N., R. Dorit, P.H. Greenwood, G. Fryer, N. Hughes, P.B.N. Jackson, H. Kawanabe, R.H. Lowe-McConnell, M. Nagoshi, A.J. Ribbink, E. Trewavas, F.

Witte & K. Yamaoka. 1985. Destruction of fisheries in Africa's lakes. Nature 315: 19–20.

Bruton, M.N. 1989. The ecological significance of alternative life-history styles. pp. 503–553. *In:* M.N. Bruton (ed.) Alternative Life-History Styles of Animals, Kluwer Academic Publishers, Dordrecht.

Bruton, M.N. 1990. The conservation of the fishes of Lake Victoria, Africa: an ecological perspective. Env. Biol. Fish. 27: 161–175.

Buth, D.G. 1984. Allozymes of the cyprinid fishes. Variation and application. pp. 561–590. *In:* B.J. Turner (ed.) Evolutionary Genetics of Fishes, Plenum Press, New York.

Coulter, G.W., B.R. Allanson, M.N. Bruton, P.H. Greenwood, R.C. Hart, P.B.N. Jackson & A.J. Ribbink. 1986. Unique qualities and special problems of the African Great Lakes. Env. Biol. Fish. 17: 161–184.

Echelle, A.A. & I. Kornfield (ed.). 1984. Evolution of fish species flocks. University of Maine at Orono Press, Orono. 257 pp.

Flegler-Balon, C. 1989. Direct and indirect development in fishes – examples of alternative life-history styles. pp. 71–100. *In:* M.N. Bruton (ed.) Alternative Life-History Styles of Animals, Kluwer Academic Publishers, Dordrecht.

Greenwood, P.H. 1981. The haplochromine fishes of the East African lakes. Kraus Intern. Publ., München. 839 pp.

Hoffman, A. 1989. Arguments on evolution. A paleontologist's perspective. Oxford University Press, New York. 274 pp.

Hoogerhoud, R.J.C., F. Witte & C.D.N. Barel. 1983. The ecological differentiation of two closely resembling *Haplochromis* species from Lake Victoria (*H. iris* and *H. hiatus*, Pisces, Cichlidae). Neth. J. Zool. 33: 283–305.

Holčík, J. & I. Bastl. 1976. Ecological effects of water level fluctuation upon the fish populations in the Danube River floodplain in Czechoslovakia. Acta Sci. nat. Brno 10: 1–46.

Holčík, J., I. Bastl, M. Ertl & M. Vranovsky. 1981. Hydrobiology and ichthyology of the Czechoslovak Danube in relation to predicted changes after the construction of the Gabčikovo-Nagymaros River Barrage System. Práce Laboratória Rybarstva a Hydrobiologie 3: 19–158.

Jantsch, E. 1980. The self-organizing universe. Scientific and human implications of the emerging paradigm of evolution. Pergamon Press, Oxford. 343 pp.

Jonsson, B., S. Skúlason, S.S. Snorrason, O.T. Sandlund, H.J. Malmquist, P.M. Jónasson, R. Gydemo & T. Lindem. 1988. Life history variation of polymorphic arctic charr in Thingvallavatn, Iceland. Can. J. Fish. Aquat. Sci. 41: 1537–1547.

Koblickaya, A.F. 1966. On spawning of freshwater fishes. Pishtchevaya Promyshlennost, Moscow. 109 pp. (In Russian).

Kryzhanovsky, S.G. 1949. Ecomorphological principles of development of carps, loaches and catfishes. Trudy Inst. Morphol. Zhiv. Severtsova 1: 5–332. (In Russian).

Løvtrup, S. 1987. Darwinism: the refutation of a myth. Croom Helm, London. 469 pp.

Magnússon, K.P. & M.M. Ferguson. 1987. Genetic analysis of four morphs of arctic charr (*Salvelinus alpinus* L.) from Thingvallavatn, Iceland. Env. Biol. Fish. 20: 67–73.

Maturana, H.R. & F.J. Varela. 1988. The tree of knowledge. The biological roots of human understanding. Shambhala, Boston. 263 pp.

McKaye, K.R., T. Kocher, P. Reinthal & I. Kornfield. 1982. A sympatric sibling species complex of *Petrotilapia* Trewavas from Lake Malawi analysed by enzyme electrophoresis (Pisces: Cichlidae). Zool. J. Linn. Soc. 76: 91–96.

McKaye, K.R., T. Kocher, P. Reinthal, R. Harrison & I. Kornfield. 1984. Genetic evidence for allopatric and sympatric differentiation among color morphs of a Lake Malawi cichlid fish. Evolution 38: 215–219.

Noakes, D.L.G., S. Skúlason & S.S. Snorrason. 1989. Alternative life-history styles in salmonine fishes with emphasis on arctic charr, *Salvelinus alpinus*. pp. 329–346. *In:* M.N. Bruton (ed.) Alternative Life-History Styles of Animals, Kluwer Academic Publishers, Dordrecht.

Ogutu-Ohwayo, R. 1990. The decline of the native fishes of lakes Victoria and Kyoga (East Africa) and the impact of introduced species, especially the Nile perch, *Lates niloticus*, and the Nile tilapia, *Oreochromis niloticus*. Env. Biol. Fish. 27: 81–96.

Popper, K.R. 1968. Conjectures and refutations. The growth of scientific knowledge. Harper & Row, New York. 417 pp.

Prigogine, I. 1980. From being to becoming. Time and complexity in the physical sciences. W.H. Freeman, San Francisco. 272 pp.

Ribbink, A.J. 1986. The species concept, sibling species and speciation. Ann. Kon. Mus. Mid. Afr., Zoöl. Wetensch. 251: 109–116.

Ribbink, A.J. 1987. African lakes and their fishes: conservation scenarios and suggestions. Env. Biol. Fish. 19: 3–26.

Ribbink, A.J., B.A. Marsh, A.C. March, A.C. Ribbink & B.J. Sharp. 1983. A preliminary survey of the cichlid fishes of rocky habitats in Lake Malawi. S. Afr. J. Zool. 18: 149–310.

Riedl, R. 1975. Die Ordnung des Lebendigen: Systembedingungen der Evolution. Paul Parey, Hamburg. 372 pp. (also as: Order in living organisms. A system analysis of evolution, John Wiley & Sons, Chichester, 1978, 313 pp.).

Sandlund, O.T., B. Jonsson, H.J. Malmquist, R. Gydemo, T. Lindem, S. Skúlason, S.S. Snorrason & P.M. Jónasson. 1987. Habitat use of arctic charr, *Salvelinus alpinus*, in Thingvallavatn, Iceland. Env. Biol. Fish. 20: 263–274.

Smirnova, E.N. 1957. Principles of development of the Kuban zährte in embryonic and larval periods of life. Trudy Inst. Morphol. Zhiv. Severtsova 20: 71–94. (In Russian).

Smirnova, E.N., M.Z. Vladimirov & P. Volskis. 1970. Development. pp. 155–290. *In:* P. Zayanchkauskas (ed.) Biology and Fisheries of *Vimba* in Europe, Mintis Press, Vilnius. (In Russian).

Wanzenböck, J., H. Kovacek & B. Herzig-Straschil. 1989. Zum Vorkommen der Gründlinge (Gattung: Gobio; Cyprinidae) im österreichischen Donauraum. Österreichs Fischerei 42: 118–128.

Witte, F. 1984. Ecological differentiation in Lake Victoria haplochromines: comparison of cichlid species flocks in African lakes. pp. 155–167. *In:* A.A. Echelle & I. Kornfield (ed.) Evolution of Fish Species Flocks, University of Maine at Orono Press, Orono.

Environmental Biology of Fishes **33**: 181–193, 1992.

Comparative microhabitat use of cyprinid larvae and juveniles in a lotic floodplain channel

Gordon H. Copp
Université Claude Bernard, Lyon I, Départment de Biologie animale et Ecologie, 43, bulv. du II novembre 1918, 69622 Villeurbanne, cedex, France
Present address: Hatfield Polytechnic, Division of Environmental and Biosciences, College Lane, Hatfield, Herts, AL10 9AB, U.K.

Received 15.1.1990 Accepted 5.9.1990

Key words: Pisces, Cyprinidae, Correspondence analysis, Electrofishing, Point abundance sampling, Upper Rhône River, Early life history, Electivity index, Predation, Resource partitioning

Synopsis

The microhabitat of six cyprinids in a braided side-channel of the Upper Rhône River, France, was studied June to September 1985, using 'point abundance sampling' by electrofishing. Correspondence analysis of the samples-by-species data and Jacob's electivity index (calculated from the availability and exploitation of eight environmental variables) revealed three groups of species: (1) *Leuciscus cephalus* and *Alburnus alburnus,* which used relatively similar microhabitat during both larval and 0+ juvenile development; (2) *Rutilus rutilus, Leuciscus leuciscus* and *Chondrostoma nasus,* which used different microhabitats as larvae but whose microhabitat overlapped notably during 0+ juvenile development; (3) and *Gobio gobio,* whose juvenile microhabitat overlapped only marginally with that of other 0+ juveniles (no data were available on larvae). Overlap in microhabitat use by 0+ juveniles increased notably during a period of reduced discharge, when the amount of available vegetal and ligneous structures decreased; while most species exploited the increased area of shallow waters, juvenile *A. alburnus* appeared to be the only species that exploited the deeper waters off the channel's steeper banks. Although not quantified, numerous field observations of predation threat, particularly during the period of reduced discharge (which reduced the amount of cover), suggest that predation risk may influence the microhabitat use of young fishes. Compared with similar species in European lakes, the overlap in microhabitat use of these fluvial cyprinids appears to be more intensive.

> Individuals of different sizes in size-structured populations often differ greatly in the use of resources and/or space... The presence or absence of such diet or habitat segregation between different size classes can greatly affect population structure and dynamics.
>
> Gilliam & Fraser (1988)

Introduction

Microhabitat use is an important aspect in the life history of fluvial fishes, particularly during early ontogeny, when relatively small body size and restricted motility render young fishes susceptible to environmental hazards such as elevated water velocities (Lightfoot & Jones 1979, Heggenes 1988) and predation (Baltz et al. 1982, Cerri & Fraser 1983, Power & Matthews 1983, Scott 1987, Rozas & Odum 1988, Robinson 1989). It is therefore suprising that most accounts of microhabitat use by cyprinid larvae and early juveniles in European rivers have been incidental to studies of growth or diet (e.g. Stankovitch 1921, Brown 1979, Carrel 1986, Scott 1987, Haberlehner 1988). The few investigations directly concerned with the microhabitat use of fish larvae and juveniles in medium-to-large European rivers have either been limited to one species (Lightfoot & Jones 1979, Copp 1990a) or addressed but one or a few environmental variables (Schröder 1979, Schiemer & Spindler 1989).

In European lakes, the repartition of spatial resources by young cyprinids is facilitated by offshore migrations in 0 + juvenile roach *Rutilus rutilus* (L.), and the vertical segregation of other species, e.g. *Leuciscus cephalus* (L.) and *Scardinius erythrophthalmus* (L.), and their sub-cohorts (Hammer 1985, Mark et al. 1987, Rheinberger et al. 1987). In smaller rivers, *R. rutilus* and dace *Leuciscus leuciscus* (L.) of a few months age are reported to move from backwaters into lotic areas of the channel (Scott 1987), though Lightfoot & Jones (1979) found *R. rutilus* to be closely associated with macrophyte beds, which considerably reduced water velocities therein. By contrast, in a relatively wider side channel of a large river, *R. rutilus* larvae and juveniles were found to avoid water currents and preferred the shallow lentic waters of weakly sloped banks (Copp 1990a). To understand the dynamics of such differences in microhabitat use by different species of fish larvae and juveniles, comparative studies of habitat segregation/co-exploitation are essential. Thus, in the present study I compare the microhabitat use of cyprinid larvae and juveniles in a braided side-channel of the Upper Rhône River, a comparison that examines the temporal and spatial dynamics, as well as the preference for, or avoidance of, eight environmental variables by *R. rutilus, L. cephalus, L. leuciscus*, nase *Chondrostoma nasus* (L.), bleak *Alburnus alburnus* (L.) and gudgeon *Gobio gobio* (L.) during their larval and early juvenile (0 +) development.

Methods and materials

The cyprinid larvae and juveniles were sampled once a week (in daylight) from early June to mid September 1985, then once again in late September (15 excursions in total), in a side-channel of the Brégnier-Brangues flood plain (Upper Rhône River, France) colloquially known as the 'Lône d'Ilon', henceforth referred to as 'Ilon' (Copp & Peňáz 1988). This side-channel, which is of braided geomorphological origin (see Petts & Foster 1985), has a surface area of about 6 ha and lies within a flood plain that was in the process of being by-passed by a hydroelectric development under construction during the study year. Just upstream of this flood plain, the Upper Rhône has a mean summer discharge of $\approx 700\,m^3\ s^{-1}$, whereas the by-passed flood plain receives $\approx 150\,m^3\ s^{-1}$ (Copp & Peňáz 1988). Ilon is approximately 40 m across at its downstream confluence with the main river and about 60 m across at its upstream end. The water temperature at Ilon fluctuated between 15 and 19° C in June 1985, remained at about 15° C through late June and early July, rose progressively to about 20° C through July until mid-August, then fluctuated between 15 and 21° C until the end of the study. The water conductivity generally fluctuated between 250 and 350 μS (Copp 1990a).

A wing dam had been constructed at Ilon's upstream end to deflect the main river's flow. This structure accentuated aggradation of Ilon's upstream end, forming an alluvial plug that in turn created a range of conditions, depending upon river discharge. Colonisation of the alluvial plug by pioneer vegetation had helped establish an island a few metres behind the wing dam. This island was separated from the channel's right (convex) bank by a 20 to 30 m wide riffle bed and a small, slightly deeper gully ($\approx$ 0.75 m) through which water could pass during slightly reduced discharge (overall river $\approx 500\,m^3\ s^{-1}$). The island was separated from the wing dam and the left (concave) bank by a relatively deep gully (≈ 1.5 m), through which water passed continuously except during extremely reduced river discharge (overall river $\approx 350\,m^3\,s^{-1}$, by-passed flood plain minimum $\approx 120\,m^3\,s^{-1}$). Downstream of the alluvial plug, the weakly-sloped convex right bank was composed of gravel changing progressively downstream to sand and silt deposits, the bank became muddier and steeper as the channel approached its downstream confluence with the river's main channel. In contrast, because of scour action, the concave left bank remained steep throughout its length, was composed of mud/silt and exposed rock embankment, and possessed a number of dead and living branches/trees lying at, or submerged below, water level.

During elevated discharge ($> 700\,m^3\,s^{-1}$), the right bank of alluvial plug became a riffle bed, with some lentic zones behind the island, behind the wing dam where it joins the left bank, and along the periphery of the convex right bank. During mean discharge, weak-to-moderate current velocities were limited to the gullies on either side of the mid-channel island. As scouring action and alluvial deposition still occurred during elevated discharge, aquatic vegetation was generally confined to the margin of the right bank and almost entirely around the mid-channel island. During extremely reduced discharge, such as occurred for about three weeks between mid-August and early September 1985, the upstream end was almost completely isolated from the main river, rendering the channel almost entirely lentic and exposing vast shallows along the right bank, opposite the long, deep gully of the concave left bank. During this period of extremely reduced discharge, the available aquatic vegetation and ligneous debris was limited to some areas on the island's periphery and to the more downstream sections of either bank of the channel.

Sampling by a two-person crew from a rubber boat was undertaken using a battery-powered, portable electrofisher equipped with a 10 cm anode for greater efficiency with fish larvae and juveniles (Copp 1989b). The samples were collected according to the 'point abundance sampling' strategy, originally described by Blondel et al. (1970) for nesting birds, adapted for adult fishes by H. Persat (Nelva et al. 1979) and for fish larvae and juveniles by Copp & Peňáz (1988). Similar sampling approaches have been applied to microhabitat studies elsewhere (Baltz et al. 1982, Bain et al. 1985, Moyle & Baltz 1985). Selection of the 25 weekly sampling points was made with a random point of the index finger (eyes closed), a valid alternative to computer-generated coordinates (D. Chessel personal communication), which would be virtually impossible to locate in the field. The anode's effective field, i.e. the sample area, was measured as $\approx 0.07\,m^2$ (Copp 1989b). Although 25 point samples were anticipated for each week, the number of samples on 12 June was limited to 15 because of equipment failure, thus a total of 365 point samples were collected over the study period (i.e. 14×25, plus 15).

At each sampling point, the fishes collected were preserved in 4% formaldehyde, after which microhabitat availability and use were evaluated via qualitative and quantitative measurements of eight environmental variables: situation (channel, weak bank, steeper bank), water depth (<0.2, 0.2–0.5, 0.5–1.0, >1.0 m), water current (absent, weak, strong), substratum (mineral/silty mud [<0.06 cm], silty sand [0.06–0.2 cm], silted gravel [0.2–2.0 cm], silted rocks [2–20 cm], indeterminable), water surface illumination (sun, mixed, shade, overcast), ligneous debris (absent, some, dense), macrophytes (absent, some, dense) and periphyton (absent, attached, detached). Water depth was measured with a graduated dip-net pole. Water current was evaluated using a small dip net with flexible 0.5 mm mesh netting; upon immersion of the hand net, no ballooning of the dip net was noted as current absent, gentle ballooning indicated a weak water velocity and rapid ballooning of the net represented a strong current. Because rapid water velocities were rarely encountered (less than 10 points from 365 points), water current is considered here as either absent or present. Similarly, dense vegetation rarely occurred and therefore macrophytes are also considered as either present or absent. Water surface illumination was evaluated as indicated above to test the common assumption that freshwater fish larvae prefer sunny locations to shaded areas.

In the laboratory, the cyprinids were identified according to Koblickaya (1981), measured using an observation microscope fitted with a calibrated micrometer lens, and categorised in three intervals with respect to crucial events in their larval and 0 + juvenile development (E.K. Balon personal communication): young larvae, first feeding to differentiation of the finfold; older larvae, differentiation of the finfold to transition to the juvenile period; and the juvenile period (as defined by Balon 1985).

The data were arranged in three matrices. Firstly, a 'fish-by-attribute' matrix (1365 × 3) contained the sample number, the developmental interval (young larvae, older larvae, juvenile) and standard length in mm of each specimen collected. From this matrix the mean standard length and associated statistics were calculated. Secondly, a 'samples-by-species' matrix (365 × 16) contained the number of young larvae, older larvae and 0 + juveniles of each species found at each sampling point. This matrix was reduced to non-null samples only (i.e. 115 × 16), used to calculate the index of dispersion (division of the variance by the mean), and then was submitted to 'correspondence analysis' (Benzecri 1969) using the multivariate package 'Macmul' (Thioulouse 1989). Correspondence analysis is particularly suitable for revealing patterns in fractional data sets, i.e. those comprised of partial or incomplete histograms (Persat & Chessel 1989), such as in the present case. Thirdly, a 'samples-by-variables' matrix (365 × 10) contained the various classes of the eight environmental variables recorded at each sampling point. Proper evaluation of microhabitat availability and use requires not only an unbiased, a priori sampling strategy (Bain & Finn 1990) but also a comprehensive interpretative tool. Thus, from the original 'samples-by-species' and the 'samples-by-variables' matrices, microhabitat preferences were calculated using Jacobs' (1974) improved version of Ivlev's electivity index:

$$D = \frac{r - p}{(r + p) - 2rp},$$

where r is the proportion of the resources used by each species (interval) and p is the proportion available in the environment. When calculating p, only weeks containing a given species interval (i.e. young larvae, older larvae, juveniles) were included. To evaluate the effect of reduced discharge on the microhabitat use of 0 + juveniles, electivities for this development interval were calculated separately for the period of mean discharge and for reduced flow. As in Bain et al. (1985), use of an *a priori* sampling strategy (e.g. 'point abundance sampling') incorporates microhabitat availability into the data through the collection of randomly-selected sampling points.

Results

Of the 365 point samples taken, 115 contained at least one of the six cyprinids addressed in the present study: *C. nasus, L. leuciscus, L. cephalus, A. alburnus, R. rutilus* and *G. gobio*. Of these, 686 were larvae and 679 were 0 + juveniles, of which the latter were generally found to be more dispersed among the samples than the former (Table 1). Only one *G. gobio* larva was collected, but the juvenile gudgeon demonstrated the highest degree of clumping. Two other cyprinids, common bream *Abramis brama* (L.) and silver bream *Blicca bjoerkna* (L.), were collected sporadically and in very low numbers. Throughout most of the summer of 1985, but particularly during the period of extremely reduced discharge, shoals of 0 + cyprinids were often observed rushing up into the shallow areas when threatened by a predator, usually observed to be perch *Perca fluviatilis* L.

Relevant axes from the 'correspondence analysis' of the reduced 'samples-by-species' matrix (115 × 16) were selected graphically (Fig. 1A), as described by Persat & Chessel (1989), who stated that 'in most cases we can graphically identify a rupture between two groups of values [dotted lines in Fig. 1A]. The first group of values corresponds to the structured part of the matrix [the initial four in the present case, Fig. 1A], and the second group, which constitutes the weakly decreasing values, indicates the ensemble of individual variations among the [species] and the samples' (my translation). Interpretation of the resulting ordinations from the 'correspondence analysis' (Figs. 1B, C) was facilitated using inertia analysis of the initial four axes (Table 2, programme by Thioulouse 1989), which cumulatively accounted for 43% of the variance, as well as the information on microhabitat electivites (Fig. 2, 3). Three groups of species were identified: (1) *L. cephalus* and *A. alburnus*, which used similar but apparently adjacent (i.e. overlapping) microhabitat during both larval and 0+ juvenile development; (2) *R. rutilus*, *L. leuciscus* and *C. nasus*, which used different microhabitats as larvae but whose microhabitats overlapped more notably during 0+ juvenile development, particularly during the three-week period of reduced river discharge; (3) and *G. gobio*, whose juvenile microhabitat overlapped only marginally with that of other 0+ juveniles (no data were available during larval development).

Despite the earlier appearance of the lithophilic (sensu Balon 1985) *L. leuciscus* and *C. nasus*, microhabitat co-exploitation by these species and the phyto-lithophilic *R. rutilus* was suggested by the proximity of the larvae in the ordination of axes one and two of the 'correspondence analysis' (Fig. 1B), overlap that could have been facilitated by the protracted spawning demonstrated in all three species (Table 1). However, spatial differences in microhabitat use of these three species were apparent from their respective electivities (white bars, Fig. 2), differences accounted for in the ordination of axes 3 and 4 (Fig. 1C). Although both young and older larvae of *R. rutilus* and *L. leuciscus* (Rr1 &

Table 1. Samples and specimen statistics for young larvae (exogenous feeding to finfold differentiation), older larvae (finfold differentiation to juvenile metamorphosis) and 0+ juvenile cyprinids (see Balon 1985) collected in the braided channel 'Ilon' during summer 1985, with the period of presence (dates) given in day/month. For samples, the number (n) of individuals from each species interval, the number of points (f) in which they occurred, the mean number and standard error (S.E.) of individuals in the 365 samples, and their index of dispersion (variance divided by mean) are given. For specimens, the mean standard length (mean SL) in mm, the standard error (S.E.) and range of lengths are given.

Species/interval	Dates	Samples					Specimens		
		n	f	mean	S.E.	I.D.	mean SL	S.E.	range(mm)
C. nasus									
young larvae	5/6–18/6	16	4	0.14	0.11	10.71	12.9	0.2	11.8–14.8
older larvae	12/6–16/7	13	6	0.11	0.05	2.29	16.4	0.4	14.6–18.8
juveniles	from 16/7	68	20	0.59	0.22	9.70	27.7	0.7	19.6–45.3
L. leuciscus									
young larvae	12/6–24.6	40	8	0.35	0.15	6.91	12.1	0.3	7.6–15.6
older larvae	12/6–16.7	74	8	0.63	0.38	25.83	15.0	0.2	10.5–18.3
juveniles	from 16/7	24	11	0.17	0.06	2.04	40.1	2.9	16.5–75.2
A. alburnus									
young larvae	22/7– 5/8	16	3	0.14	0.11	10.84	8.4	0.2	6.6–10.0
older larvae	22/7–27/8	58	5	0.51	0.34	25.52	10.4	0.1	9.4–14.0
juveniles	from 16/7	31	8	0.28	0.14	8.30	22.5	1.1	14.8–36.1
L. cephalus									
young larvae	5/6– 5/8	38	12	0.37	0.17	9.24	11.5	0.3	8.1–14.7
older larvae	5/6–27/8	60	15	0.52	0.18	7.20	12.2	0.2	9.9–17.1
juveniles	from 5/8	30	12	0.26	0.11	4.85	25.7	2.0	14.6–56.1
Rutilus rutilus									
young larvae	24/6 12/8	82	18	0.68	0.29	13.91	9.9	0.2	6.9–13.8
older larvea	18/6– 5/8	275	27	2.53	0.88	35.22	14.3	0.1	10.4–20.3
juveniles	from 16/7	342	53	3.09	0.49	8.90	25.3	0.3	13.7–39.1
Gobio gobio									
juveniles	from 5/8	184	12	1.61	0.91	59.03	23.8	0.3	16.0–40.2

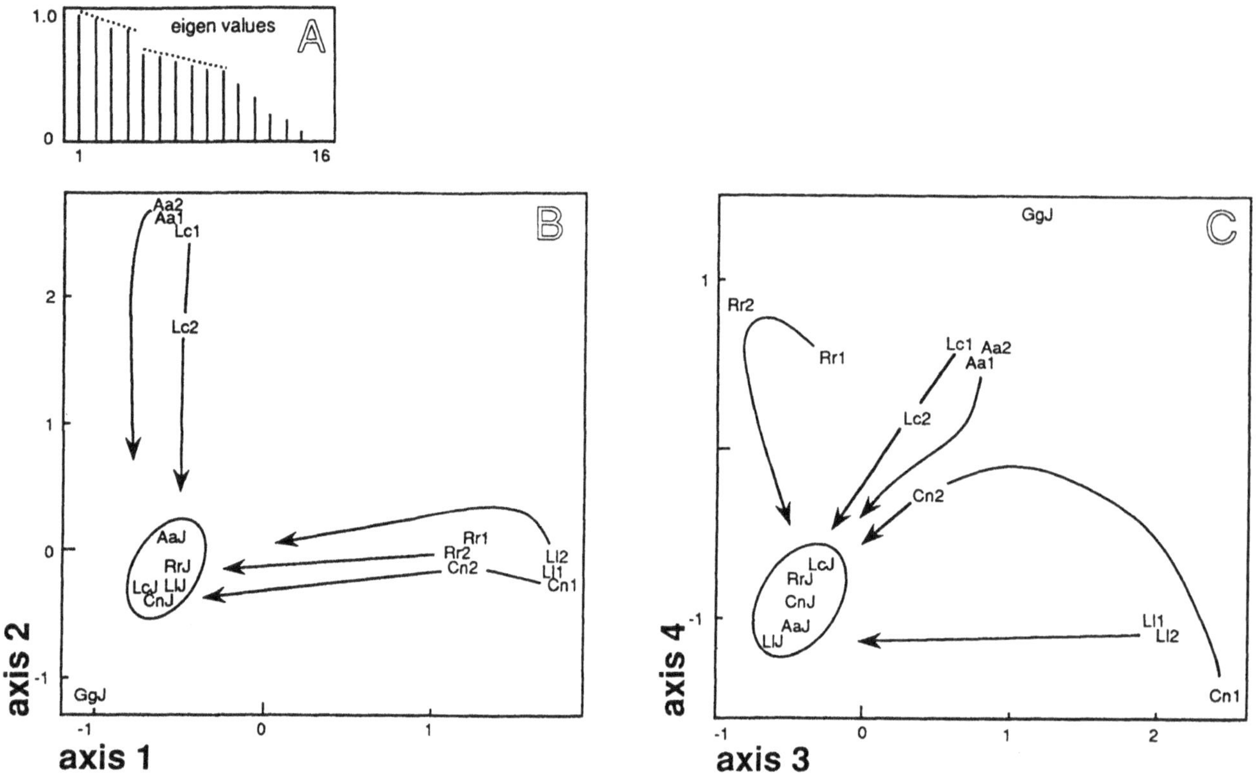

Fig. 1. 'Correspondence analysis' of the 'samples-by-species' matrix (115 × 16). A - Graphical selection of relevant eigen values, four in this case, by identification of the rupture in decreasing values as described by Persat & Chessel (1989). B - Ordination of axes one and two, C - ordination of axes two and three, revealing three groups of species: (1) *R. rutilus* (Rr), *C. nasus* (Cn) and *L. leuciscus* (Ll), which have little larval co-exploitation of spatial resources (suffixes: 1 = young larvae, 2 = older larvae), but more as juveniles (suffix: J = 0 + juveniles); (2) *A. alburnus* (Aa) and *L. cephalus* (Lc), which demonstrate overlap during both larval and juvenile development; (3) *G. gobio* (Gg.), which maintained some spatial distinction from other juveniles but for which no data on larvae were available.

Rr2, Ll1 & Ll2 respectively in Fig. 1) preferred lentic waters over weakly sloped banks of mineral/silty mud or silty sand (Fig. 2), the young *L. leuciscus* larvae inhabited a range of water depths, as did the young *C. nasus* (Cn1), and demonstrated a preference for macrophytes and attached periphyton (D = +0.8), but avoided ligneous debris (e.g. branches and areas without periphyton (D = − 1.0). In contrast, young *R. rutilus* larvae preferred waters of 0.5–1.0 m deep with dense debris and/or macrophytes. The young *C. nasus* larvae also preferred dense debris but differed from both *R. rutilus* and *L. leuciscus* in its avoidance of macrophytes and periphyton (D = − 1.0 for presence, D = + 0.6 in absence) as well as its preference for the silted gravel bottoms of the channel (Fig. 2). Young larvae of all three species demonstrated little preference for directly illuminated (i.e. sunny) waters (D = 0 to + 0.3).

As older larvae (black bars, Fig. 2), *R. rutilus* (Rr2 in Fig. 1) moved into the moderate deep waters (0.2–0.5 m) preferred by *L. leuciscus* larvae (both Ll1 & Ll2), but *R. rutilus* exploited areas of mineral/silty mud with detached periphyton (D = + 0.6 in presence, D = − 1.0 in absence) and macrophytes, whereas *L. leuciscus* preferred silted gravel with attached periphyton (D = + 0.8 in presence, D = − 1.0 in absence) and no aquatic vegetation. The older *C. nasus* larvae (Cn2) also moved to shallower waters (<0.2 m) but rather than over weak banks they preferred the steeper banks of mineral/silty mud without periphyton (D = − 1.0 for presence, D = + 0.5 in absence) though with some ligneous debris and macrophytes

(Fig. 2). Preference in all three species for sun illuminated areas increased only marginally (D values of + 0.3 to + 0.6), with only older *R. rutilus* larvae avoiding shaded waters (D = − 1.0).

Despite a seven-week difference (Table 1) in the appearance of *L. cephalus* larvae (Lc1, Lc2) and *A. alburnus* larvae (Aa1, Aa2), protracted spawning in *L. cephalus* resulted in similar microhabitat use by these two phyto-lithophilic species during larval development, which is evident in both ordinations of the 'correspondence analysis' (Fig. 1). The microhabitat electivities of *A. alburnus* and *L. cephalus* larvae (both young and older) were essentially the same: lentic waters of 0.2–0.5 m depth over silted gravel with macrophytes and some ligneous debris (Fig. 2). Both species avoided waters shaded from the sun (D = 1.0), but only *A. alburnus* demonstrated a preference for directly illuminated waters (D = 0.6 and 0.7). Because the sixteen specimens of young *A. alburnus* larvae occurred in only three samples (Table 1), but concurrently with *L. cephalus*, their apparent preference for steeper banks and sunny areas should be considered tentative.

After completion of juvenile metamorphosis, the cyprinids demonstrated notable spatial overlap (Fig. 1), with only the immigrating *G. gobio* (GgJ in Figs. 1B & 1C) showing notable segregation. With the near absence of *G. gobio* larvae during the entire summer, I assume that they migrated in from upstream, a phenomen observed elsewhere on the Upper Rhône (Copp & Cellot 1988). With respect to microhabitat electivities of juveniles during normal discharge (white bars, Fig. 3), most species preferred lentic, shallow waters (< 0.2 m in all but *R. rutilus*, which preferred 0.2–0.5 m deep) over miniral/silty mud or silted gravel. However, while *C. nasus* and *L. leuciscus* demonstrated some preference for areas with macrophytes, both *L. cephalus* and *R. rutilus* avoided these and showed some preference for ligneous debris, which was avoided by *C. nasus*, *L. leuciscus* and *A. alburnus*

Table 2. Inertia analysis (programme by Thioulouse 1989) establishing the absolute and relative inertia (from 10 000), i.e. contribution, of each species interval to the first four axes of the 'correspondence analysis' (Fig. 1). Major contributors for each axis are underlined.

Species	axis 1		axis 2		axis 3		axis 4	
	absolute	relative	absolute	relative	absolute	relative	absolute	relative
C. nasus								
young larvae	344	560	11	17	763	<u>1104</u>	318	451
older larvae	121	289	4	9	13	27	12	25
juveniles	263	481	115	202	152	248	539	860
L. leuciscus								
young larvae	838	<u>2067</u>	18	42	1233	<u>2709</u>	388	835
older larvae	1504	<u>2481</u>	37	58	2442	<u>3587</u>	832	<u>1198</u>
juveniles	77	289	21	77	66	223	228	753
A. alburnus								
young larvae	53	261	859	<u>4045</u>	62	272	38	165
older larvae	199	334	3293	<u>5280</u>	243	362	151	220
juveniles	100	159	0	1	89	126	356	494
L. cephalus								
young larvae	97	156	1990	<u>3053</u>	112	159	121	169
older larvae	141	278	1347	<u>2542</u>	38	67	9	16
juveniles	139	245	32	55	36	56	175	271
R. rutilus								
young larvae	836	<u>1293</u>	0	0	55	76	167	226
older larvae	2557	<u>3721</u>	16	23	2148	<u>2782</u>	1639	<u>2081</u>
juveniles	962	<u>1706</u>	108	184	776	<u>1224</u>	2099	<u>3247</u>
G. gobio								
juveniles	1769	<u>2216</u>	2148	<u>2579</u>	1773	<u>1977</u>	2926	<u>3199</u>

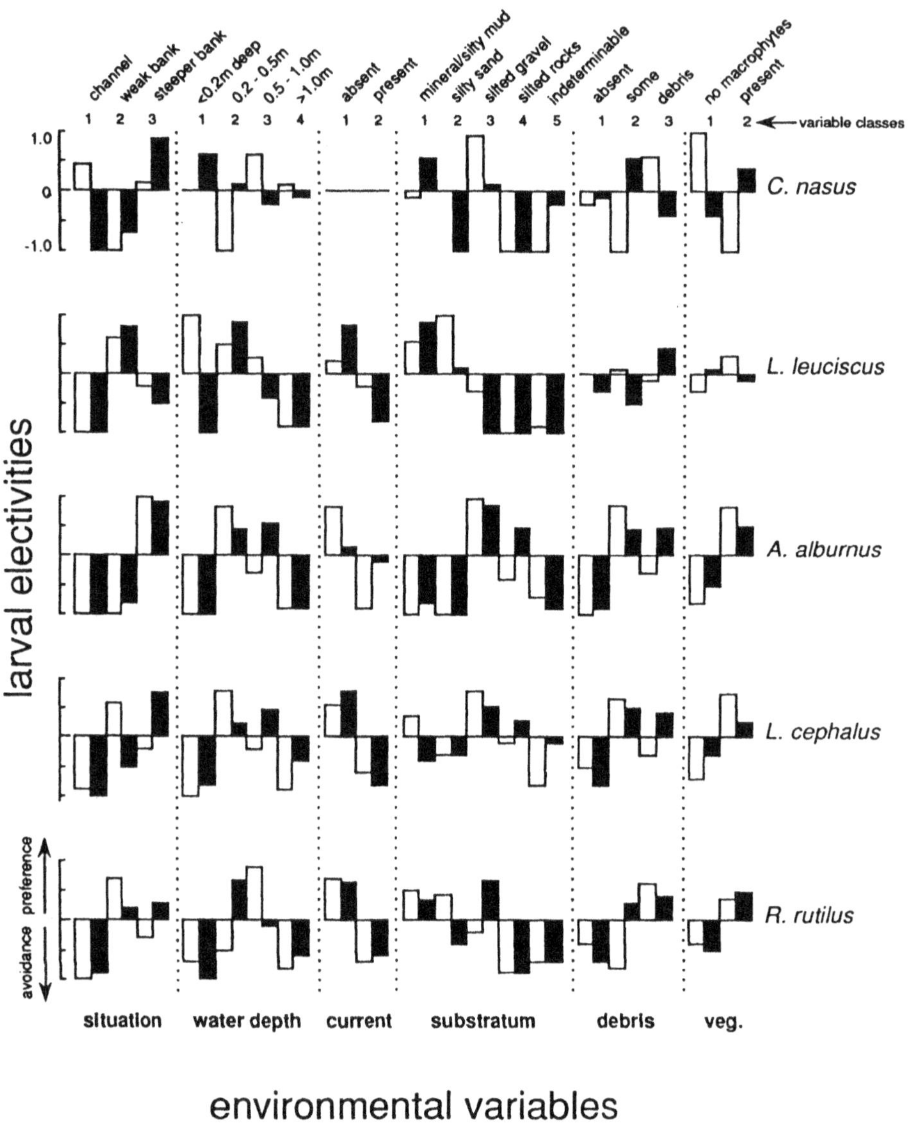

Fig. 2. Electivity index D from Jacob's (1974) formula for cyprinid young larvae and older lavae in Ilon during 1985. The white bars correspond to young larvae (suffix '1' in Fig. 1B, C), and black bars to older larvae (suffix '2' in Fig. 1B, C).

(Fig. 3). Only *A. alburnus* juveniles demonstrated any preference for directly illuminated waters (D = +0.8), and little or no preference for attached or detached periphyton was evident (D values of −1.0 to 0.4); in fact, both *L. cephalus* and *A. alburnus* juveniles demonstrated an avoidance of periphyton areas (D = 1.0 in absence, D = −1.0 in presence).

During reduced discharge, most species modified their microhabitat to some extent (black bars, Fig. 3) except the *G. gobio*, whose continued inhabitation of shallow weak banks (Fig. 3) was reflected in a moderate preference for illuminated waters (D = +0.7 in sun, D = −1.0 in shade). Both *L. cephalus* and *R. rutilus* juveniles moved to either weak or steeper banks of silty sand (also silted gravel in *L. cephalus*) with a strong preference for macrophytes. Although *R. rutilus* moderately preferred dense ligneous debris and shaded or mixed (shade and sun) areas (D = +0.4 and

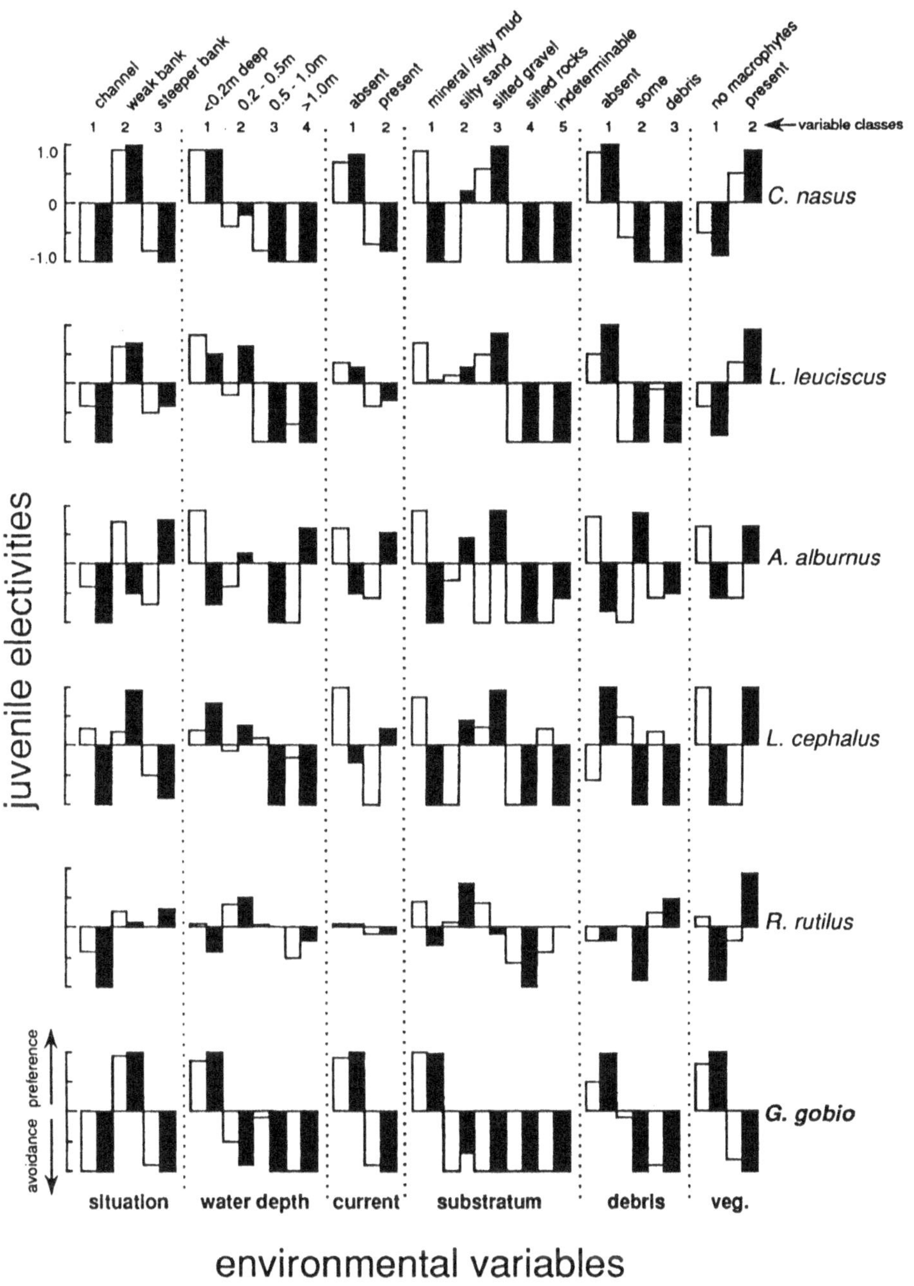

Fig. 3. Electivity index D from Jacob's (1974) formula for cyprinid 0 + juveniles (suffix 'J' in Fig. 1B, C) during mean discharge (white bars) and extremely reduced discharge (black bars) in Ilon during 1985.

+ 0.5), *L. cephalus* avoided debris and shade (D = − 1.0) as well as demonstrating a weak preference for water current. Juvenile *A. alburnus* also showed a mild preference for water current, but differed from the other species in its strong preference for steeper banks at >1.0 m depth with macrophytes, some ligneous debris (Fig. 3) and mixed illumination (D = + 1.0). Juvenile *L. leuciscus* and *C. nasus* co-exploited a very similar microhabitat during low discharge, lentic waters of 0.2 m depth (also 0.2–0.5 m for *L. leuciscus*) over silted gravel with macrophytes but avoided ligneous debris. However, while juvenile *C. nasus* showed a preference for directly illuminated waters (D = + 0.9) and avoided attached periphyton (D = + 0.8 in absence, D = − 1.0 in presence), *L. leuciscus* were indifferent to both illumination (D = + 0.2 in sun, D = + 0.6 in shade) and periphyton (D ≈ 0 for all classes).

Discussion

> It would be interesting to know if the distribution differences apparent in the size classes of many species are indicative of a general trend in fishes or of sampling emphasis.
>
> G.S. Helfman (1978)

The young cyprinids of Ilon, a braided channel of the Upper Rhône River, exhibited considerable overlap in microhabitat use (Fig. 2, 3), particularly during a period of reduced river discharge. The drastic drop in water level experienced by the 0 + juveniles as a result of this decrease in flow severely limited the amount of littoral vegetation and branches, structures that small fishes (e.g. < 70 mm) use for feeding and as protective refuge against predation (Arnold 1974, Helfman 1978, Lightfoot & Jones 1979, Northcott 1979, Schröder 1979, Kriksunov et al. 1985, Haberlehner 1988, Rozas & Odum 1988). Similar associations between marine fishes and organic structures, principally reefs, are well known (Helfman 1978, Howard 1989).

Commencement of shoaling behaviour and feeding migrations away from the littoral zone of lakes, particularly in turbid waters (Bohl 1980), often coincide with changes in morphology/physiology (Mark et al. 1989) and in diet (Hartmann 1983). In the present study (Fig. 2, 3), only juvenile *A. alburnus* demonstrated any appreciable movement away from the channel's littoral (i.e. banks), and this coincided with a preference for aquatic vegetation (previously avoided) in lotic waters along Ilon's steeper banks. The preference of most young cyprinids in Ilon for lentic, shallower waters at weakly sloped banks corroborates Schiemer & Spindler's (1989) observations of young rheophilic cyprinids in the Austrian Danube, suggesting that shallow littoral areas may provide safer daytime refuge from predation, as postulated by Bohl (1980). This contrasts Scott's (1987) observations in the River Frome, a small chalk stream of Southern England, where juvenile *L. leuciscus* moved into the elevated water velocities of the channel after inhabiting lentic areas for the first two months of life. The avoidance by *R. rutilus* larvae and juveniles of Ilon's deeper channel waters differs from their reported microhabitat dynamics in lakes (Rheinberger et al. 1987), where they apparently undergo a migration to deeper waters (≈ 1.0 m) a few weeks after hatching, whereas the microhabitat of *L. cephalus* larvae and juveniles appears to be the same under both lacustrine (Rheinberger et al. 1987) and fluvial conditions: near physical structures (vegetation or branches) in the relatively shallow littoral areas.

The preference demonstrated here for shallow waters and protective structures (Fig. 2, 3) is presumed to be an anti-predatory action (Manteifel et al. 1978) employed in response to the drop in discharge. Such predation was not empirically tested in the present study, but predator threat was frequently observed, particularly during the period of reduced discharge, when *P. fluviatilis* would chase shoals of 0 + cyprinids up into the shallows of Ilon's weakly-sloped banks. Young cyprinids such as *R. rutilus* and *Leucaspius delineatus* (L.) respond to predator threat and reduced light levels (< 0.1 lux) by moving to the substratum (Manteifel et al. 1978), a reaction also observed in *R. rutilus* of a Danubian backwater (Haberlehner 1988). This anti-predator response may well be employed by other small cyprinids (Fig. 3; Haberlehner 1988), which are generally more susceptible to predation because of their 'highly visible' shoaling behaviour and their lack of spines or barbs (Bohl 1980, Robinson 1989).

Littoral macrophytes in river channels reduce water velocities, thus creating lentic areas favourable to young fishes such as *R. rutilus* (Lightfoot & Jones 1979). In Ilon, macrophytes were rare in mid-channel and mainly restricted to the bank margins, however all of the cyprinids except *G. gobio* preferred areas with macrophytes during some if not most of their development in the first summer of life (Fig. 2, 3). In small North American cyprinids, a similar pattern of microhabitat selection has been observed; larvae also tend to frequent shallow, lentic areas with protective structures, whereas 0+ juveniles may exploit similar microhabitats but are known to venture out into deeper waters, particularly those in proximity to submerged vegetation (Keast 1978, Halyk & Balon 1983, Power 1987).

Contrary to the colloquial belief that cyprinid larvae and juveniles prefer sunny shallow waters (e.g. Stankovitch 1921), most cyprinids did not demonstrate a preference for sunny areas. Indeed, the weak preferences shown in some species for sunny areas, and the stronger preference in larvae of *A. alburnus* (see also Schröder 1979), are probably coincidental to their preferences for shallow (> 0.5 m), weakly-sloped banks. In Ilon, riparian cover generally stood well back from the weakly-sloped banks (i.e. the right bank and island), which were thus directly exposed to the sun throughout most of the day. Nonetheless, young fishes probably benefit from this coincidence, being that the illuminated shallows are often notably warmer than those of the adjacent deeper waters (Copp 1990b). Jacob's (1974) formula for electivities appeared to account well for such relations between available and exploited microhabitat. In a recent study limited to the growth and microhabitat of *R. rutilus* larvae and early juveniles in Ilon (Copp 1990a, b), chi-square analysis in general indicated the same microhabitat preferences but did not account for numerical differences in species (or interval) abundance between the point samples as does Jacob's electivity index D.

In conclusion, cyprinids of Ilon demonstrated notable spatial segregation during their early development in 1985, despite the co-occurrence of similar life-history intervals from different species caused by protracted spawning in *R. rutilus, C. nasus, L. cephalus* and *L. leuciscus*. Although the microhabitats of *L. cephalus* and *A. alburnus* larvae were very similar, those of *R. rutilus, L. leuciscus* and *C. nasus* larvae overlapped only partially (Fig. 1B, C, 2). After transformation to the juvenile period, the cyprinids continued to demonstrate some repartition of spatial resources (Fig. 3), despite the strong interspecific associations observed in Ilon (Fig. 1B, C) and elsewhere on the French Upper Rhône (Copp 1989a). The overlap in their microhabitat increased during a period of reduced discharge, which restricted the amount of available vegetal/ligneous structures and increased the amount of shallow waters around the island and the alluvial plug. While most species exploited some portion of these vast shallows, juvenile *A. alburnus* appeared to be the only species that exploited the deeper waters off the channel's steeper banks.

The numerous unquantified events of predation threat observed in the field, supported by the general preference of 0+ cyprinids for shallows and protective structures, suggest that predation risk may influence the microhabitat use of young fishes, contrary to conclusions derived elsewhere from studies of microhabitat use in adult fishes (e.g. Grossman & Freeman 1987). Although the present study provided considerable information on the spatial and temporal dynamics of microhabitat use in 0+ cyprinids, future investigations should address both the available and exploited nutritional resources, as well as the extent of predation risk, in order to clarify: the extent of resource partitioning or co-exploitation in fluvial fishes during their larval and juvenile development, the extent to which predation risk influences this repartition of resources, and the similarities or differences in resource repartitioning by 0+ cyprinids in lacustrine and fluvial conditions.

Acknowledgements

I would like to express my gratitude to R.H.K. Mann and anonymous referees for their critical review of the manuscript, to P. Leroy for field assistance, and to A.L. Roux and the PIREN-Rhône vallée fluviale of the French Scientific Research Council (CNRS) for research assistance and funding. I am grateful to the Freshwater Biological Association for providing research time to complete this study and travel funds to attend the symposium. Finally, I thank D. Chessel and J. Thioulouse for providing the 'ADE.Base' multivariate and graphics software package.

References cited

Arnold, G.P. 1974. Rheotropism in fishes. Biol. Rev. 49: 515–576.

Bain, M.B. & J.T. Finn. 1990. Analysis of microhabitat use by fish: investigator effect and investigator bias. Rivers (in press).

Bain, M.B., J.T. Finn & H.E. Booke. 1985. A quantitative method for sampling riverine microhabitats by electrofishing. N. Amer. J. Fish. Manag. 5: 489–493.

Balon, E.K. 1985. Early life histories of fishes: new developmental, ecological and evolutionary perspectives. Dev, Env. Biol. Fish. 5, Dr Junk Publishers, Dordrecht. 280 pp.

Baltz, D.M., P.B. Boyle & N.J. Knight. 1982. Competitative interactions between benthic stream fishes, riffle sculpin *Cottus gulosus,* and speckled dace, *Phinichthys osculus.* Can. J. Fish. Aquat. Sci. 39: 1502–1511.

Benzecri, J.P. 1969. Statistical analysis as a tool to make patterns emerge from data. pp. 35–60, *In:* S. Watannabe (ed.) Methodologies of Pattern Recognition, Academic Press, New York.

Blondel, J., Ferry, C. & B. Frochot. 1970. La méthode des indices ponctuels d'abondance (I.P.A.) ou des relevés d'avifaune par 'Stations d'écoute'. Alauda 38: 55–71.

Bohl, E. 1980. Diel pattern of pelagic distribution and feeding in planktivorous fish. Oecolgica 44: 368–375.

Brown, D.J.A. 1979. The distribution and growth of juvenile cyprinid fishes in rivers receiving power station cooling water discharge. pp. 217–229. *In:* Proc. 1st Brit. Freshwat. Fish. Conf., Liverpool.

Carrel, G. 1986. Characterisation physico-chimique du Haut-Rhône français et de des annexes: incidences sur la croissance des populations d'alevins. Thèse de Doctorat, Université Claude Bernard, Lyon. 185 pp.

Cerri, R.D. & D.F. Fraser. 1983. Predation and risk in foraging minnows: balancing conflicting demands. Amer. Nat. 121: 552–561.

Copp, G.H. 1989a. The habitat diversity and fish reproductive function of floodplain ecosystems. Env. Biol. Fish. 26: 1–26.

Copp, G.H. 1989b. Electrofishing for fish larvae and 0 + juveniles: equipment modifications for increased efficiency with short fishes. Aqua. Fish. Manag. 20: 177–186.

Copp, G.H. 1990a. Shifts in the microhabitat of larval and juvenile roach *Rutilus rutilus* (L.) in a floodplain channel. J. Fish Biol. 36: 683–692.

Copp, G.H. 1990b. Recognition of cohorts and growth of larval and juvenile roach *Rutilus rutilus* (L.), using size-class ordination of developmental steps. J. Fish Biol. (in press).

Copp, G.H. & B. Cellot. 1988. Drift of embryonic and larval fishes, especially *Lepomis gibbosus* (L.), in the Upper Rhône River. J. Freshwat. Ecol. 4: 419–424.

Copp, G.H. & M. Peňáz. 1988. Ecology of fish spawning and nursery zones in the flood plain, using a new sampling approach. Hydrobiologia 169: 209–224.

Gilliam, J.F. & D.F. Fraser. 1988. Resource depletion and habitat segregation by competitors under predation hazard. pp. 173–184. *In:* B. Ebenman & L. Persson (ed.) Size-Structured Populations, Springer-Verlag, Berlin.

Grossmann, G.D. & M.C. Freeman. 1987. Microhabitat use in a stream fish assemblage. J. Zool., London 212: 151–176.

Haberlehner, E. 1988. Comparative analysis of feeding and schooling behaviour of the Cyprinidae *Alburnus alburnus* (L., 1758), *Rutilus rutilus* (L., 1758), and *Scardinius erythrophthalmus* (L., 1758) in a backwater of the Danube near Vienna. Int. Revue ges. Hydrobiol. 73: 537–546.

Halyk, L.C. & E.K. Balon 1983. Structure and ecological production of the fish taxocene of a small floodplain system. Can. J. Zool. 61: 2446–2464.

Hammer, C. 1985. Feeding behaviour of roach *(Rutilus rutilus)* larvae and the fry of perch (*Perca fluviatilis*) in Lake Lankau. Arch. Hydrobiol. 103: 61–74.

Hartmann, J. 1983. Two feeding strategies of young fishes. Arch. Hydrobiol. 96: 496–509.

Heggenes, J. 1988. Effects of short-term flow fluctuations on displacement of, and habitat use by, brown trout in a small stream. Trans. Amer. Fish. Soc. 117: 336–344.

Helfman, G.S. 1978. Patterns of community structure in fishes: summary and overview. Env. Biol. Fish. 3: 129–148.

Howard, R.H. 1989. The structure of a nearshore fish community of Western Australia: diel patterns and the habitat role of limestone reefs. Env. Biol. Fish. 24: 93–104.

Jacobs, J. 1974. Quantitative measurement of food selection. Oecologia 14: 413–417.

Koblickaya, A.P. 1981. Key for indentifying young freshwater fishes. Food Industry Publishing House, Moscow. 208 pp. (In Russian).

Kriksunov, E.A., A.M. Boldyrev, T.I. Kuga & M.I. Shatunovskiy. 1985. Some ecological aspects of young roach, *Rutilus rutilus,* and bream, *Abramis brama,* in the littoral zone of Mozhaysk Reservoir. J. Ichthyol. 25: 11–17.

Lightfoot, G.W. & N.V. Jones 1979. The relationship between the size of 0 group roach (*Rutilus rutilus* (L.)), their swimming capabilities, and their distribution in a river. pp. 230–236. *In:* Proc. 1st Brit. Freshwater Fish. Conf., Liverpool.

Manteifel, B.P., I.I. Girsa & D.S. Pavlov. 1978. On rhythms of

fish behaviour. pp. 215–224. *In:* J.E. Thrope (ed.) On Rhythms of Fish Behaviour, Academic Press, New York.
Mark, W., R. Hofer & W. Wieser. 1987. Diet spectra and resource partitioning in the larvae and juveniles of three species and six cohorts of cyprinids from a subalpine lake. Oecologia 71: 388–396.
Mark, W., W. Wieser & C. Hohenauer. 1989. Interactions between developmental processes, growth, and food selection in the larvae and juveniles of *Rutilus rutilus* (L.) (Cyprinidae). Oecologia 78: 330–337.
Moyle, P.B. & D.M. Baltz. 1985. Microhabitat use by an assemblage of California stream fishes: developing criteria for instream flow determinations. Trans. Amer. Fish. Soc. 114: 695–704.
Nelva, A., H. Persat & D. Chessel. 1979. Une nouvelle méthode d'étude des peuplements ichtyologiques dans les grands cours d'eau par échantillonnage ponctuel d'abondance. C.R. Acad. Sci. Paris. t. 289. Série D: 1295–1298.
Northcott, D. 1979. The importance of aquatic macrophytes in the provision of crustacian zooplankton food for young roach. pp. 123–134. Proc. 1st Brit. Freshwater Fish. Conf., Liverpool.
Persat, H. & D. Chessel. 1989. Typologie de distributions en classes de taille: intérêt dans l'étude des populations de poissons et d'invertebrés. Acta Oecologica, Oecol. Gen. 10: 175–195.
Persson, L. 1988. Asymetrics in competition and predatory interactions in fish populations. pp. 203–218. *In:* B. Ebenman & L. Persson (ed.) Size-Structured Populations, Springer-Verlag, Berlin.
Petts, G. & I. Foster. 1985. Rivers and landscape. Edward Arnold Publishers, London. 274 pp.
Power, M.E. 1987. Predator avoidance by grazing fishes in temperate streams: importance of stream depth and prey size. pp. 333–351. *In:* W.C. Kerfoot & A. Sih (ed.) Predation: Direct and Indirect Impacts on Aquatic Communities, University Press New England, Hanover.
Power, M.E. & W.J. Matthews. 1983. Algae-grazing minnows (*Campostoma animalum*), piscivorous bass (*Micropterus* spp.), and the distribution of attached algae in a small prairie-margin stream. Oecologia 60: 328–332.
Rheinberger, V., R. Hofer & W. Wieser. 1987. Growth and habitat separation in eight cohorts of three species of cyprinids in a subalpine lake. Env. Biol. Fishes. 18: 209–217.
Robinson, C.L.K. 1989. Laboratory survival of four prey in the presence of northern pike. Can. J. Zool. 67: 418–420.
Rozas, L.P. & W.E. Odum. 1988. Occupation of submerged aquatic vegetation by fishes: testing the roles of food and refuge. Oecologia 77: 101–106.
Schiemer, F. & T. Spindler. 1989. Endangered fish species of the Danube River in Austria. Reg. Rivers: Res. & Manag. 4: 397–407.
Schröder, T. 1979. Aspekte der Ökologie von Frühentwicklungsstadien einiger Fischarten in Altrhein und Labor. Diplomarbiet, W. Goethe Universit., Frankfurt/Main. 100 pp.
Scott, A. 1987. Prey selection by juvenile cyprinids from running water. Freshwat. Biol. 17: 129–142.
Stankovitch, S. 1921. Etude sur la morphologie et la nutrition des alevins de poissons cyprinidés. Thèse de Doctorat d'Etat, Grenoble. 182 pp.
Thioulouse, J. 1989. Statistical analysis and graphical display of multivariate data on the Macintosh. CABIOS 55: 287–292.

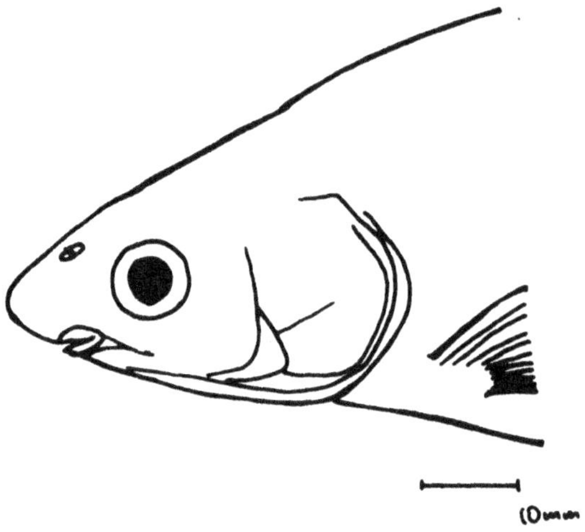
10mm

Environmental Biology of Fishes **33**: 195–205, 1992.

The role of predation and competition in determining the distribution of common bream, roach and white bream in Dutch eutrophic lakes

Eddy H.R.R. Lammens, Aafje Frank-Landman, Peter J. McGillavry & Bauke Vlink
Limnological Institute, Tjeukemeer Laboratory, De Akkers 47, 8536 VD, Oosterzee, The Netherlands

Received 12.3.1990 Accepted 2.11.1990

Key words: Cyprinids, Size structured populations, Pikeperch, Smelt, Feeding efficiency, Predator-prey-relationships

Synopsis

Most of the lakes in The Netherlands are turbid and without vegetation. This is regarded as the result of increasing eutrophication within the last decades. Under these conditions common bream, roach, and white bream are the most common cyprinids. In six shallow (1–3 m), wind exposed lakes the abundance of common bream and roach was linked to the abundance of pikeperch; bream dominated when pikeperch was abundant, but when the latter species was rare, roach was dominant and the biomass of bream was reduced. The biomass of white bream was always relatively low. In lake Tjeukemeer the distribution of roach < 20 cm fork length (FL) and bream was also related to the distribution of pikeperch. Only roach > 20 cm FL managed to coexist with pikeperch in the open water area, whereas roach < 20 cm was confined to the littoral zone where pikeperch was nearly absent. Bream occurred mainly in the open water and avoided the littoral zone where it competed with roach. White bream occupied an intermediate position, occurring in relatively low density both in the littoral zone and in open water. The importance of predation and competition in determining the distribution and abundance of roach and common bream in the eutrophic lakes of The Netherlands is discussed.

Introduction

Most fish communities in The Netherlands are dominated by cyprinids, percids and smelt. Acipenserids, salmonids, clupeids and coregonids are poorly represented, although they were abundant at the beginning of this century (Drimmelen 1987). In particular, eutrophication has brought about a change in the environment for fishes by creating turbidity and causing the replacement of macrophytes by cyanophytes (Hosper & Jagtman 1990). Cyprinids are now the most common group of fish, represented by 17 reproducing species (Nijssen & de Groot 1987). The three most common species are the common bream, *Abramis brama* (L.), roach, *Rutilus rutilus* (L.), and white bream, *Blicca bjoerkna* (L.), whereas the other species exhibit only a marginal distribution, but can be locally very abundant. In overgrown waters, tench, *Tinca tinca* (L.), and rudd, *Scardinius erythrophthalmus* (L.), are quite common, whereas crucian carp *Carassius carassius* (L.) occurs in lesser numbers (Nijssen & de Groot 1987, Lammens 1986, Pieczinska 1976). Because the overgrown areas are becoming scarce, tench, rudd and crucian carp are decreasing drastically. Roach and white bream are most flexible and can live in both overgrown and open habitats.

An 18-year study in Tjeukemeer showed that the density and size distribution of common bream, roach and white bream varied in relation to the

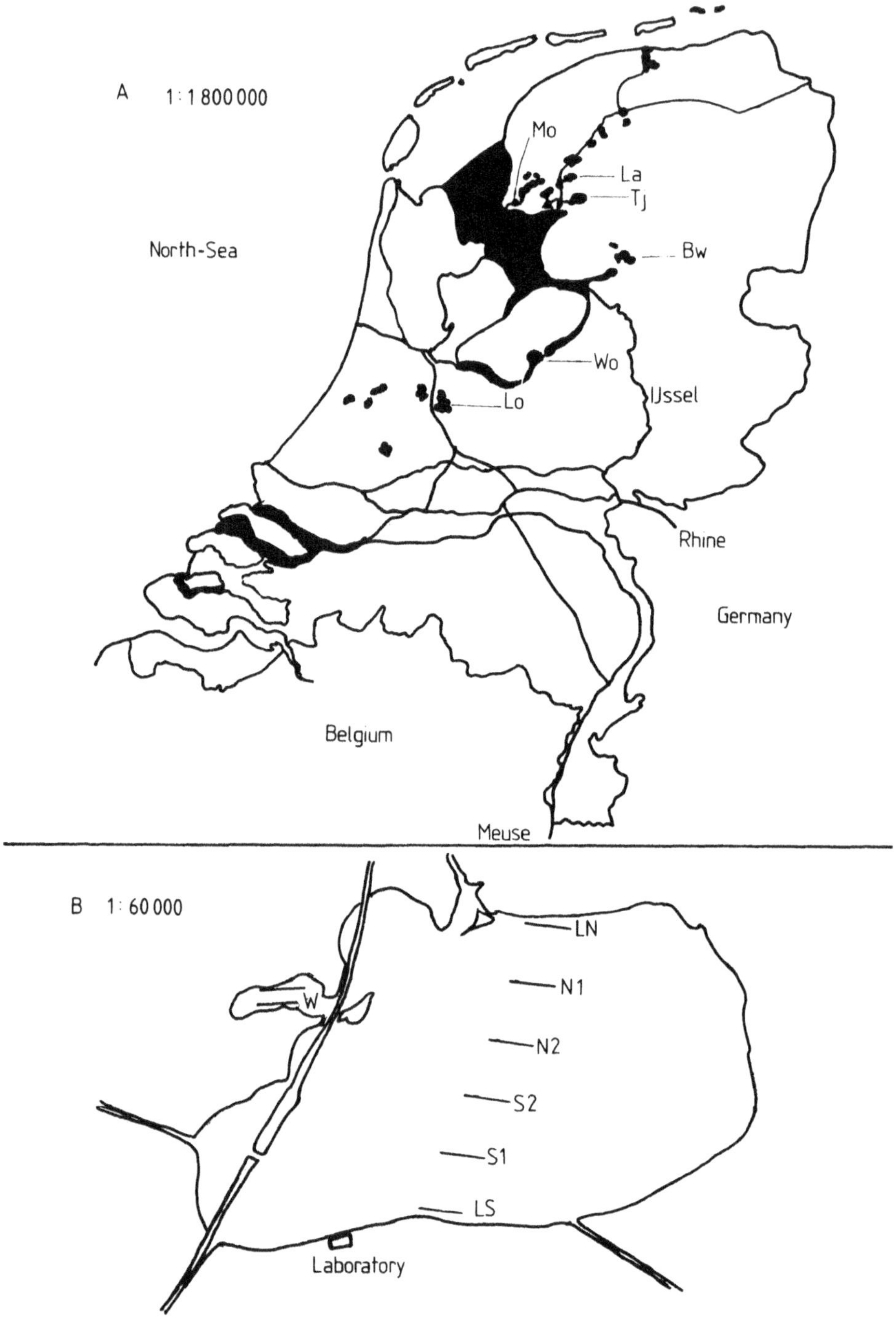

Fig. 1. A – Location of the six lakes investigated in The Netherlands (Tj = Tjeukemeer, Lo = Loosdrecht, La = Langweerder Wielen, Bw = Beulaker Wijde, Mo = Morra, Wo = Wolderwijd). B – Location of the localities in and near lake Tjeukemeer (W = two localities in 'Het Wiel', LS = littoral zone south, LN = littoral zone north, S1 = 1 km from south shore, N1 = 1 km from north shore, S2 = 2 km from south shore, N2 = 2 km from north shore).

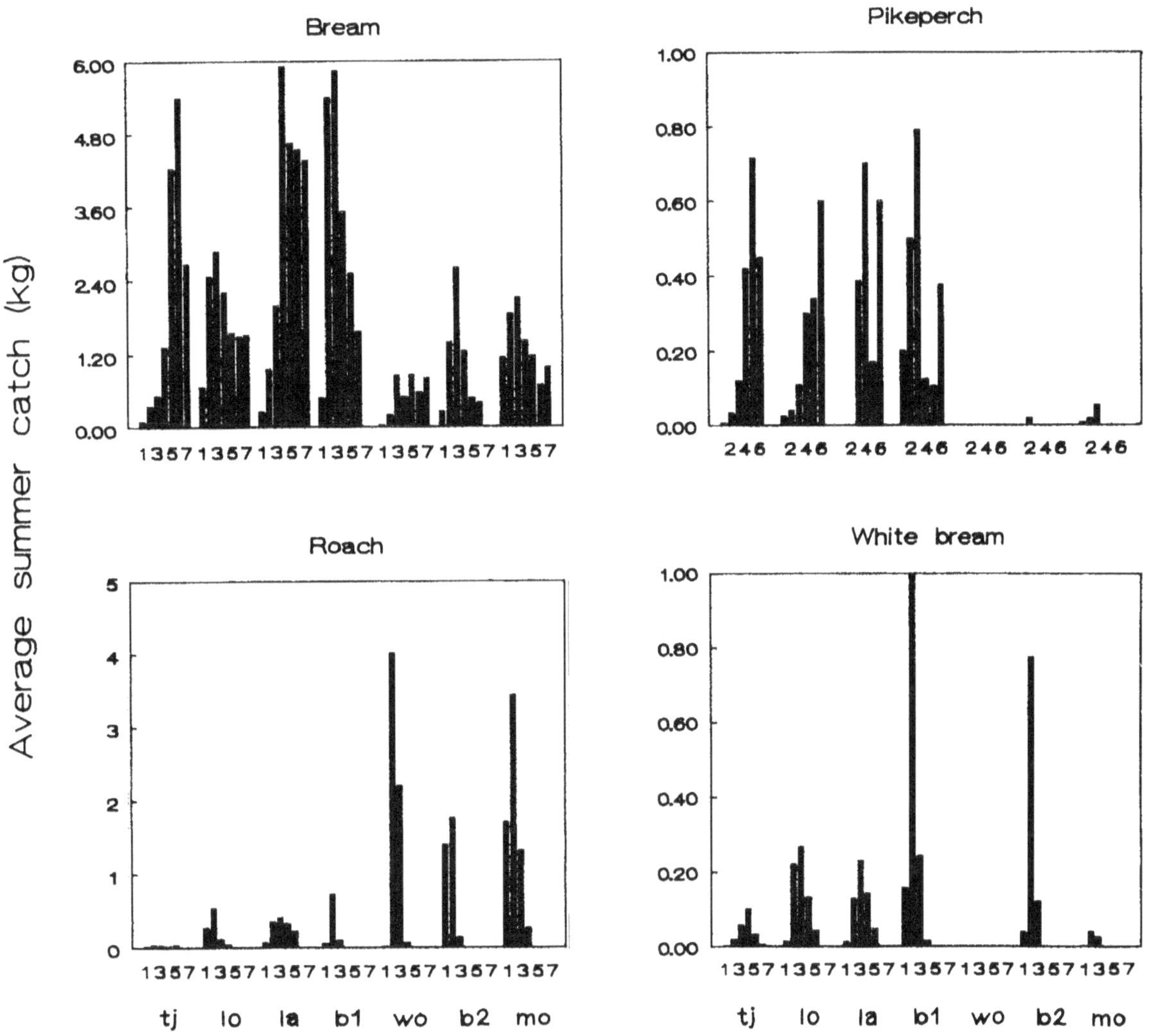

Fig. 2. Size distribution of four fish species caught during 10 min trawling and expressed in biomass per length-class (1 = 5–10 cm, 2 = 10–15 cm etc,; for pikeperch 1 = 10–20 cm, 2 = 20–30 cm etc.) for seven lakes (tj = Tjeukemeer, lo = Loosdrecht, la = Langweerder Wielen, b1 = Beulaker Wijde, first two years, wo = Wolderwijd, b2 = Beulaker Wijd, last three years, mo = Morra). Figures represent averages of five years, with three catches per year.

density and size distribution of the pikeperch, *Stizostedion lucioperca* (L.), population (Lammens et al. 1990). To test our hypothesis that the biomass and size distribution of cyprinids are related to the abundance of pikeperch, we compared the fish communities of six lakes with varying pikeperch populations. In addition, the distribution of cyprinids in Tjeukemeer was compared with that of pikeperch.

Study areas

Six lakes differing with respect to size (200–3000 ha) and fishery management (Fig. 1A) were studied: Langweerder Wielen (222 ha), Loosdrechtse Plassen (900 ha) and Tjeukemeer (2150 ha) were similar in depth (1–2 m), transparency (30–40 cm Secchi-disc) and fishery management (eel fishery with fykes and no gill-net fishing); Morra (234 ha), Beulaker Wijde (950 ha) and Wolderwijd (3000 ha) were of similar transparency (30–40 cm). Only in Beulaker Wijde was there an intensive gill-net fish-

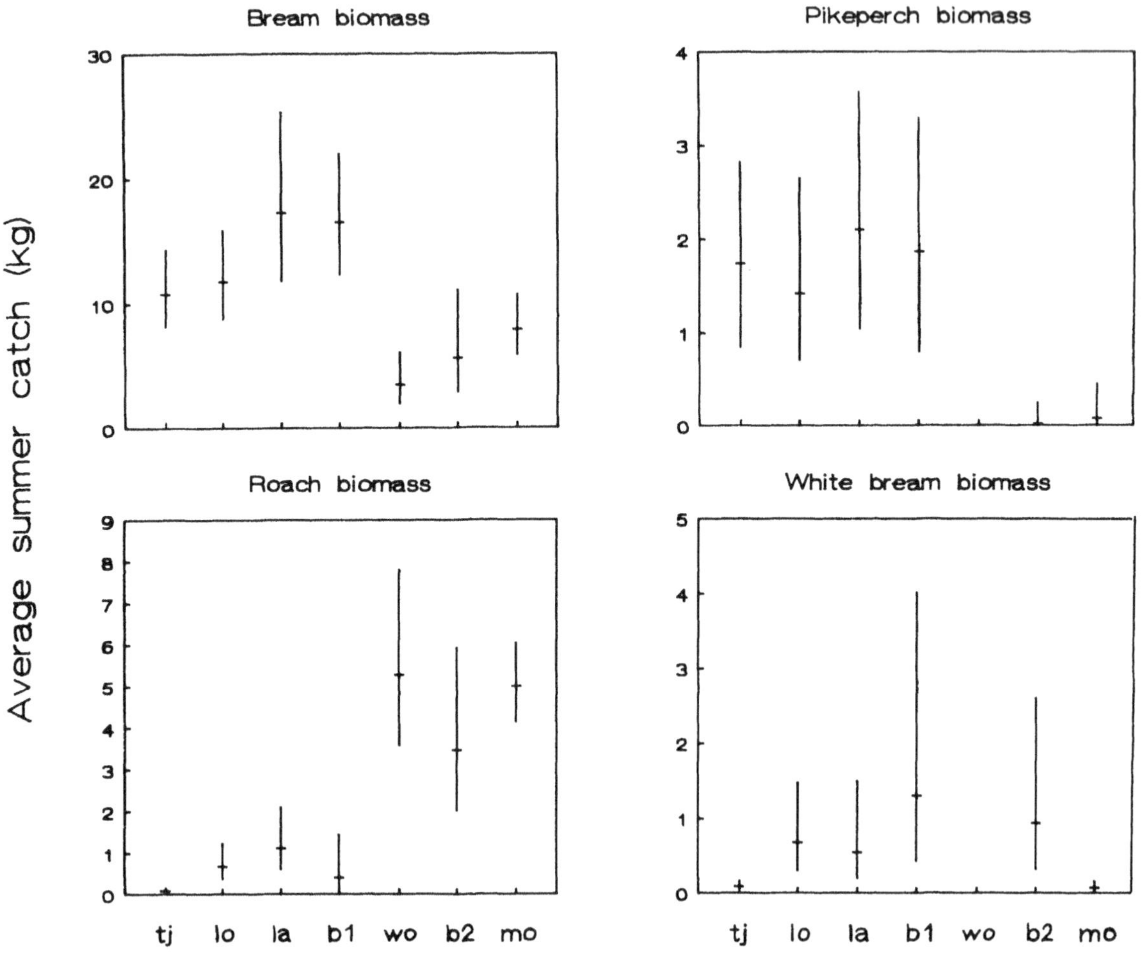

Fig. 3. The catch of four fish species expressed in kg per 10 min trawling in seven lakes (tj = Tjeukemeer, lo = Loosdrecht, la = Langweerder Wielen, b1 = Beulaker Wijde, first two years, wo = Wolderwijd, b2 = Beulaker Wijde, last three years, mo = Morra). Figures represent an average of five years, with three catches per year, the bars indicating means and ranges.

ery and in Wolderwijd and Morra a relatively large area was shallower than 1 m.

The shore line of all these lakes is poorly developed, the only vegetation being some reeds, except for Wolderwijd where fields of *Potamogeton perfoliatus* are present. The most important secondary

Table 1. Correlation coefficients (* = $p < 0.005$) between average length of common bream (LB), roach (LR), white bream (LWB), biomass of the population of bream (BB), roach (BR), white bream (BWB), pikeperch (BS), and lake surface area (SUR).

	LB	LR	LWB	BB	BR	BWB	BS
LR	0.819*						
LWB	0.195	− 0.195					
BB	0.126	0.095	0.528				
BR	− 0.042	0.221	0.463	− 0.619			
BWB	− 0.474	− 0.536	0.133	0.615	− 0.646		
BS	0.166	0.036	0.570	0.896*	− 0.844	0.620	
SUR	0.675	0.546	− 0.228	− 451	0.099	− 0.524	− 0.186

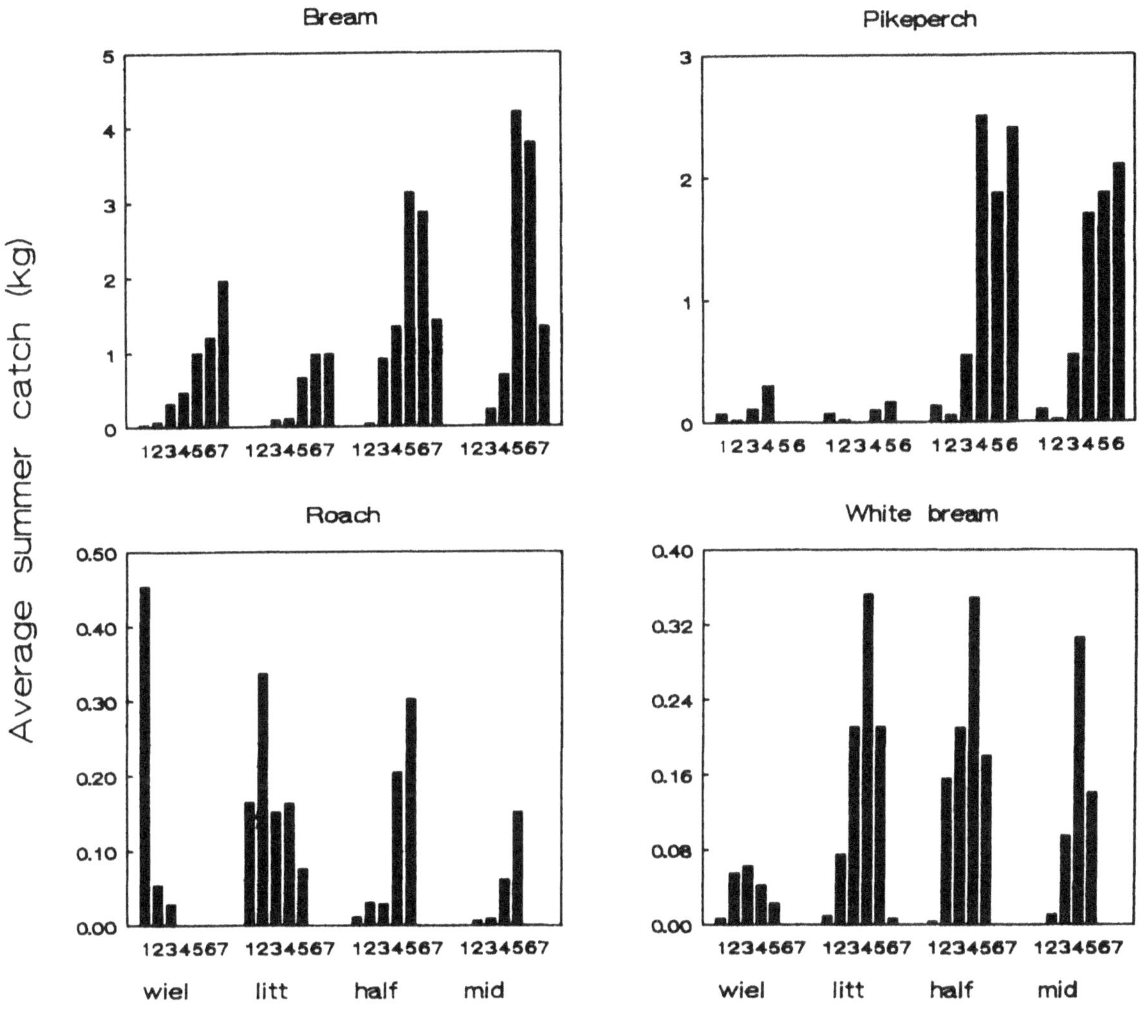

Fig. 4. The size distribution of four fish species caught overnight with 100 m gill-nets (10, 20, 35, 50 and 65 mm mesh size knot to knot) and expressed in biomass per length-class (1 = 5–10 cm, 2 = 10–15 cm etc.; for pikeperch 1 = 10–20 cm, 2 = 20–30 cm etc.) for four locations in and near Tjeukemeer (wiel = two localities in 'Het Wiel', litt = two sites in the south and north littoral zone, half = 1 km from the south and north shore, mid = 2 km from south and north shore). Figures represent averages of two years, with three catches per year.

producers are cladocerans, copepods (Vijverberg & Richter 1982a, b), and chironomids (Beattie 1982). As the lakes are shallow and continuously mixed by wind, zooplankton is almost homogeneously distributed (de Nie et al. 1978). These lakes receive water from the river Rhine during the summer, when evaporation is higher than precipitation, and from the surrounding polders in winter, when precipitation is greater than evaporation (Fig. 1A). The man-made hydrological regime maintains the constant water level necessary for the surrounding agricultural area. For chemistry, see Haan (1982) and Haan & Moed (1984), and for hydrology, Leenen (1982).

Materials and methods

Fish collection by trawling

Fish were caught between May and September 1984–1988 with a small mesh (5.5 mm cod end) 5 m beam-trawl at five localities in each lake every one or two months. Standard hauls were made with a

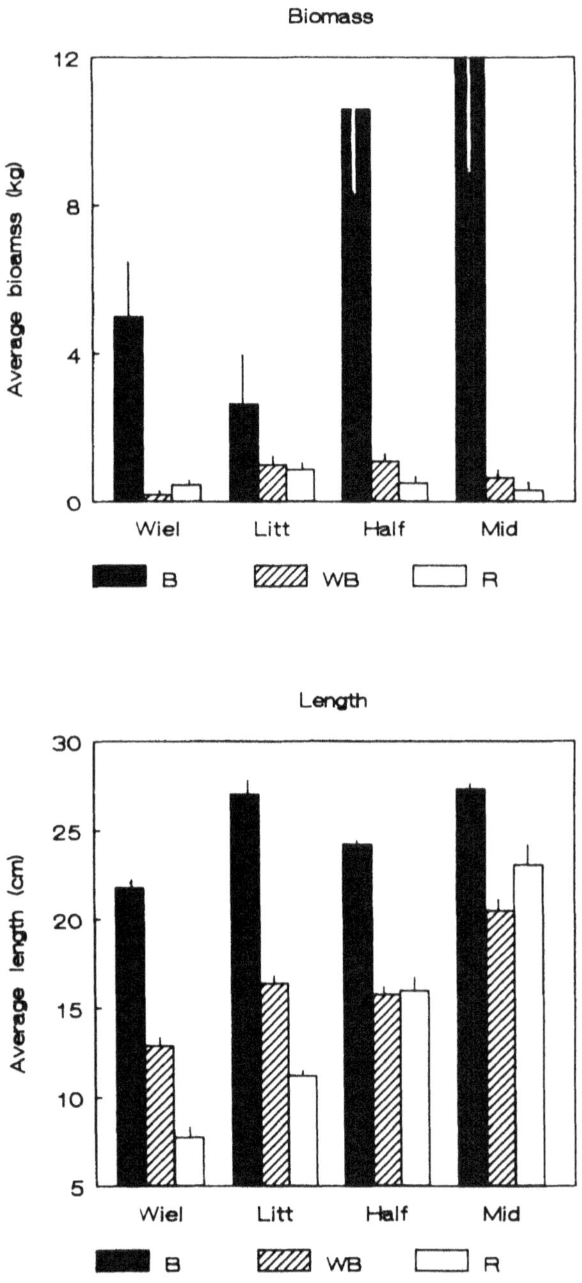

Fig. 5. Average biomass and average size of bream (B), white bream (WB) and roach (R) caught in gill-nets set overnight (10, 20, 35, 50 and 65 mm mesh size, knot to knot) of four locations in and near Tjeukemeer (wiel = two sites in 'Het Wiel', litt = two sites in the south and north littoral zone, half = 1 km from the south and north shore, mid = 2 km from south and north shore). Figures represent averages of two years and three catches per year. Bars indicate S.E.

speed of 1 m sec^{-1} for 10 min. Average length frequencies (fork length, FL in cm) and catch per unit effort (c.p.u.e.) were determined for each species. Length frequencies were transformed to biomass per catch, using an average length-weight relationship for summer months. Young-of-the-year were not considered in the calculations since they were not caught effectively until the end of summer. In these lakes, smelt, *Osmerus eperlanus* (L.), was the only fish species of which the first year class dominated the population.

Distribution of the fish species in Tjeukemeer

In 1988 and 1989 the distribution of fish was determined using gill-nets from June to September. Mesh-sizes were 10, 20, 35, 50 and 65 mm from knot to knot, the retained fish ranging in size from 5 to 75 cm fork length. Every month the nets were placed at six localities in lake Tjeukemeer, in a South-North transect at 1 km intervals, and also at two localities in a small, very shallow (0.5 m) lake, Het Wiel, opening into lake Tjeukemeer (Fig. 1B).

Table 2. Levels of significance of differences in biomass of common bream, roach, white bream and pikeperch at four sites in and near lake Tjeukemeer (W = 'Het Wiel', L = littoral zone of north and south combined, 1 = 1 km site of north and south combined, 2 = 2 km site of north and south combined (Kruskall-Wallis test).

	L	1	2
Common bream			
W	0.06	0.08	0.02
L		0.001	0.003
1			0.36
Roach			
W	0.05	0.16	0.21
L		0.16	0.02
1			0.36
White bream			
W	0.002	0.003	0.008
L		0.77	0.28
1			0.21
Pikeperch			
W	0.25	0.001	0.001
L		0.001	0.001
1			0.57

At each locality a total length of 500 m net was set, consisting of randomly arranged nets, 50 m long and 2 m high, of five different mesh sizes. The nets were set at 1600 h and collected after 16 hours. Catch effort at each locality and date were similar. Since only two localities could be monitored per sampling date, four sampling days were needed to monitor the whole S-N transect of Tjeukemeer, as well as Het Wiel, each month. Fish were grouped in length-classes of 5 or 10 cm, depending on the species and its size range.

Results

Variation between lakes

Apart from some very rare catches of ide, *Leuciscus idus* (L.), bleak, *Alburnus alburnus* (L.), and carp, *Cyprinus caprio* (L.), common bream, roach and white bream were the only cyprinids in the trawl catches of the six lakes. Roach and white bream stocks consisted mainly of small individuals, up to 25 cm FL. In the bream stocks specimens measured up to 40 cm FL (Fig. 2). The average biomass of bream catches was significantly higher than that of roach and white bream (Kruskall-Wallis, $p < 0.0033$ and < 0.0001; Fig. 3). The average biomass of roach catches was significantly higher than that of white bream (Kruskall-Wallis, $p < 0.06$; Fig. 3).

Within the stocks of each species considerable differences in size and biomass distribution occurred (Fig. 2, 3), but no correlation was found

Table 3. Correlation coefficients (* = $p < 0.05$) of the biomass of bream > 20 cm (B2), roach > 20 cm (R2), white bream > 20 cm (WB2), bream < 20 cm (B1), roach < 20 cm (R1), white bream < 20 cm (WB2) and pikeperch (SN).

	B2	R2	WB2	B1	R1	WB1
R2	0.450					
WB2	0.219	0.804				
B1	0.452	0.637	0.054			
R1	−0.993*	−0.549	−0.311	−0.501		
WB1	−0.156	0.797	0.630	0.536	0.045	
SN	0.939*	0.728	0.460	0.612	−0.972*	0.191

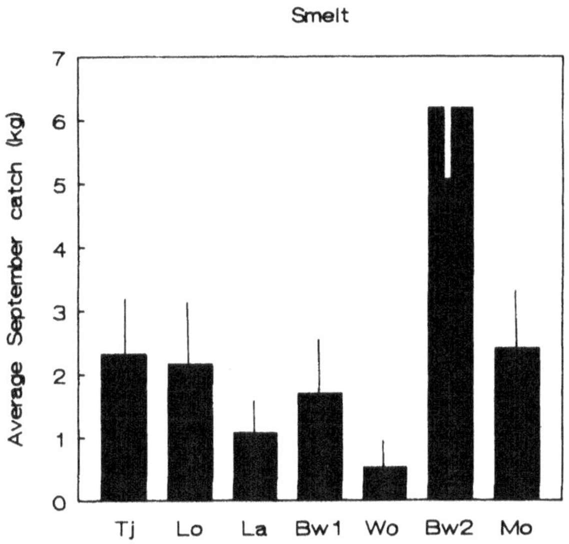

Fig. 6. The smelt catches expressed in kg per 10 min of trawling in September and averaged for five years in seven lakes (tj = Tjeukemeer, lo = Loosdrecht, la = Langweerder Wielen, bw1 = Beulaker Wijde, first two years, wo = Wolderwijd, bw2 = Beulaker Wijde, last three years, mo = Morra). Bars indicate S.E.

between average size of the fish and biomass (Table 1).

The biomass of bream was positively, that of roach negatively, correlated with the biomass of pikeperch (see also Fig. 3). The average size of roach and bream were significantly correlated with each other but not with other variables. Lake area was not correlated with the biomass of any of the fish stocks (Table 1).

Smelt were present in all lakes and constituted, with roach, the highest proportion of the biomass of fish < 20 cm FL. Most of the smelt stock was young-of-the-year < 10 cm FL. In the lakes where pikeperch were abundant, smelt were the most abundant prey-fish (Fig. 2, 6). In these lakes the catches in September did not differ significantly from each other (Kruskall-Wallis, $p > 0.05$). Only the smelt in Wolderwijd deviated significantly (Kruskall-Wallis, $p < 0.05$) from that of other lakes, and only in this lake did the smelt stock consist of three year classes.

Distribution within Tjeukemeer

At all localities in the lake, bream biomass was greater than that of the other species (Fig. 4, 5). However, size distribution and total biomass per locality varied. In open water, biomass was significantly higher than in the littoral zone and in 'Het Wiel' (Table 2, Fig. 5). Catches were lowest in the littoral and highest at the 2 km locality. Although the biomass of the 1 km and 2 km catches did not differ significantly, the length distribution at the 1 km site consisted of significantly more small bream and fewer large specimens (Kolmogorov-Smirnov test, $p < 0.05$). The 'Het Wiel' and the 1 km locality included the smallest specimens, the length distributions of which were not significantly different (Kolmogorov-Smirnov test, $p > 0.05$). The littoral zone and the middle of the lake were avoided by small individuals and preferred by larger individuals (Fig. 4).

The biomass of roach was lowest in 'Het Wiel' and in the middle of the lake (Table 2, Fig. 4). Length distributions at these sites, however, were completely different: in 'Het Wiel' small specimens dominated, whereas in the middle of the lake large fish dominated. The littoral zone was also inhabited by significantly smaller specimens than the 1 and 2 km localities, but littoral zone roach were significantly larger than the roach in 'Het Wiel'. The average size of roach showed a steep gradient from 'Het Wiel' towards the middle of the lake, whereas biomass was highest in the littoral zone (Fig. 5).

Biomass of white bream was lowest in 'Het Wiel' and highest in the littoral zone and at the 1 km locality (Fig. 4). It was not significantly different at the littoral, 1 km, and 2 km localities (Table 2). However, the littoral zone harboured significantly (K-S test, $p < 0.013$) smaller specimens than the open water sites, but those of 'Het Wiel' were even smaller (K-S test, $p < 0.001$). On average, the largest specimens were found in the middle of the lake and the smallest in 'Het Wiel', but the gradient was not as steep as for roach and the maximal biomass occurred closer to the 1 km locality (Fig. 5). At the 1 km and 2 km localities white bream were more abundant than roach, but roach were more abundant in the littoral zone and in 'Het Wiel'. The abundance of pikeperch in the littoral zone and in 'Het Wiel' was very low and limited to smaller specimens. Biomass at the 1 and 2 km localities was 10–20 times higher than in the littoral and in 'Het Wiel' (Fig. 4).

The abundance of small (< 20 cm FL) roach was negatively correlated both with the abundance of pikeperch and bream > 20 cm FL, but not with that of small bream and white bream (Table 3).

Discussion

A large part of the inland waters in The Netherlands are shallow lakes that have their origin in peat digging. Most of these lakes are severely eutrophic, have lost all submerged vegetation, and are very turbid (Hosper & Jagtman 1990); bream, roach and white bream are the dominant cyprinids (Lammens 1986). In lake Tjeukemeer, Lammens et al. (1990) found that the biomass and size structure of the bream, white bream, and roach populations changed after termination of the pikeperch gill-net fishery. The proportion of the pikeperch stock > 50 cm FL increased dramatically, altering the predation pressure on the cyprinids. In order to test the hypothesis that the structure of cyprinid communities is closely related to the abundance of pikeperch, we compared the cyprinid communities of six lakes, and within lake Tjeukemeer, with the densities of pikeperch.

In these fish communities the biomass of common bream was invariably found to be very high, varying from 40 to 80% of the total fish biomass. Roach stocks were large only in lakes where the biomass of pikeperch was very low and in these lakes the biomass of bream was lower than in lakes with a low biomass of roach. The same tendency was found within Tjeukemeer: small roach were restricted to the littoral zone of the lake where pikeperch were nearly absent. In the open water zone only roach > 20 cm FL, barely vulnerable to predation (Popova & Sytina 1977, Willemsen 1977), occurred together with pikeperch. An inverse relationship was found not only between pikeperch and roach but also between roach and bream, pointing to the effects of both predation and competition.

The interaction between roach and pikeperch

Pikeperch feeding on prey of about 1/3 of its body length (Popova & Sytina 1977, Willemsen 1977) was the most abundant predator in the lakes investigated. Slender and pelagic prey, such as smelt and roach, are preferred to benthic prey with a deep body, such as common bream, ruffe, and white bream. Since roach grow more slowly than bream (Geyer 1939, Hartley 1947, Lammens 1982, Lammens et al. 1987) their vulnerability is of longer duration, and since they are more slender than bream they are more vulnerable at similar length. Bream of > 15 cm FL (> 3 years old) are almost invulnerable to predation by pikeperch, whereas this is only achieved by roach of 20 cm length (> 7 years old) (Densen unpublished results). Thus, the negative correlation between the biomass of roach and that of pikeperch is most probably a predation effect.

Abundance of bream and white bream < 20 cm FL was not negatively related to the abundance of pikeperch. These length classes of bream and white bream even avoided the littoral zone. Only bream < 5 cm FL preferred this zone, migrating, as they grew larger, to more open areas somewhere between the centre of the lake and the littoral zone (Lammens et al. 1990). The largest bream preferred the central part of the lake. Most probably the high abundance of pikeperch keeps the density of small fish low, leaving more food available to the larger fish.

A similar distribution pattern was found by Werner et al. (1983) for the bluegill, *Lepomis macrochira*, in relation to the distribution of its main predator, largemouth bass, *Micropterus salmoides*. Only large bluegill coexisted with largemouth bass in open water, the smaller, vulnerable individuals remaining in the predator-free littoral zone. In their later studies, Werner & Gilliam (1984) demonstrated that foraging in the littoral zone was less profitable for the small individuals.

The interaction between common bream and roach

The abundance of roach and bream was always negatively correlated, both in individual lakes and at different localities of lake Tjeukemeer. If, in the absence of pikeperch, roach had the opportunity to develop a large stock, the bream stock was always much lower. When in one of the lakes, Beulaker Wijde, the pikeperch stock collapsed, the roach stock recovered, and the bream stock decreased drastically. Both roach and bream can utilize zooplankton and chironomids, but when fish of similar length are compared bream are more efficient (Lammens et al. 1987). However, the roach stocks were mainly formed by small individuals which are efficient predators on zooplankton (Cryer et al. 1986). Bream > 20 cm FL can utilize zooplankton large enough to be retained by their gill-rakers (Lammens et al. 1987), but if more efficient zooplanktivores are abundant, then bream switch to chironomids. The same switch was found to occur in Tjeukemeer when the recruitment of smelt was successful (Lammens et al. 1985) and this species exploited zooplankton just as successfully as young roach (Cryer et al. 1986). In consequence, the food source left for bream was chironomids, which can support only a much lower fish biomass (Beattie 1982). Thus an increase in the biomass of roach or smelt will lower the carrying capacity of the bream stock, and the negative relation between bream and roach most probably resulted from food competition in the absence of pikeperch. In the presence of pikeperch, bream are at an advantage, due to the preferred predation of pikeperch on roach. This type of interaction between species has been called "apparent competition" by Holt (1977).

The intermediate position of white bream

White bream were not dominant in any of the 6 lakes, and their distribution within lake Tjeukemeer was intermediate to that of roach and common bream. White bream are morphologically similar to common bream, but their growth pattern more closely resembles that of roach. Thus white bream seem less vulnerable to predation than roach. However, since white bream spawn a month later than roach (at the end of June), young white bream are more vulnerable to young pikeperch (Mooy unpublished results). MacArthur & Levins (1967) postulated a minimal difference in the utilization pattern of food resources of coexisting species. Bream are the most efficient predators on chironomids (Lammens et al. 1987), whereas roach prey most efficiently on molluscs and can also utilize detritus and algae (Persson 1983). Both can feed on zooplankton, the small specimens most efficiently (Lammens & Hoogenboezem 1991). Because the diet and feeding efficiency of white bream are somewhere in between those of bream and roach (Lammens et al. 1987, Lammens & Hoogenboezem 1991) their feeding opportunities may be limited when both bream and roach are abundant.

Pikeperch, smelt and roach

In Tjeukemeer the distribution of pikeperch is restricted to the open area where they feed mainly on smelt. Smelt are abundant in the open water area and avoid the littoral zone (Lammens et al. 1990). In contrast to pike, pikeperch is not an ambush predator but chases its prey, catching it tail first (Grimm 1981). Since deep-bodied fish cannot be swallowed that way (unpublished observations), the predator has to turn them around in the mouth which is easy for pike but difficult for pikeperch (Hoogland et al. 1957). Thus the latter species should prefer feeding on slender fish. Because pikeperch require open water to pursue their prey, roach may have better survival chances in the littoral zone. In contrast to roach, smelt remain in the open water zone, where they are very vulnerable to predation. There are, however, two compensating mechanisms counteracting extinction: smelt already begin to reproduce after completion of their first year, and even if the smelt stock is totally extinguished through predation, it can be replaced by immigration from lake Ijsselmeer. Water from this large lake (Fig. 1A) is let into the other lakes to maintain the water level during summer; this leads to passive transport of smelt larvae (Densen & Vijverberg 1982) which support a larger pikeperch stock (Densen 1985).

In Morra and Wolderwijd, however, the development of the pikeperch stock may be poor because of large shallow areas in which foraging of this species is probably inefficient. In Beulaker Wijde, the pikeperch stock collapsed, probably due to a combination of overexploitation by gill-net fishery and very poor recruitment. Both smelt and roach became very abundant and the bream stock was reduced to 25% of its original biomass.

Werner et al. (1983) demonstrated the difference in distribution of bluegills in the presence and absence of predators. We did not conduct experiments, but circumstantial evidence that pikeperch is a very important factor determining the distribution of cyprinids in turbid open waters appears very convincing.

Acknowledgements

The authors are very grateful for the help they received from the skippers Koos Swart, Steven Visser and Bertus Lemsma, from Wim van Densen, Ruud Knijn, Wolf Mooij, Wim Hoogenboezem, Diederik ter Laak Poot, Maarten Boersma and Anja van der Honig, who cooperated in this monitoring programme and contributed substantially by their ideas and cheerful company. Joy and Wolfgang Wieser critically reviewed the manuscript.

References cited

Beattie, D.M. 1982. Distribution and production of the larval chironomids populations in Tjeukemeer. Hydrobiologia 95: 287–306.

Cryer, M., G. Peirson & C.R. Townsend. 1986. Reciprocal interactions between roach, *Rutilus rutilus*, and zooplankton in a small lake: prey dynamics and fish growth and recruitment. Limnol. Oceanogr. 3: 1022–1038.

Densen, W.L.T. van. 1985. Piscivory and the development of bimodality in the size distribution of O+ pikeperch (*Stizostedion lucioperca* L.). J. Appl. Ichthyol. 3: 119–131.

Densen, W.L.T. van & J. Vijverberg. 1982. The relations between 0+ fish density, zooplankton size and vulnerability of pike-perch, *Stizostedion lucioperca*, to angling in the Frisian lakes. Hydrobiologia 95: 321–336.

Drimmelen, D.E. van 1987. Description of the Dutch riverine and inland fishery until the middle of the 20th century. O.V.B. Nieuwegein. 128 pp. (In Dutch).

Geyer, F. 1939. Alter und Wachstum der wichtigsten Cypriniden Ostholsteinische Seen. Arch. Hydrobiol. 34: 543–644.

Grimm, M.P. 1981. The composition of northern pike (*Esox lucius L.*) populations in four shallow waters in the Netherlands, with special reference to factors influencing O+ pike biomass. Fish. Manag. 12: 61–79.

Haan, H. de. 1982. Physico-chemical environment in Tjeukemeer with special reference to speciation of algal nutrients. Hydrobiologia 95: 205–221.

Haan, H. de & J.R. Moed. 1984. Phosphorus, nitrogen and chlorophyll – a concentrations in a typical Dutch polder lake, Tjeukemeer in relation to its water regime between 1968 and 1982. Wat. Sci. Technol. 17: 733–743.

Hartley, P.H.T. 1947. The natural history of some British freshwater fishes. Proc. Zool. Soc. Lond. 117: 129–206.

Holt, R.D. 1977. Predation, apparent competition, and the structure of prey communities. Theor. Pop. Biol. 12: 197–229.

Hoogland, R., D. Morris & N. Tinbergen. 1957. The spines of sticklebacks (*Gasterosteus* and *Pygosteus*) as a means of defense against predators (*Perca* and *Esox*. Behaviour 10: 207–236.

Hosper, S.J. & E. Jagtman 1990. Biomanipulation, an additional tool to nutrient control for the restoration of eutrophic shallow lakes in the Netherlands. Hydrobiologia 201: 523–539.

Lammens, E.H.R.R. 1982. Growth, condition and gonad development of bream (*Abramis brama*) in relation to its feeding conditions in Tjeukemeer. Hydrobiologia 95: 311–321.

Lammens, E.H.R.R. 1986. Interactions between fishes and the structure of fish communities in Dutch shallow eutrophic lakes. Ph.D. Thesis, University Wageningen, Wageningen. 100 pp.

Lammens, E.H.R.R., H.W. de Nie, J. Vijverberg & W.L.T. van Densen. 1985. Resource partitioning and niche shifts of bream (*Abramis brama*) and eel (*Anguilla anguilla*) mediated by predation of smelt (*Osmerus eperlanus*) on *Daphnia hyalina*. Can. J. Fish. Aquat. Sci. 42: 1342–1351.

Lammens, E.H.R.R., J. Geursen & P.J. McGillavry. 1987. Diet shifts, feeding efficiency and coexistence of bream *Abramis brama*, roach *Rutilus rutilus* and white bream *Blicca bjoerkna* in eutrophicated lakes. Proc. V. Congr. Europ. Ichtyol. Stockholm 1985: 153–162.

Lammens, E.H.R.R., W.L.T. van Densen & R. Knijn. 1990. The fish community structure in Tjeukemeer in relation to fishery and habitat utilization. J. Fish Biol. (in press).

Lammens, E.H.R.R. & W. Hoogenboezem. 1991. Diets and feeding behaviour. pp. 353–376. *In:* I.J. Winfield & J.S. Nelson (ed.) Cyprinid Fishes: Systematics, Biology and Exploitation, Chapman & Hall, London.

Leenen, J.D. 1982. Hydrology of Tjeukemeer. Hydrobiologia 95: 199–203.

MacArthur, R. & R. Levins. 1967. The limiting similarity, convergence, and divergence of coexisting species. Amer. Nat. 101: 377–384.

Nie, H.W. de, H.J. Bromley & J. Vijverberg. 1978. Distribution patterns of zooplankton in Tjeukemeer, The Netherlands. J. Plankt. Res. 2: 317–334.

Nijssen, H. & S.J. de Groot. 1987. The fishes of The Netherlands. Stichting Uitgeverij van de Koninklijke Nederlandse Natuurhistorische Vereniging, Utrecht. 223 pp. (In Dutch).

Persson, L. 1983. Effects of intra- and interspecific competition on dynamics and size structure of a perch *Perca fluviatilus* and a roach *Rutilus rutilus* population. Oikos 41: 126–132.

Pieczinska, E. 1976. Problems of littoral lake ecology. Wyd. Nauka, Warsaw. 238 pp.

Popova, O.A. & L.A. Sytina. 1977. Food and feeding relations of Eurasian perch (*Perca fluviatilus*) and pike-perch (*Stizostedion lucioperca*) in various waters of the USSR. J. Fish. Res. Board Can. 34: 1559–1570.

Vijverberg, J. & A.F. Richter. 1982a. Population dynamics and production of *Acanthocyclops robustus* Sars and *Mesocyclops leuckarti* Claus. Hydrobiologia 95: 261–274.

Vijverberg, J. & A.F. Richter. 1982b. Population dynamics and production of *Daphnia hyalina* Leydig and *Daphnia cucullata* Sars in Tjeukemeer. Hydrobiologia 95: 235–259.

Werner, E.E. & J.F. Gilliam. 1984. The ontogenetic niche and species interactions in size-structured populations. Ann. Rev. Ecol. Syst. 15: 393–425.

Werner, E.E. & D.J. Hall. 1988. Ontogenetic habitat shifts in blue-gill: the foraging rate – predation risk trade-off. Ecology 69: 1352–1366.

Werner, E.E., J.F. Gilliam, D.J. Hall & G.G. Mittelbach. 1983. An experimental test of the effects of predation risk on habitat use in fish. Ecology 64: 1540–1548.

Willemsen, J. 1977. Population dynamics of percids in lake IJssel and some smaller lakes in The Netherlands. J. Fish. Res. Board Can. 34: 1710–1719.

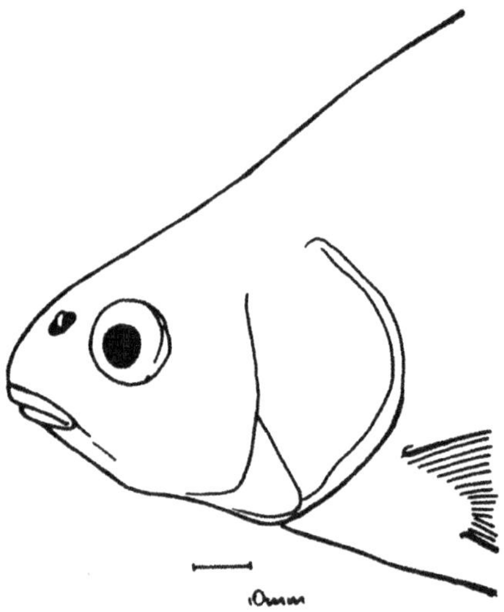
10mm

Environmental Biology of Fishes **33**: 207–214, 1992.

Interactions between the roach, *Rutilus rutilus*, and waterfowl populations of Lough Neagh, Northern Ireland

Ian J. Winfield[1], Denise K. Winfield & Catherine M. Tobin[2]
University of Ulster Freshwater Laboratory, Traad Point, Ballyronan, Co. Londonderry BT45 6LR, Northern Ireland, U.K.
[1] *Present address: Institute of Freshwater Ecology, Windermere Laboratory, Ambleside, Cumbria LA22 0LP, U.K.*
[2] *Present address: School of Applied Sciences, Wolverhampton Polytechnic, Wulfruna Street, Wolverhampton WV1 1SB, U.K.*

Received 15.1.1990 Accepted 25.10.1990

Key words: Cyprinidae, Tufted duck, Great crested grebe, Competition, Benthos, *Ligula intestinalis*

Synopsis

Following the introduction of roach, *Rutilus rutilus*, to a large eutrophic lake in ca. 1973, a subsequent increase in the abundance of this cyprinid through the 1970s was accompanied by a decline in the numbers of one of the lake's most abundant overwintering waterfowl, the tufted duck, *Aythya fuligula*, and an increase in overwintering piscivorous great crested grebes, *Podiceps cristatus*. We suggest that these contrasting trends are causally related and that competition for benthos and increased prey availability are the mechanisms responsible for the changes in the tufted duck and grebe populations respectively. In agreement with these hypotheses, a reduction in the roach population during the mid 1980s was accompanied by a recovery of tufted ducks and a decline of grebes.

Introduction

Following a long history of relative neglect, fish have now been fully incorporated within limnology and their important role in the functioning of freshwater ecosystems widely acknowledged (see Persson et al. 1988). Members of the family Cyprinidae often have particularly strong influences on other aspects of lake ecosystems (Andersson 1984, Winfield & Townsend 1991), especially the zooplankton (Andersson et al. 1978, Cryer et al. 1986) and benthos (Andersson et al. 1978, Lammens et al. 1985). While the influences of cyprinids and other fishes on zooplankton and their subsequent indirect or cascading effects have been well studied (for extensive reviews see Zaret 1980 and Lazzaro 1987), their effects on benthic macroinvertebrates have received relatively little attention. Such interactions are, however, potentially important as they raise the possibility of distant competition (sensu Hurlbert et al. 1986) between cyprinids and a group of animals which have been rarely considered in limnological studies, the waterfowl.

There is evidence from several sources that such competitive interactions do occur, at least during the breeding season of waterfowl when their food requirements are high. For example, Eriksson (1979) found that the distribution of fledged goldeneye, *Bucephala clangula* (L.), a benthivorous diving duck, was negatively correlated with the abundance of fish, mainly the cyprinid roach, *Rutilus rutilus* L., and the non-cyprinid perch, *Perca fluviatilis* L. The mallard, *Anas platyrhynchos* (L.), an omnivorous dabbling duck, has also been shown to suffer from the presence of fish including the cyprinids roach and rudd, *Scardinius erythrophthalmus* L. (Pehrsson 1984), and bream, *Abramis brama* L. (Hill et al. 1987). In contrast to these studies concerned with competitive effects of fish on the breeding success of waterfowl, we know of only one study where a long-term influence was detected at the population level. In recent years, the numbers of waterfowl, including tufted duck, *Aythya fuligula* (L.), on several increasingly eutrophic lakes in southern Sweden have declined significantly and Andersson (1981) has suggested that a synchronous increase in the abundance of cyprinids, principally roach and bream, is at least partly responsible.

In this paper, we investigate possible competitive interactions between the roach and overwintering tufted duck populations of Lough Neagh in Northern Ireland by an analysis of long-term trends in population abundances. Information was also gathered over the same period on the abundance of piscivorous great crested grebes, *Podiceps cristatus* (L.), a common predator of small roach (Cramp & Simmons 1977). Lough Neagh is particularly suitable for such a study because the roach has only recently been introduced to this water body and, in contrast to the lakes studied by Andersson (1981), the potentially confounding effects of severely increasing eutrophication are not present (Gibson 1986).

Materials and methods

Study area

Lough Neagh is the largest lake in the British Isles (surface area 387 km^2), although it is relatively shallow (mean depth ca. 9 m). As a result, stratification and consequent gradients in oxygen concentration are rare despite the highly eutrophic state of the lake (Gibson 1986). The benthos is thus accessible to both diving ducks and the lake's fish community which in addition to roach contains significant populations of eels, *Anguilla anguilla* L., perch, pollan, *Coregonus autumnalis pollan* Thompson, bream, and gudgeon, *Gobio gobio* L. (Winfield et al. 1991). Lough Neagh also supports internationally important (sensu Smart 1976) numbers of overwintering waterfowl including tufted duck, pochard, *Aythya ferina* (L.), goldeneye, scaup, *Aythya marila* (L.), and great crested grebes (Winfield et al. 1989).

Roach monitoring

Sampling was carried out on one day each year between 8 May and 2 June from 1985 to 1989 inclusive in Ballyronan Bay, a shallow (approximate depth 3 m) sandy bay in the north-west part of the lake (Irish Grid Ref. H 955 865) typical of the inshore summer habitat of the roach. Samples were taken using a bottom trawl, which effectively fished the entire water column in this shallow part of the lake, with a 6 mm bar mesh cod-end which ensured that all fish above 40–50 mm in length were adequately sampled. A series of five 5 min trawls was taken on each sampling date and all fish captured were taken the laboratory for identification, enumeration and measurement. In addition, scales were removed from a subsample of the roach for ageing.

For the purposes of the present paper, calculations of the mean numbers of roach taken each year allowed the abundance of the population to be monitored in terms of changes in catch per unit effort (c.p.u.e.).

Waterfowl monitoring

Due to its importance for overwintering waterfowl, Lough Neagh is included in a series of counts covering the whole of the U.K. organised by the Wildfowl and Wetlands Trust, Slimbridge. These National Wildfowl Counts are made in the middle of each month from September to March inclusive, although counts on Lough Neagh have not been complete in terms of either winters or months within winters.

In the 1960s and 1970s and up to the winter of

1980–1981, counts were carried out by the Northern Ireland Ornithologists' Club, but since the winter of 1985–1986 they have been undertaken by a team of observers largely drawn from the Countryside and Wildlife Branch of the Department of the Environment (Northern Ireland), Craigavon Borough Council, and the Royal Society for the Protection of Birds. Counting is performed by at least 12 shore-based observers using binoculars and telescopes, with the entire 125 km shoreline of the lake being covered on the same day.

Data were thus collected on the numbers of tufted duck and great crested grebes during the winters of 1964–1965, 1965–1966, 1966–1967, 1967–1968, 1968–1969, 1977–1978, 1979–1980, 1980–1981, 1985–1986, 1986–1987, 1987–1988 and 1988–1989. However, as stated above, coverage from September through to March was not complete for every winter. In order to reduce the effects of varying inter-winter sampling effort, the present analysis only considers counts from the important months of November, December and January, although not even these months were consistently counted each year.

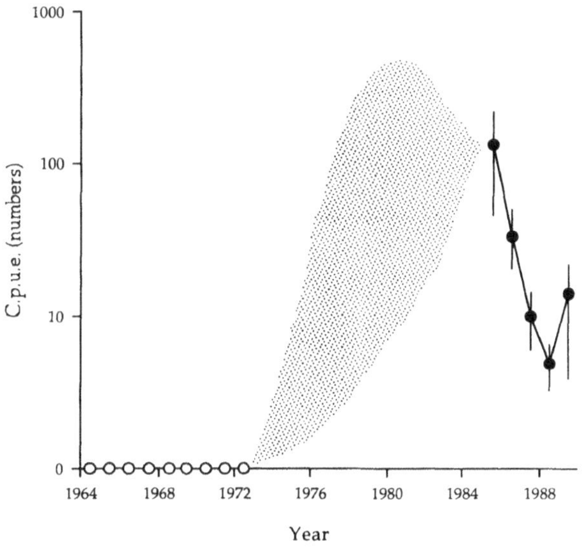

Fig. 1. Long-term trends in the abundance of roach in terms of catch-per-unit-effort (individuals per trawl) during the period 1964 to 1989. Values from 1985 onwards are given as closed symbols and error bars indicating means ± S.E. Assumed c.p.u.e.s. of zero up to 1972 and the likely range of values for the period between 1973 and 1984 are shown as open symbols and a stippled area respectively.

Results

Roach population

The abundance of roach in Lough Neagh has changed greatly during the 1980s as assessed by trends in mean c.p.u.e. by numbers (Fig. 1). From an average catch of over 100 roach in 1985, catches declined consistently until 1988 when a mean of less than 10 individuals was recorded. However, this trend was reversed in 1989 when the mean catch increased slightly. Furthermore, the latter sampling probably underestimated the relative abundance of roach because in 1989, in contrast to earlier years, the population was largely composed of small fish near the lower limit of size retention of the trawl (see below). Changes in roach biomass followed the same trends as the numerical abundance (I.J.W. and C.M.T. unpublished data).

Prior to their first detection in the early 1970s (Cragg-Hine 1973), it can safely be assumed that the roach was absent from Lough Neagh. Although no scientific records of the Lough Neagh fish community composition of this period have been published, the failure of the extensive commercial fisheries of the lake to capture a single roach is strong evidence for its continued absence during this time. This phase is represented in Fig. 1 as a series of assumed zero levels of c.p.u.e. from 1964 through to 1972. In the following period of 13 years the roach increased quickly and greatly in population size, although again the absence of any scientific records from this period means that the precise timing and magnitude of the increase are unknown.

Between 1985 and 1988, the declining roach population showed significant changes in size and age structure (Fig. 2). In 1985 the population was dominated by fish of approximately 2 years of age and 90 mm in length. This year class continued to dominate through to 1988, and thus the modal size class of the population reflected the growth of these fish. As a result, in 1988 the much reduced roach population was composed largely of fish approximately 5 years old and 150 mm in length. Through this

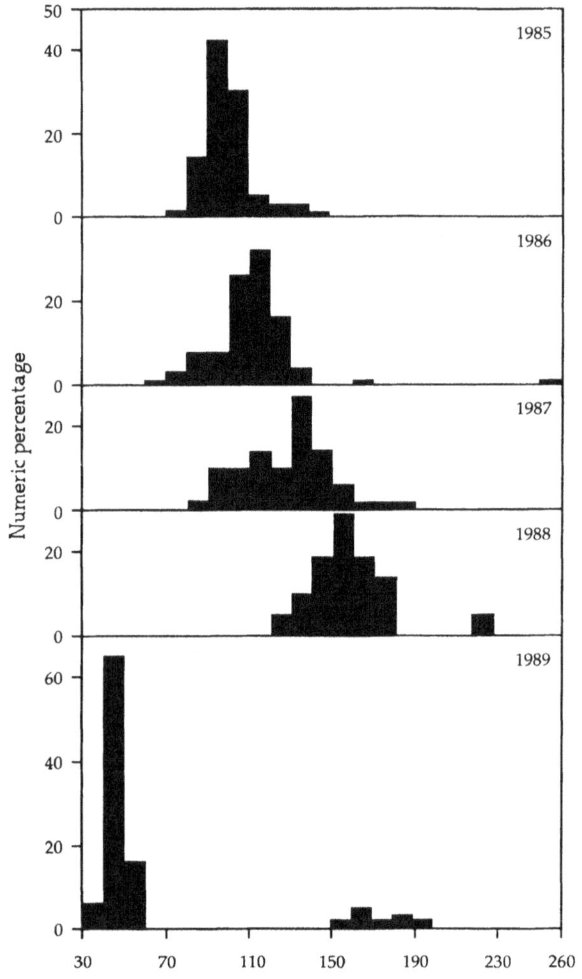

Fig. 2. Changes in the size structure of the roach population during the period 1985 to 1989.

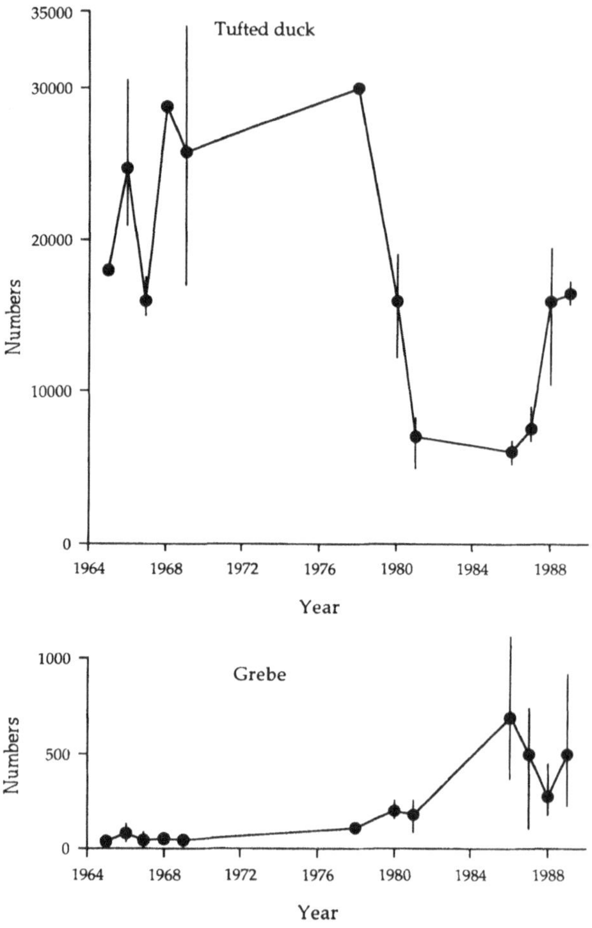

Fig. 3. Long-term trends in the numbers of overwintering tufted duck and great crested grebe during the period 1964 to 1989. Points represent the mean counts of November, December and January (or means of the remaining months if counting was incomplete) of each winter counted, while the error bars indicate the ranges of values.

entire period there was very little recruitment of young roach. In 1989 this situation changed completely with the emergence of a new very strong year class, and thus the population became dominated by approximately 1 year old fish of 40 mm in length.

Waterfowl populations

Both tufted duck and great crested grebe have shown significant changes in their overwintering stocks on Lough Neagh over the last 25 years (Fig. 3).

Following a period of stability through the 1960s and 1970s when mean numbers of tufted duck varied between 16084 and 29622, in the winter of 1980–1981 the average value was just 7053 birds. Similarly low levels were observed during the winters of 1985–1986 and 1986–1987 when average numbers of tufted duck were 5933 and 7871 respectively. Indeed, during this period the counts during individual months dropped as low as 5265 and never exceeded 8943. In contrast, during the subsequent winters of 1987–1988 and 1988–1989 the numbers of tufted duck returned to the levels of the

1960s with averages of 15877 and 16397 individuals respectively.

Like the tufted duck, the numbers of great crested grebes overwintering on Lough Neagh were stable through the 1960s and 1970s when between 1964–1965 and 1977–1978 the mean counts varied from 16 to 38 birds. Numbers began to increase in the late 1970s such that by the winters of 1979–1980 and 1980–1981 average counts were 197 and 181 birds respectively, and counts reached a peak in 1985–1986 with an average value of 671 grebes. Numbers then fell steadily during the next two winters to a low of 273 grebes in 1987–1988, before recovering in the winter of 1988–1989 to produce an average count of 505 birds.

The increase in abundance of roach during the late 1970s and early 1980s coincided with the increase of grebes and the decrease of tufted duck (Fig. 1, 3). Furthermore, the decline of roach in the mid to late 1980s was accompanied by a decline in grebes and a recovery of tufted duck to levels seen before the arrival of this cyprinid. The very recent observed recovery of yearling roach in the summer of 1989 (which means that underyearlings must have been abundant during the previous winter) was accompanied by an increase in great crested grebes during the winter of 1988–1989 but not by a reduction in the numbers of tufted duck. The c.p.u.e. of roach in a given summer is significantly correlated with the average numbers of tufted duck and grebe present during the following winter, although with contrasting signs (Spearman's Rank Correlation Coefficients: roach and tufted duck, $r_s = -0.86$, $N = 9$, $p < 0.01$; roach and great crested grebe, $r_s = +0.86$, $N = 9$, $p < 0.01$). Thus, lower numbers of overwintering tufted duck are associated with higher numbers of roach, while the converse is true for grebes (Fig. 4).

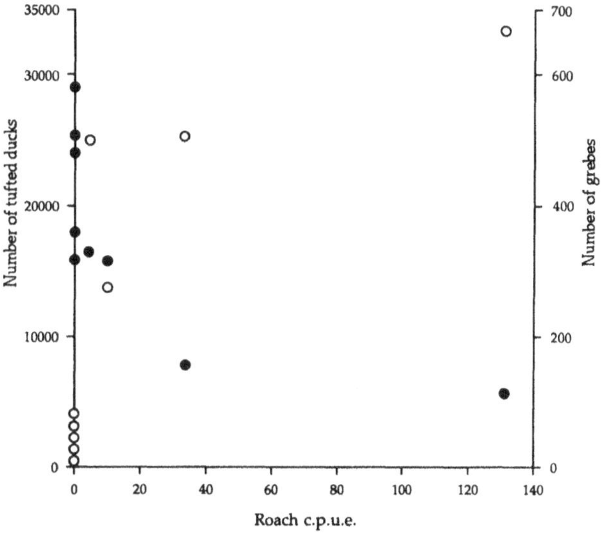

Fig. 4. The relationships between the mean c.p.u.e. of roach in a given summer and the mean numbers of tufted duck (dots) and great crested grebe (circles) present during the following winter. Points are included from the pre-1973 period when the c.p.u.e. of roach was assumed to be zero.

Discussion

Following its introduction in the early 1970s, the roach has shown extreme variations in abundance in Lough Neagh. The increase in roach during the mid to late 1970s, and possibly early 1980s, was simply the result of its introduction: severely increasing eutrophication, which has been associated with increases in roach populations elsewhere (Johansson & Persson 1986), was not involved. The mechanism of the decrease in roach during the mid 1980s is more difficult to establish although evidence indicates that a major factor had been poor recruitment due to parasitic infestation by the tapeworm *Ligula intestinalis* (L.) (C.M.T. unpublished data), possibly involving the increased numbers of great crested grebes as final hosts.

The observed changes in the numbers of tufted duck and grebes overwintering on Lough Neagh may be the result of one or more of three classes of factors. Firstly, it is arguable that they are the artifacts of changes in counting practices over the last 25 years, although Winfield et al. (1989) concluded that this was unlikely. Secondly, the changes may be due to factors operating not on Lough Neagh but elsewhere on the breeding and summering grounds of the waterfowl. However, this explanation is also unlikely because, in the case of the tufted duck, individuals overwintering on Lough Neagh come from a variety of European breeding grounds (evidenced by ringing returns held by the British Trust for Ornithology, Tring) which are unlikely to have all suffered simultaneous poor recruitment, particularly as consistent increases in the numbers of tufted duck have been recorded elsewhere in the British Isles and Europe (Owen et al. 1986). In the case of the grebe, a significant proportion of the local overwintering stock breed on Lough Neagh and are thus unaffected by changing conditions elsewhere. The final possibility, which we consider most likely, is that the trends are real and are the result of changes in local conditions. A major ecological event that has occurred in Lough Neagh synchronous with these changes in overwintering waterfowl has been the arrival of the roach. Furthermore, Andersson (1981) suggested three ways in which cyprinids such as roach may influence duck populations, i.e. by an indirect reduction of higher plants, by an increase in eutrophication resulting from cascading effects through the zooplankton, and finally by direct competition for invertebrate food supplies. As macrophytes were already rare in the eutrophic Lough Neagh before the roach introduction, only the latter mechanism can be operative in our study.

The contention that fish and waterfowl may undergo exploitative competition for benthos is relatively new. Early circumstantial evidence was produced by Eriksson (1979) who found that the distribution among lakes of fledged goldeneye was negatively correlated with the abundance of fish, mainly roach and perch. Subsequently, Eadie & Keast (1982) have found dietary evidence to support the competition hypothesis for goldeneye and *Perca* spp., while Pehrsson (1984) and Hill et al. (1987) have found experimental evidence for competitive interactions between fish and breeding mallard. In all of these instances, the observed competition was asymmetrical with the waterfowl being outcompeted by the fish. Molluscs are often important in the diets of both adult roach and tufted duck (Lammens & Hoogenboezem 1991, Cramp & Simmons 1977), and recent increases in the abundance of tufted duck on mainland Europe have been ascribed at least in part to the continental spread of the zebra mussel, *Dreissena polymorpha* (Pallas) (Owen et al. 1986). While this bivalve mollusc is absent from Lough Neagh, initial studies of the local diet of tufted duck indicate that other species of molluscs are important (D.K.W. unpublished data) and so competition for this prey type may be responsible for the trends observed locally. The large numbers of young roach in the recovering population presently in the lake are still largely zooplanktivorous and too small to feed on molluscs. Consequently, if the competition hypothesis is correct we would not expect the numbers of tufted duck to decrease again yet, although we predict that they will do so in the near future as the roach grow and become benthivorous. The investigation of tufted duck and roach diets, and the availability of molluscs, during periods of both high and low roach abundance would produce further evidence for or against the competition hypothesis.

In general, the investigation of competitive interactions between fish and waterfowl is extremely difficult. In a study of competition between fish (*Orestias* spp.) and waterfowl (Chilean flamingos, *Phoenicopterus chilensis*) for plankton rather than benthos, Hurlbert et al. (1986) remarked that such investigations are also neglected for intellectual reasons, specifically the general neglect of competitive interactions between taxonomically distant animals. In addition, the long generation times and mobile nature of waterfowl stocks preclude the use of experimental techniques with the exception of short-term duckling survival investigations. As a result, it is not surprising that, to the best of our knowledge, the study of Andersson (1981), is the only previous documentation of possible fish-waterfowl competitive interactions manifesting themselves in population changes over time.

The trends in overwintering numbers of piscivorous great crested grebes on Lough Neagh, i.e. an initial increase followed by a decrease and finally a recovery during the last winter, largely mirror those of the tufted duck. We suggest that these trends reflect changes in the abundance of suitable prey in the form of young roach. While the first grebe increase on the lake occurred some years before the arrival of the roach, it is likely that significant numbers of grebes were migrating onto Lough Neagh from nearby lakes and rivers where this cyprinid was already well established. The recovery in roach numbers seen over the last year has been accompanied by an immediate revival in the grebe population, although no response was shown by the tufted duck. We suggest that this is because while underyearling roach in Lough Neagh are too small to compete with tufted duck, they are large enough to be consumed by grebes. As is the case for tufted duck, the next stage in this investigation should be the determination of grebe diet composition, although this is extremely difficult on practical and conservation grounds.

Andersson (1981) remarked that his article was based on a restricted number of investigations. Similar limitations, particularly with respect to data availability and the absence of experimental manipulation, apply to the fish-flamingo-plankton work of Hurlbert et al. (1986) and to the present paper, reflecting in large part the inherent difficulties of fish-waterfowl studies. Nevertheless, such broad investigations are essential if the role of fish in freshwater ecosystems is to be properly appreciated in all of its complexities and manifestations. They will also ultimately make significant contributions to the conservation biology of both waterfowl (Andersson 1982, Wright & Street 1985, Kerekes 1990) and fish (Kennedy & Greer 1988).

Acknowledgements

We would like to thank Matt Quinn, Colin Bean and Caron Montgomery for their help in the roach monitoring which was itself made possible by the kind co-operation of the Department of Agriculture (Northern Ireland), the Fisheries Conservancy Board, the Honourable the Irish Society, and the Lough Neagh Fishermen's Co-operative. We are also indebted to all of the people who have been involved in the Lough Neagh Monthly Wildfowl Counts over the last 25 years, including members of the Northern Ireland Ornithologists' Club, Craigavon Borough Council, the Department of the Environment (Northern Ireland), the Royal Society for the Protection of Birds, and particularly Bob Davidson. We thank the N.I.O.C. for making all of their count data available to us, and the D.o.E. for allowing publication of work partly carried out under contract to them. Colin Bean made useful comments on an earlier version of this manuscript, as did two anonymous referees on a later version. Part of the roach monitoring was carried out while C.M.T. was in receipt of a studentship from the Department of Fisheries, Dublin.

References cited

Andersson, A. 1982. Competition between fishes and waterfowl. pp. 248–249. *In:* D.A. Scott (ed.) Managing Wetlands and their Birds, International Waterfowl Research Bureau, Slimbridge.

Andersson, G. 1981. Influence of fish on waterfowl in lakes. Anser 20: 21–34.

Andersson, G. 1984. The role of fish in lake ecosystems – and in limnology. Nordisk Limnologsymposium, Oslo: 189–197.

Andersson, G., H. Berggren, G. Cronberg & C. Gelin. 1978. Effects of planktivorous and benthivorous fish on organisms and water chemistry in eutrophic lakes. Hydrobiologia 59: 9–15.

Cragg-Hine, D. 1973. Coarse fish and fishery management in Northern Ireland. Proc. Brit. Coarse Fish. Conf. (6th), Univ. Liverpool: 52–59.

Cramp, S. & K.E.L. Simmons. 1977. The birds of the western Palearctic, Vol. 1. Oxford University Press, Oxford. 720 pp.

Cryer, M., G. Peirson & C.R. Townsend. 1986. Reciprocal interactions between roach, *Rutilus rutilus*, and zooplankton in a small lake: prey dynamics and fish growth and recruitment. Limnol. Oceanogr. 31: 1022–1038.

Eadie, J. McA. & A. Keast. 1982. Do goldeneye and perch compete for food? Oecologia 55: 225–230.

Eriksson, M.O.G. 1979. Competition between freshwater fish and goldeneye *Bucephala clangula* (L.) for common prey. Oecologia 41: 99–107.

Gibson, C.E. 1986. Preliminary results on phytoplankton reduction in Lough Neagh – assessing the effect against a background of change. Hydrobiological Bulletin 20: 173–182.

Hill, D., R. Wright & M. Street. 1987. Survival of mallard ducklings *Anas platyrhynchos* and competition with fish for invertebrates on a flooded gravel quarry in England. Ibis 129: 159–167.

Hurlbert, S.H., W. Loayza & T. Moreno. 1986. Fish-flamingo-plankton interactions in the Peruvian Andes. Limnol. Oceanogr. 31: 457–468.

Johansson, L. & L. Persson. 1986. Planktivorous fish. pp. 237–266. *In:* B. Riemann & M. Sondergaard (ed.) Carbon Dynamics in Eutrophic Temperate Lakes, Elsevier Science, Amsterdam.

Kennedy, G.J.A. & J.E. Greer. 1988. Predation by cormorants (*Phalacrocorax carbo* (L.)) on the salmonid populations of an Irish river. Aquaculture and Fisheries Management 19: 159–170.

Kerekes, J.J. 1990. Possible correlation of summer loon population with the trophic state of a water body. Proceedings

XIV Congress of the International Association of Theoretical and Applied Limnology, Munich (in press).
Lammens, E.H.R.R., H.W. de Nie & J. Vijverberg. 1985. Resource partitioning and niche shifts of bream (*Abramis brama*) and eel (*Anguilla anguilla*) mediated by predation of smelt (*Osmerus eperlanus*) on *Daphnia hyalina*. Can. J. Fish. Aquat. Sci. 42: 1342–1351.
Lammens, E.H.R.R. & W. Hoogenboezem 1991. Diets and feeding behaviour. pp. 353–376. *In:* I.J. Winfield & J.S. Nelson (ed.) Cyprinid Fishes: Systematics, Biology and Exploitation, Chapman & Hall, London.
Lazzaro, X. 1987. A review of planktivorous fishes: their evolution, feeding behaviours, selectivities, and impacts. Hydrobiologia 146: 97–167.
Owen, M., G.L. Atkinson-Willes & D.G. Salmon. 1986. Wildfowl in Great Britain. Cambridge University Press, Cambridge. 613 pp.
Pehrsson, O. 1984. Relationships of food to spatial and temporal breeding strategies of mallard in Sweden. J. Wildlife Manag. 48: 322–339.
Persson, L., G. Andersson, S.F. Hamrin & L. Johansson. 1988. Predator regulation and primary production along the productivity gradient of temperate lake ecosystems. pp. 45–65. *In:* S.D. Carpenter (ed.) Complex Interactions in Lake Communities, Springer-Verlag, New York.
Smart, M. (ed.) 1976. Proceedings of the international conference on the conservation of wetlands and waterfowl, Heiligenhafen 1974. International Waterfowl Research Bureau, Slimbridge. 492 pp.
Winfield, D.K., R.D. Davidson & I.J. Winfield. 1989. Long-term trends (1965–1988) in the numbers of waterfowl overwintering on Lough Neagh and Lough Beg, Northern Ireland. Irish Birds 4: 19–42.
Winfield, I.J. & C.R. Townsend. 1991. The role of cyprinids in ecosystems. pp. 552–571. *In:* I.J. Winfield & J.S. Nelson (ed.) Cyprinid Fishes: Systematics, Biology and Exploitation, Chapman & Hall, London.
Winfield, I.J., C.M. Tobin & C.R. Montgomery. 1991. Ecological studies of the fish populations of Lough Neagh, Northern Ireland: a prelude to fisheries management. Proceedings of an Institute of Fisheries Management (Northern Ireland Branch) Study Course, University of Ulster (in press).
Wright, R. & M. Street. 1985. The influence of fish on the survival of wildfowl broods. Game Conservancy Annual Review 1985: 77–80.
Zaret, T.M. 1980. Predation and freshwater communities. Yale University Press, New Haven. 187 pp.

Environmental Biology of Fishes **33**: 215–221, 1992.

Seasonal feeding activity and ontogenetic dietary shifts in crucian carp, *Carassius carassius*

Olli-Pekka Penttinen & Ismo J. Holopainen
Department of Biology, University of Joensuu, P.O. Box 111, SF-80101 Joensuu, Finland

Received 15.1.1990 Accepted 2.11.1990

Key words: Alkaline phosphatase, Gut contents, Cyprinidae, Habitat, Anoxia, Forest pond, Finland

Synopsis

A pronounced seasonal cycle in the activity of the intestinal alkaline phosphatase indicated a 6 month pause in feeding by crucian carp in an anoxic pond during winter. Zero enzyme activity was not reached until ice cover of the pond was complete, although even in the preceding two months no food was found in the guts of the fish. The summer diets of four size classes of fish consisted of varying proportions of planktonic and benthic arthropods, including Odonata which were consumed exclusively by the largest size class of crucian carp. Chironomid larvae were consumed even by the smallest fish investigated, although planktonic Cladocera were preferred by this size class. Benthic Cladocera were almost totally absent in the diet of fish > 10 cm. Habitat segregation between size classes developed synchronously with the shift to benthic food. The stunting of crucian carp populations in small ponds is interpreted as resulting from limited resources.

Introduction

The physiology of crucian carp has received considerable attention, mainly due to its unique tolerance of anoxia (e.g. Blažka 1960, Holopainen & Hyvärinen 1985). Exceptionally large glycogen reserves are stored in liver and muscle tissue in late summer for the demands of anaerobic metabolism during the winter (Hyvärinen et al. 1985). The pathways of anaerobic metabolism appear to be the same as in goldfish, *Carassius auratus*, with a minimum of lactate production, but with ethanol and CO_2 as the most important end-products (Shoubridge & Hochachka 1981, Holopainen et al. 1986).

Because of these adaptations, crucian carp is often the only fish species present in shallow, seasonally anoxic ponds in northern and eastern Europe. In a benign environment, with multispecies fish communities, only a few but large crucian carp may be found, whereas in seasonally anoxic ponds crucian carp is likely to be the only species present, forming stunted, high-density populations dominated by small individuals (Piironen & Holopainen 1988).

Seasonality in the feeding activity of crucian carp has been studied by Prejs (1973). In summer, crucian carp show the common cyprinid feeding mode, with high plasticity in prey selection. The species has been classified as a benthophage (e.g. Ciepielewski 1985) with a preference for molluscs, chironomid larvae and sediment-dwelling chydorids. Zooplankton may form a substantial part of the diet, with the addition of algae and macrophytes (Uspenskaja 1953, Prejs 1973).

Ontogenetic dietary shifts are common in size-structured fish populations (Werner & Gilliam 1984), including many cyprinids (Lammens et al. 1987, Mark et al. 1987). Size-related shifts in feeding efficiency of crucian carp have been demonstrated in the laboratory (Paszkowski et al. 1989), but the data on diets are based mainly on adult fish and size specific changes have not previously been documented.

The aim of this study is to follow seasonality of feeding in crucian carp as indicated by the activity of an intestinal enzyme, and to examine possible changes in diet with changes in body size.

Study area and methods

Fish examined in this study came from a small (1.5 ha), shallow (max depth 1.6 m), forest lake in eastern Finland (62° 41' N, 29° 41' E). A well-vegetated main pond was connected to a shallower (max 1 m) side pond containing crucian carp as the only fish species. A detailed description of the pond is given in Holopainen & Pitkänen (1985).

Fish were trapped in the small side pond between the beginning of April and the middle of November 1988. The pond was visited every 2 weeks in the spring and autumn, and monthly during the summer. Water temperature and dissolved oxygen concentration were measured on each trapping date. Adult fish (10–11 cm) were captured in traps set for 24 h. Captured fish ($n = 3$ on each occasion) were transported alive to the laboratory (10 km from the pond) and killed within 15 minutes of removal from the traps.

A small part of the anterior end of the gut, between the oesophagus and the bile duct, was excised. The tissue sample (60–100 mg) was weighed, homogenized in distilled water, and the activity of intestinal alkaline phosphatase (E.C.3.1.3.1) was measured by means of the Bessey-Lowry-Brock method (Toro & Ackermann 1975). This method uses p-nitrophenol as a substrate at pH 10.1. The incubation temperature was 25° C. Liberated nitrophenol creates an intense yellow colour when the solution is made strongly alkaline after incubation. The colour is then measured spectrophotometrically (wavelength 410 nm) and converted to micromoles of nitrophenol formed per hour per gram of tissue.

Material for diet analysis came from fish captured in the main pond. The numbers of fish handled are given in Figure 2.

Crucian carp is a portional spawner with up to three spawning periods between late May and early August (Holopainen & Pitkänen 1985). In 1987 the first young were captured by hand-net two weeks after the first observed hatching (June 26), then sampled weekly up to mid-September. During the study, young-of-the-year (YOY) grew from a total length of < 1 cm to almost 3 cm.

Juveniles (3.5–6 cm, 1+ age class) and adult (6–16 cm) fish were captured by traps 100 × 80 × 50 cm, with a 50 cm guiding fence and 5 mm green plastic mesh. Juveniles were collected at least three times a month, larger fish once a month. Traps were placed randomly within the pond and set for < 2 h. Young were always collected between 0900 and 1000 h, larger fish between 1000 and 1300 h.

The fish were killed and their total length was measured. Young were immersed in 4% formalin; larger fish were dissected and the gut was preserved in 70% alcohol. The anterior third of the gut was analyzed. Gut contents were removed and placed in a petri dish, detritus was separated and its volume measured. All prey items were identified to the lowest taxonomic group possible. Prey items were counted, the body length of crustaceans and the width of the head capsule of chironomid larvae measured. In order to estimate more precisely the nutritional value, prey numbers were transformed into carbon weights by use of length-weight relationships based, for chironomids, on measurements with a Carlo Erba CHN analyzer or, for zooplankton, on species-specific carbon values (M. Viljanen unpublished).

Diet was analyzed in terms of the mean percentage of total carbon weight of the gut contents (excluding detritus) contributed by individual prey taxa. Taxa were later combined into six groups: benthic Cladocera, pelagic Cladocera, Copepoda, Chironomidae (including larvae, pupae and Ceratopogonidae), Odonata larvae, and other zoobenthos.

Results

Seasonal feeding activity

No intestinal alkaline phosphatase activity was

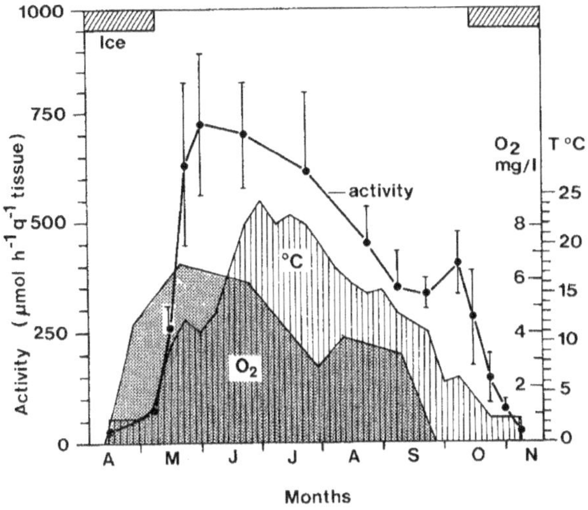

Fig. 1. Seasonal variation in intestinal alkaline phosphatase activity measured at 25° C (dots), oxygen content of water (shaded) and water temperature (vertical lines). Each point of alkaline phosphatase activity is a mean of three fish; vertical bars indicate standard deviation.

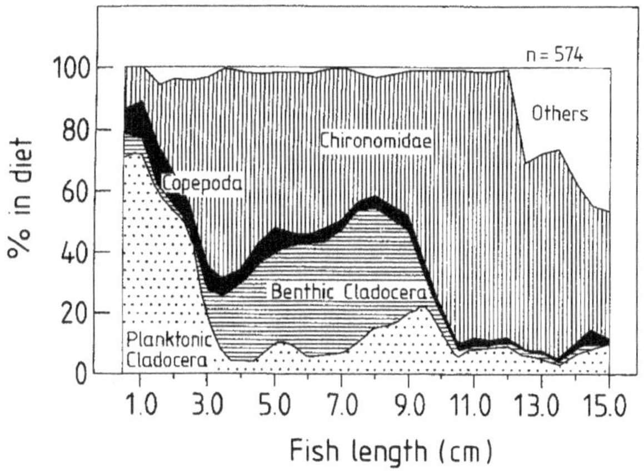

Fig. 2. The size-specific percentage composition of diet of crucian carp over the whole feeding season in 1987. Percentages are calculated from carbon weights.

found in winter (Fig. 1). After the ice melted in spring, activity increased rapidly and stayed high from late May to late July. From August through October a lower level of activity was maintained. In late September, fish entered an inactive phase with decreased mobility and feeding, as reflected by a decline in the catch and by the presence of fish with empty guts. Enzyme activity decreased in synchrony with water temperature, reaching a low level with ice formation in late October, and zero activity in early November.

Diet of crucian carp

Four size classes of crucian carp could be distinguished on the basis of gut contents. The food of all length classes (1–3 cm, 3–10 cm, 10–13 cm and > 13 cm) consisted of zooplankton, benthic Cladocera, and chironomid larvae, in varying proportions. Odonata, the largest benthic prey, were present in the diet of the largest fish only (Fig. 4).

The diet of small fish (< 3 cm; Fig. 3A, B) consisted mainly of planktonic Cladocera (especially *Bosmina longispina* and *B. longirostris*), but included Copepoda, benthic Cladocera and chironomid larvae even in the smallest fish. The relative abundance of chironomid larvae in the diet of fish > 2 cm increased during autumn (Fig. 3B).

In early June, the diet of the 3.5–6 cm size class consisted almost entirely of chironomid larvae, but later other prey taxa were included (Fig. 3C, D). Copepods and planktonic Cladocera were only occasionally eaten. Besides Chironomidae, the benthic Cladocera (mainly *Chydorus sphaericus*) appeared in the diet in mid-June and were eaten through summer and autumn (Fig. 3D). Among chironomid larvae, the most important groups were Orthocladiinae and Chironominae, especially *Tanytarsus*.

The diet of large fish (6–10 cm, see Fig. 4A) was very similar to that of the previous size class. In the inshore habitats (fish caught in traps 0.5–1.5 m from the shoreline), the diet consisted of benthic Cladocera (*Chydorus sphaericus* and *Alonella nana* in June and the larger *Alona affinis* and *Camptocercus rectirostris* in September). In July-August chironomid larvae were equally important. In the offshore habitat (traps more than 5 m from the shoreline) benthic Cladocera were replaced by planktonic forms, like *Bosmina longispina, Diaphanosoma brachyurum*, and *Limnosida frontosa*, as well as the calanoid *Eudiaptomus* spp. The diet of fish > 10 cm (Fig. 4B) consisted of chironomid larvae and/or the planktonic crustaceans listed above. Differences in diet between inshore and offshore

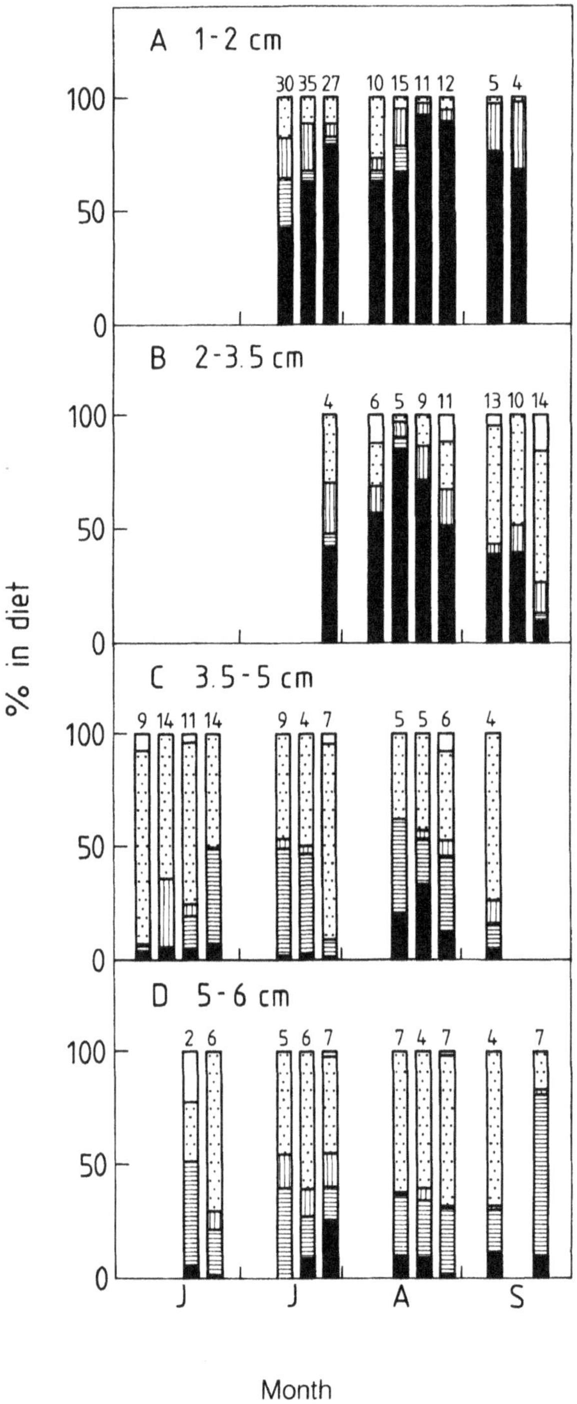

Fig. 3. Diet of four size classes of young crucian carp in summer 1987. Percentages are calculated from carbon weights. Numbers above the bars indicate the number of guts analysed. For explanations on food items see Figure 4.

areas were not as obvious in this size class as they were in fish > 13 cm. Large Odonata larvae (*Libellula quadrimaculata*) occurred in the diet of fish > 13 cm, mainly from the inshore samples (Fig. 4C). Notable was the almost total absence of benthic Cladocera in the diet of fish > 10 cm. Occasionally hundreds of them were found in the gut and are included in the data, although it is possible that these cladocerans may actually have represented the gut contents of predaceous Tanypodinae, especially *Procladius*, the most common prey of crucian carp > 10 cm.

Discussion

Dynamics of alkaline phosphatase

The function of the enzyme alkaline phosphatase in fish intestines is not fully understood. A role in active transport of phosphate and metabolites in general as well as involvement of the alkaline phosphatase-CaATPase complex in calcium absorption have been proposed (Wassermann & Fullmer 1983).

In the present study, enzyme activity was used as an indicator of digestive activity and thus of feeding. Empty guts, together with the apparent immobility of fish in winter, support our interpretation of enzymatic patterns. The total inactivation of this enzyme indicates the loss of digestive activity in the anoxic environment.

In crucian carp, June-July is the period of growth and reproduction, whereas in August-September winter reserves are built up (e.g. Holopainen et al. 1988). The two levels of alkaline phosphatase activity in the open-water period coincide with these two phases. The decrease in enzyme activity towards autumn may also be partly due to the diminished metabolic demands at the lower temperatures and oxygen contents.

The activity of alkaline phosphatase remained relatively high in September-October, although no food was found in the gut and there was a remarkable decrease in the mobility of the fish. The high level of activity of a digestive enzyme suggests that the fish maintain their ability to process food until complete ice cover. This could be of advantage if conditions suddenly became more favourable.

In roach, *Rutilus rutilus*, a similar seasonal pattern in the activity of another digestive enzyme, amylase, has been shown to occur (Hofer 1979). Kuzmina (1988) demonstrated a clear relationship between the seasonal dynamics of total amylase, saccharase, and alkaline phosphatase activity, and feeding behaviour, of five fish species including two cyprinids, roach and bream. Pike, *Esox lucius*, a typical predator, which feeds all year round shows much less seasonal variation than roach and the benthophagous common bream, *Abramis brama*, which both exhibit significantly lower levels of enzymatic activity in winter (7–23% of the values in summer). Bream and roach stop feeding at temperatures below 7° C (Kuzmina 1988) and show seasonal changes in glycogen content (Valtonen et al. 1983).

Higher activity of alkaline phosphatase in summer was reported by Gelman et al. (1984) for pikeperch, *Stizostedion lucioperca*, whereas bream displayed essentially no seasonal differences. These patterns correlate with differences in the biology of the two species.

In all studies so far reported the lowest activity values in winter stayed well above zero. No data are available on enzymatic activities of crucian carp in a normoxic winter environment. Prejs (1973) reported a summer peak in feeding intensity (the amount of food in the gut in relation to fish weight) of crucian carp in Lake Warniak. This lake experiences a short period of lack of oxygen under ice in spring, but Prejs' material was collected during the open water period only.

Ontogenetic dietary shift

Changes in diet and/or habitat during ontogeny are common in fish and often associated with size-specific differences in ability to use resources from different habitat types (Werner & Gilliam 1984).

In cyprinids, the most significant change in diet, i.e. the shift from phytoplankton and rotifers to

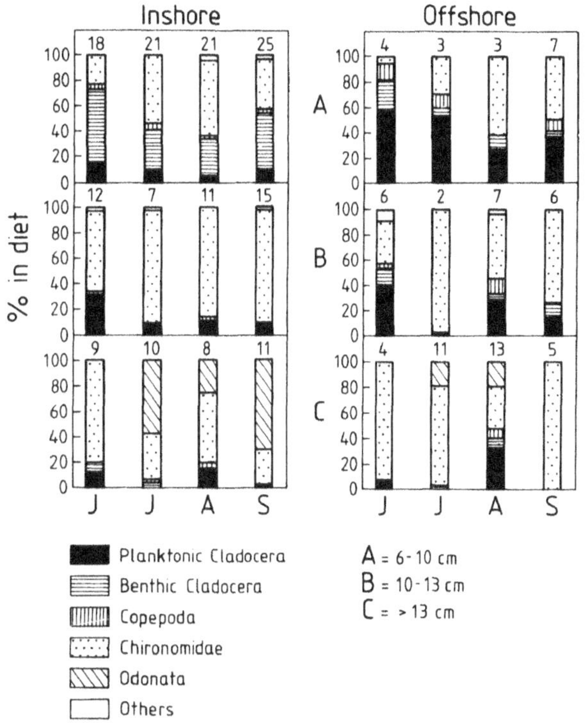

Fig. 4. Diet of three size classes of adult crucian carp in the two habitats in summer 1987. Numbers above the bars indicate the number of guts analysed.

crustaceans and chironomid larvae, takes place at an early juvenile phase when the fish have reached a length of ca. 2 cm (Adzhimudarov 1972, Hammer 1985, Mark et al. 1987, 1989). During the juvenile period, ecological diversification continues as the fish go through structural and functional reorganization; when they have reached the adult period cyprinids display high interspecific variation in their use of resources.

At lengths < 3 cm, the diet of crucian carp was found to resemble the general cyprinid pattern proposed by Mark et al. (1987). The only notable variation was the almost complete absence of rotifers in the diet, although this group was the most abundant in zooplankton samples in the pond (Mitronen unpublished data).

In crucian carp the transition to an adult diet occurred at lengths of 3 cm (Fig. 2). Habitat segregation between size classes developed together with the shift to benthic food. Crucian carp < 3 cm inhabited only vegetated inshore areas. Juveniles and adults < 10 cm still preferred inshore habitats, particularly in parts of the pond that supported limited growth of floating-leaved macrophytes (Holopainen et al. 1991). However, these juveniles and adults did feed in deeper water on sediment-dwelling chironomid larvae and chydorids. Chydorids (except for *Chydorus sphaericus*) were not present in zooplankton samples (Mitronen unpublished data), but they were commonly found in the guts of fish caught in the inshore area (Fig. 4). The crustacean zooplankton, both inshore and offshore, was dominated by *Bosmina longispina, Eudiaptomus* spp. and *Bosmina longirostris*. Because the zoobenthos was not sampled quantitatively, the abundance of benthic Cladocera and Chironomidae is not known. Qualitative analysis indicates an absence of molluscs and scarcity of Oligochaeta, although remains of oligochaetes were occasionally found in guts.

Fish < 10 cm preferred small benthic prey, the preference for larger prey items increased with the size of the fish. Crucian carp > 10 cm fed mainly on chironomid larvae, but occasionally large planktonic crustaceans were also eaten. Paszkowski et al. (1989) examined handling times and return rate-body size relations for crucian carp in the laboratory with standardized food items. It was demonstrated that return rates (dry mass ingested per unit handling time) increased with fish size and decreased with increasing prey size, especially in small fish. Larger fish had high return rates with intermediate sized prey and were better able to use much larger prey (cf. Odonata in the diet of fish > 13 cm) than small fish. The absence of benthic Cladocera in the diet of fish > 10 cm is partly explained by the dominance of this size-class in offshore rather than inshore habitats (see Holopainen et al. 1991). However, the fact that Chydoridae were absent even in fish > 10 cm caught inshore probably reflects behavioral and morphological changes accompanying growth in these fish. No studies on ontogenetic changes in the morphology of crucian carp are available. In many cyprinids the efficiency in catching smaller prey decreases with growth as a result of the decreasing powers of retention of the gill-raker apparatus (Mikheev 1984, Lammens et al. 1987). Lammens et al. (1987) have shown that in the diet of common bream, roach and white bream, *Blicca bjoerkna*, the proportion of small Cladocera and Copepoda decreased with increasing fish length, as gradually more chironomids and/or snails were eaten. Morphologically, these changes can be related to the structure of the gill-raker apparatus, mouth protrusability, and the strength of pharyngeal teeth. Also the general shape of larger fish contributed to the decreased efficiency of filtering small Cladocera and Copepoda, and increased efficiency of collecting larger organisms from the bottom.

In many small ponds only a minority of crucian carps reach the length necessary for effective prey handling. Populations are characteristically dominated by fish < 10 cm, whose diet breadth is limited and which feed effectively on small prey only (Paszkowski et al. 1989).

A subject for further studies is the stunting of crucian carp populations in small ponds and its possible dependence on resource limitations (e.g. paucity of benthic Cladocera) during the juvenile period. Developmental bottlenecks in fish populations with a clear ontogenetic dietary shift have previously been shown to cause stunting (see e.g. Persson 1987, Persson & Greenberg 1990).

Acknowledgements

We wish to thank H. Hyvärinen for help with the enzyme analysis and comments on the manuscript. W.M. Tonn, C.A. Paszkowski, both members of the Ruutana-team, and W. Wieser also commented on the manuscript. We are most grateful for their help.

References cited

Adzhimudarov, K.A. 1972. The food of juvenile carp (*Cyprinus carpio* (L.)) in early development stanzas in bodies of water of the Arakum (Terek river delta). J. Ichthyol. 12: 981–986.
Blažka, B. 1960. On the biology of the crucian carp (*Carassius carassius* L. morpha humilis Heckel). Zool. Zhurnal 9: 1384–1389. (In Russian).
Ciepielewski, W. 1985. Experimentally increased fish stock in

the pond type Lake Warniak. Changes of ichthyofauna between 1970 and 1983. Ekol. Pol. 33: 37–59.

Gelman, A., S. Mokady & U. Cogan. 1984. The effect of seasonal changes on the activity of intestinal alkaline phosphatase of pike perch, *Lucioperca lucioperca* and bream, *Abramis brama*. J. Fish Biol. 25: 207–212.

Hammer, C. 1985: Feeding behaviour of roach (*Rutilus rutilus*) larvae and the fry of perch (*Perca fluviatilis*) in Lake Lankau. Arch. Hydrobiol. 96: 496–509.

Hofer, R. 1979. The adaptations of digestive enzymes to temperature, season and diet in roach, *Rutilus rutilus* and rudd *Scardinius erythrophthalmus* L. 1. Amylase. J. Fish Biol. 14: 565–572.

Holopainen, I.J. & H. Hyvärinen. 1985. Ecology and physiology of crucian carp (*Carassius carassius* (L.)) in small Finnish ponds with anoxic conditions in winter. Verh. Internat. Verein. Limnol. 22: 2566–2570.

Holopainen, I.J. & A.K. Pitkänen. 1985. Population size and structure of crucian carp (*Carassius carassius* (L.)) in two small, natural ponds in Eastern Finland. Ann. Zool. Fennici 22: 397–406.

Holopainen, I.J., H. Hyvärinen & J. Piironen. 1986. Anaerobic wintering of crucian carp (*Carassius carassius* (L) – II. Metabolic products. Comp. Biochem. Physiol. 83A: 239–242.

Holopainen, I.J., W.M. Tonn, C.A. Paszkowski & A.K. Pitkänen. 1988. Habitat use, diel activity, and growth of crucian carp in a manipulated pond. Verh. Internat. Verein. Limnol. 23: 1743–1750.

Holopainen, I.J., W.M. Tonn & C.A. Paszkowski. 1991. Ecological responses of crucian carp populations to predation by perch in a manipulated pond. Verh. Int. Ver. Limnol. 24: (in press).

Hyvärinen, H., I.J. Holopainen & J. Piironen. 1985. Anaerobic wintering of crucian carp (*Carassius carassius* L.) – I. Annual dynamics of glycogen reserves in nature. Comp. Biochem. Physiol. 82A: 797–406.

Kuzmina, V.V. 1988. Seasonal dynamics in the activity of some carbohydrase and alkaline phosphotase enzymes in the intestinal mucosa of fish. J. Ichthyol. 28: 156–161.

Lammens, E.H.R.R., J. Geursen & P.J. McGillavry. 1987. Diet shifts, feeding efficiency and coexistence of bream (*Abramis brama*), roach (*Rutilus rutilus*) and white bream *Blicca bjoerkna*) in hypertrophic lakes. pp. 153–162. Proceedings V Congress of European Ichthyologists, Stockholm.

Mark, W., R. Hofer & W. Wieser. 1987. Diet spectra and resource partitioning in the larvae and juveniles of three species and six cohorts of cyprinids from a subalpine lake. Oecologia 71: 388–396.

Mark, W., W. Wieser & C. Hohenauer. 1989. Interactions between developmental processes, growth, and food selection in the larvae and juvenile of *Rutilus rutilus* (L) (Cyprinidae). Oecologia 78: 330–337.

Mikheev, V.N. 1984. Prey size and food selectivity in young fishes. J. Ichthyol. 24: 66–76.

Paszkowski, C.A., W.M. Tonn & I.J. Holopainen. 1989. An experimental study of body size and food size relations in crucian carp, *Carassius carassius*. Env. Biol. Fish. 24: 275–286.

Persson, L. 1987. The effects of resource availability and distribution on size class interactions in perch, *Perca fluviatilis*. Oikos 48: 148–160.

Persson, L. & L.A. Greenberg. 1990. Juvenile competitive bottlenecks: the perch (*Perca fluviatilis*)-roach (*Rutilus rutilus*) interactions. Ecology 71: 44–56.

Piironen, J. & I.J. Holopainen. 1988. Length structure and reproductive potential of crucian carp (*Carassius carassius* (L.)) populations in some small forest ponds. Ann. Zool. Fennici 25: 203–208.

Prejs, A. 1973. Experimentally increased fish stock in the pond type Lake Warniak. IV. Feeding of introduced and autochthonous non-predatory fish. Ekol. Pol. 21: 465–503.

Shoubridge, E.A. & P.W. Hochachka. 1981. The origin and significance of metabolic carbon dioxide production in the anoxic goldfish. Molecular Physiology 1: 315–338.

Toro, G. & P.G. Ackermann. 1975. Practical clinical chemistry. Little, Brown and Company, Boston. 779 pp.

Uspenskaja, V.D. 1953. Feeding of crucian carp in lakes of river Kljzma. Trudy Vsesojuzn Gidrobiol. Obshtch. 5: 347–364. (In Russian).

Valtonen, T., H. Hyvärinen & S. Lampelo. 1983. The effects of winter and summer temperature acclimation on some carbohydrate metabolism parameters in fish. Aquilo Ser. Zool. 22: 99–108.

Wasserman, R.H. & C.S. Fullmer. 1983. Calcium transport proteins, calcium absorption and vitamin D. Ann. Rev. Physiol. 45: 375–390.

Werner, E.E. & J.F. Gilliam. 1984. The ontogenetic niche and species interactions in size-structured populations. Ann. Rev. Ecol. Syst. 15: 393–426.

Environmental Biology of Fishes **33**: 223–227, 1992.

Epilogue: food and feeding, ecomorphology, energy assimilation and conversion in cyprinids

Fritz Schiemer[1] & Wolfgang Wieser[2]
[1] *Institut für Zoologie, Universität Wien, Althanstrasse 14, A-1090 Wien, Austria*
[2] *Institut für Zoologie, Universität Innsbruck, Technikerstrasse 25, A-6020 Innsbruck, Austria*

The roach, *Rutilus rutilus,* may be used as an example for illustrating the scope of the mechanistic and comparative approach to community ecology alluded to in the Introduction. This species has been dealt with, in one way or the other, in 14 of the 21 papers collected in this special issue. How does the information conveyed in these papers relate to the functional role of the roach in European lake and stream ecosystems, and in what way might a deeper knowledge of the reaction norms of this and other species influence the construction of models of population ecology and biological interactions?

Overall the roach is the most successful generalist fish in Central European freshwater habitats, occurring in all types of lotic and lentic water bodies in which the summer temperature attains mean levels of at least 15° C. How can this success be explained? What are the biological attributes of a successful generalist within a large group of similar species?

Biological success may be evaluated in terms of a species' ability to:
(1) find and incorporate food,
(2) assimilate energy,
(3) use energy in processes related to fitness, especially growth (in the early life phase) and reproduction, and
(4) reduce predation pressure.

Acquisition and incorporation of food

Ecological flexibility

Success in acquiring a rich and varied diet depends to a large degree on the extent to which an animal is capable of exploiting the full range of resources offered by the environment. In this respect the roach shows extraordinary flexibility in the choice of both food and microhabitats. For example, the ecologically uniform backwaters of the Danube are inhabited by a highly diverse and abundant fish fauna, among which there are specialists for Anflug (*Alburnus alburnus*), zoobenthos (*Abramis brama*), and zooplankton (*A. ballerus*). Under these conditions the roach occurs almost exclusively in the littoral zone, feeding on macrophytes, Aufwuchs, detritus, and molluscs (Schiemer, Dudzinski & Zweimüller unpublished). On the other hand, in lakes with low transparency in which zooplanktivorous specialists are absent, the roach may lead a predominantly pelagic life, feeding largely on zooplankton (Persson 1986).

Contrary to other species with which it may coexist, the roach is a facultative rather than an obligate 'schooler', often joining other species to form aggregations or loose feeding associations. From field observations Haberlehner (1988) concluded that 'the variable distribution of this species in different zones and the wide range of feeding techniques indicate flexible behavioural patterns or a great capacity for learning and imitation'. Along the same line, the intraspecific diversity of food choice, i.e. the wide range of individual food spectra within a population, may reflect this species' flexible behavior pattern. In the backwaters of the Danube, for example, intraspecific diversity of food choice proved to be much higher in roach compared to bream, bleak, and blue bream (Schiemer, Dudzinski & Zweimüller unpublished).

Conforming to its status as a generalist, the roach is also a eurythermal species, but contrary to the opinion sometimes held about poikilothermic species of this kind (Schlieper 1950) the rate of its energy metabolism does not show any sign of being temperature compensated. On the other hand, important aspects of its behaviour and metabolism are closely linked to the reproductive cycle which itself is controlled by season (Koch et al., Koch & Wieser 1983). Ionic regulation presents an exception to this generalization since in fish acclimated to 5° C the ion pumps of the kidney and the gills are much more active, and energy consumption of the kidney is much greater, than in warm-acclimated fish measured at the same temperature (Schwarzbaum et al. 1990). In consequence, this particular physiological function shows perfect temperature compensation, which implies that, relative to total turnover, ionic regulation requires more metabolic energy at low than at high environmental temperatures. However, on the organismic level the roach seems to be able to make up for the increased energy demand of this function at low temperature, probably by spending less energy on locomotion (Forstner & Wieser 1990). Behavioural flexibility thus may be a component of strategies allowing exploitation of a wide range of thermal environments.

Feeding behavior

In a comparative study it was found that food searching-cyprinids obey species-specific decision rules. In contrast to *Vimba elongata, Chalcalburnus chalcoides,* and *Abramis brama,* the roach relied almost exclusively on directed search methods, even at reduced visibility at which the other species switched to undirected sampling styles. Roach proved to be more successful than bream and vimba in discriminating prey (chironomid larvae) from a coarse sediment substratum (Uiblein). On the other hand, the visual ability for detecting planktonic prey was clearly lower in roach than in bleak. Lower reactive distances together with a lower cruising speed when searching for food (and thus lower prey encounter probabilities) render the roach an inferior planktivorous fish compared to bleak (Wanzenböck & Schiemer 1989). In consequence, the roach requires higher prey concentrations for attaining a sufficiently high consumption rate in the planktonic feeding mode (Schiemer et al. 1989). However, comparisons of this kind are valid only for a specified set of environmental conditions. For example, as Persson (1986) has shown, the foraging ability of roach increases with temperature. In Swedish lakes its swimming speed and capture success (with the phantom midge, *Chaoborus,* as prey) increased dramatically above 18° C, surpassing the performance of perch, *Perca fluviatilis,* which is the superior competitor at lower temperature.

Ecomorphological attributes

Sensory performance and neural integration – On the basis of the density and distribution of external taste buds a rank of life styles, from benthivory to open water planktivory, has been established, the two extremes being characterized by the highest and the lowest densities of taste buds, respectively (Gomahr et al.). In this series, encompassing 10 species, the roach occupied position no. 6, i.e. almost exactly in the middle of the range. The same seems to be true for the sense of vision. On the basis of a quantitative histological comparison, Zaunreiter et al. (1991) concluded that with respect to several characters of the optic tract and the retina, such as number of fibres, fibre diameter, ratio of large versus small fibres, photoreceptor density across the retina, and receptor cell-interneuron ratios, roach occupied an intermediate position among European cyprinids.

A comparison of the relative volumes of 17 primary sensory and higher brain areas in 28 mid-European cyprinids revealed four major groups of brain types: the generalized basic type, the abramine or advanced generalized, the chemosensory, and the octavo-lateralis brain (Kotrschal & Palzenberger). Two major pathways of cyprinid evolutionary and ecological radiation were defined, one leading from the basic cyprinid towards octavo-lateralis dominated midwater and surface planktivores, the other towards chemosensory dominated benthivores. The roach belongs to the most generalized group with a basic cyprinid type of brain. Among the brains of the cyprinid species investigated that of the roach would have represented the 'arithmetic mean' (Kotrschal & Junger 1988).

During development from larva to juvenile to adult, the relative size of the *tectum opticum* decreased, but, contrary to other species of cyprinids, the relative size of the remaining sensory lobes of the brain did not change in the roach (Brandstätter & Kotrschal 1989). This signals the switch from an optically dominated larval brain to a more generalized and balanced juvenile and adult brain.

Muscle and locomotion

The swimming performances of various cyprinid species examined are similar in the larval period. For example, in a tunnel respirometer the larvae and juveniles of roach proved just as powerful and efficient in short-term swimming exercises as the larvae and juveniles of the Danube bleak, one of the most effectively swimming cyprinid species (Kaufmann 1990). In the adult period, however, structure and functional properties of the swimming muscles exhibit distinct interspecific differences. Sänger (this volume) ranked 11 species of cyprinids according to lipid content, density of mitochondria and myofibrils in both red and white muscle fibres. The roach belonged to the species with low values of mitochondrial and myofibrillar density, thus confirming its status as a non-specialist.

Assimilation of energy

The roach is capable of rapid and sensitive adjustment of the activity of its digestive enzymes to temperature and diet. The activity of amylase and proteases in the gut increased dramatically with increasing temperature, and hence energy demand, as well as with the amount of animal food ingested (Hofer 1979a, b). The digestive system of the closely related rudd, *Scardinius erythrophthalmus,* is more inflexible, being less affected by environmental factors. Thus in a comparative study of these two species co-existing in a subalpine lake, protease activity in the gut was always higher in the roach than in the rudd, irrespective of the composition of the diet (Hofer 1979b).

As far as the relative length of the intestine is concerned, the roach occupies an intermediate position. In a list comprised 22 cyprinid species, relative gut length ranged from 0.78 in the predatory Sichling, *Pelecus cultratus,* to 2.05 in the grazing nase, *Chondrostoma nasus*. The relative gut length of the roach was 1.25, indicative of an omnivorous feeding type (Junger et al. 1989).

The respiratory system of the roach exhibits remarkable phenotypic plasticity. A comparison of populations from alpine lakes and from the Danube backwaters revealed significant differences in the number of the secondary gill lamellae forming the respiratory surface (Palzenberger & Pohla unpublished).

Conversion of energy

The larvae and juveniles of roach are fairly efficient in converting zooplankton food into growth. At high rations they grew faster than the bleak (Keckeis & Schiemer) and the Danube bleak (Wieser et al. 1988). At low rations, however, roach, bleak and blue bream, *Abramis ballerus,* grew at the same, slower rate. As far as general growth performance is concerned, roach appears to occupy an intermediate position between bleak, Danube bleak, and silver bream on the one hand, barbel, nase, tench and carp on the other hand. From a broad comparison of literature data it emerged that at three years of age, bleak weigh 5–10 g, silver bream less than 25 g, roach and rudd up to 40 g, but barbel, nase, tench and carp more than 100 g. This sequence agrees with experimental data on maximal growth performance at 20° C and high food densities during the initial growth period, specific daily growth rates ranging from about 12% in bleak (Keckeis & Schiemer), 18–20% in roach (Schiemer et al. 1989, Wieser & Medgyesy 1990a, b), to 50% in carp (Bryant & Matty 1981).

The problem of energy allocation to reproduction has not been addressed in this volume, but with regard to a maximum gonadosomatic index (about 20%), egg size, and size specific fecundity, roach resemble other cyprinids, such as *Scardinius erythrophthalmus, Blicca bjoerkna,* and *A. brama*. As in other species eutrophication of the environment leads to an increase of fecundity and to a decrease of egg size.

Conclusion

The success of the roach may reflect the increasing eutrophication of European lakes and streams which not only ensures a rich and varied menu for the fish fauna, but by causing increasing uniformity in aquatic habitats also facilitates the spread of the successful generalist. The data summarized above link the success of the roach to:

(1) its flexibility with regard to food choice and microhabitat use, which reflects its great capacity for learning and imitation;
(2) the generalized and balanced architecture of its brain, which is an expression of the balanced representation of all major sensory modalities;
(3) the non-specialized nature of several functions, like swimming performance and digestive ability; and
(4) the moderately high efficiency with which food energy is converted into growth during the larval and early juvenile life period.

Additional features may be relevant for explaining the present position of roach in European freshwater communities but are difficult to assess due to lack of comparative data. Such features could be:

- the use of a combination of compensating and non-compensating metabolic styles which enables the roach to cope with most thermal regimes in the temperate regions of Europe, possibly better than the 'champion compensators' (Guderley & Blier 1988) among cyprinids: goldfish, crucian carp and carp.
- the fact that roach seem to be able to adjust the amount of digestive enzymes secreted into the gut more rapidly and more sensitively to environmental temperature and diet than another, closely related, species, the rudd.
- the high efficiency of the larvae and juveniles of roach in catching zooplankton in dim light, and even in the dark (Wieser & Medgyesy 1990a). Is it this ability which gives small roach an advantage over bream in turbid water (Lammens et al.)?

Several questions can be formulated with regard to the ecological scenarios for the success of *R. rutilus;* for example:

- is the success of this species related to conditions of low ecological stability or, conversely, is the roach outcompeted by other species if environmental conditions remained stable for a long time?
- is the roach particularly successful only when interspecific competition is low?
- is the roach more efficient than other species in exploiting high food concentrations and is this the reason why it thrives under eutrophic conditions?

A more detailed knowledge of the reaction norm of the roach would be of assistance in validating these questions.

The mechanistic, autecological and comparative approach is only one of several pathways leading into the mainland of ecology. There can be no doubt that each organism is part of a network of genotypically and phenotypically determined relationships with the environment; in order to provide a sounder basis for generalizations on the community level a more detailed analysis of such relationships is essential.

References cited

Brandstätter, R. & K. Kotrschal. 1989. Life history of roach, *Rutilus rutilus* (Cyprinidae, Teleostei). Brain, Beh. Evol. 34: 35–42.

Bryant, P.L. & A.J. Matty. 1981. Adaptation of carp (*Cyprinus carpio*) larvae to artificial diets. I. optimum feeding rate and adaptation age for a commercial diet. Aquaculture 23: 275–286.

Forstner, H. & W. Wieser. 1990. Patterns of routine swimming and metabolic rate in juvenile cyprinids at three temperatures: analysis with a respirometer-activity-monitoring system. J. comp. Physiol. B 160: 71–76.

Guderley, H. & P. Blier. 1988. Thermal acclimation in fish:

conservative and labile properties of swimming muscle. Can. J. Zool. 66: 1105–1115.

Haberlehner, E. 1988. Comparative analysis of feeding and schooling behaviour of the Cyprinidae *Alburnus alburnus* (L., 1758), *Rutilus rutilus* (L., 1758), and *Scardinius erythrophthalmus* (L., 1758) in a backwater of the Danube near Vienna. Int. Rev. ges. Hydrobiol. 73: 537–546.

Hofer, H. 1979a. The adaptation of digestive enzymes to temperature, season and diet in roach, *Rutilus rutilus* L. and rudd, *Scardinius erythrophthalmus* L.: Amylase. J. Fish Biol. 14: 565–572.

Hofer, H. 1979b. The adaptation of digestive enzymes to temperature, season and diet in roach, *Rutilus rutilus* and rudd, *Scardinius erythrophthalmus:* Proteases. J. Fish Biol. 15: 373–379.

Junger, H., K. Kotrschal & A. Goldschmid. 1989. Comparative morphology and ecomorphology of the gut in European cyprinids (Teleostei). J. Fish Biol. 34: 315–326.

Kaufmann, R. 1990. Respiratory cost of swimming in larval and juvenile cyprinids. J. exp. Biol. 150: 343–366.

Koch, F. & W. Wieser. 1983. Partitioning of energy in fish: can reduction in swimming activity compensate for the cost of reproduction? J. exp. Biol. 107: 141–146.

Kotrschal, K. & H. Junger. 1988. Patterns of brain morphology in Mid-European Cyprinidae (Pisces, Teleostei): a quantitative histological study. J. Hirnforsch. 29: 341–352.

Persson, L. 1986. Temperature-induced shift in foraging ability in two fish species, roach (*Rutilus rutilus*) and perch (*Perca fluviatilis*): implications for coexistence between poikilotherms. J. Anim. Ecol. 55: 829–839.

Schiemer, F., H. Keckeis & J. Wanzenböck. 1989. Foraging in cyprinids during early development. Pol. Arch. Hydrobiol. 36: 467–474.

Schlieper, C. 1950. Temperaturbezogene Regulationen des Grundumsatzes bei wechselwarmen Tieren. Biol. Zbl. 69: 216–226.

Schwarzbaum, P., W. Wieser & H. Niederstätter. 1990. Contrasting effects of temperature acclimation on mechanisms of ionic regulation in a eurythermic and a stenothermic species of freshwater fish (*Rutilus rutilus* and *Salvelinus alpinus*). Comp. Biochem. Physiol. 98A: 483–489.

Wanzenböck, J. & F. Schiemer. 1989. Prey detection in cyprinids during early development. Can. J. Fish. aquat. Sci. 46: 995–1001.

Wieser, W., H. Forstner, N. Medgyesy & S. Hinterleitner. 1988. To switch or not to switch: partitioning of energy between growth and activity in larval cyprinids (Cyprinidae: Teleostei). Funct. Ecol. 2: 499–507.

Wieser, W. & N. Medgyesy. 1990a. Aerobic maximum for growth in the larvae and juveniles of a cyprinid fish, *Rutilus rutilus* (L.): implications for energy budgeting in small poikilotherms. Funct. Ecol. 4: 233–242.

Wieser, W. & N. Medgyesy. 1990b. Cost and efficiency of growth in the larvae of two species of fish with widely differing metabolic rates. Proc. Roy. Soc. Lond. B 242: 51–56.

Zaunreiter, M., H. Junger & K. Kotrschal. 1991. Retinal morphology of cyprinid fishes: a quantitative histological study on ontogenetic changes and interspecific variation. Vision Res. 31: 383–394.

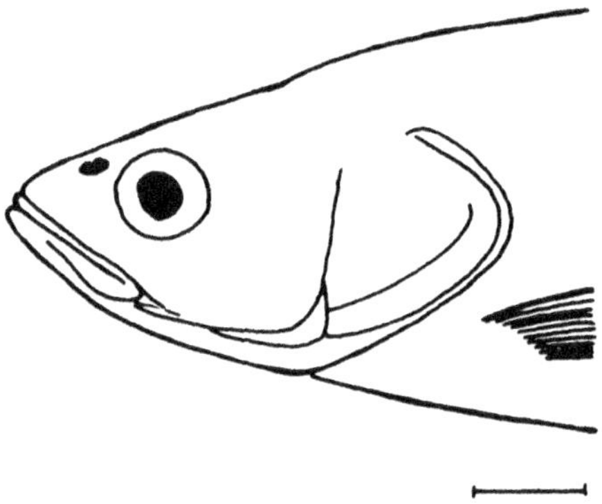

Species and subject index

GPSR Compliance

The European Union's (EU) General Product Safety Regulation (GPSR) is a set of rules that requires consumer products to be safe and our obligations to ensure this.

If you have any concerns about our products, you can contact us on ProductSafety@springernature.com

In case Publisher is established outside the EU, the EU authorized representative is:

Springer Nature Customer Service Center GmbH
Europaplatz 3
69115 Heidelberg, Germany

Batch number: 10868238

Printed by Printforce, the Netherlands